AF248620

Functionalized Nanomaterials for the Management of Microbial Infection

Functionalized Nanomaterials for the Management of Microbial Infection
A Strategy to Address Microbial Drug Resistance

Edited by

Rabah Boukherroub
*CNRS, Lille University of Science and Technology,
Villeneuve d'Ascq, France*

Sabine Szunerits
*Lille University of Science and Technology,
Villeneuve d'Ascq, France*

Djamel Drider
*Lille University of Science and Technology,
Villeneuve d'Ascq, France*

ELSEVIER

AMSTERDAM • BOSTON • HEIDELBERG • LONDON • NEW YORK • OXFORD
PARIS • SAN DIEGO • SAN FRANCISCO • SINGAPORE • SYDNEY • TOKYO

Elsevier
Radarweg 29, PO Box 211, 1000 AE Amsterdam, Netherlands
The Boulevard, Langford Lane, Kidlington, Oxford, OX5 1GB, United Kingdom
50 Hampshire Street, 5th Floor, Cambridge, MA 02139, United States

British Library Cataloguing-in-Publication Data
A catalogue record for this book is available from the British Library

Library of Congress Cataloging-in-Publication Data
A catalog record for this book is available from the Library of Congress

ISBN: 978-0-323-41625-2

For Information on all Elsevier publications
visit our website at https://www.elsevier.com

Publisher: Matthew Deans
Acquisition Editor: Simon Holt
Editorial Project Manager: Sabrina Webber
Production Project Manager: Caroline Johnson
Designer: Greg Harris

Typeset by MPS Limited, Chennai, India

Contents

List of Contributors

A. Alvarez-Ordóñez
University of León, León, Spain

P.C. Balaure
University Politehnica of Bucharest, Bucharest, Romania

Y. Belguesmia
Lille University of Science and Technology, Villeneuve d'Ascq, France

R. Boukherroub
CNRS, Lille University of Science and Technology, Villeneuve d'Ascq, France

K. Braeckmans
Ghent University, Ghent, Belgium; Centre for Nano- and Biophotonics, Ghent, Belgium

T. Coenye
Ghent University, Ghent, Belgium

L.M.T. Dicks
Stellenbosch University, Stellenbosch, South Africa

D. Drider
Lille University of Science and Technology, Villeneuve d'Ascq, France

A. Friedman
George Washington School of Medicine and Health Sciences, Washington, DC, United States

D. Gudovan
University Politehnica of Bucharest, Bucharest, Romania

I. Gudovan
University Politehnica of Bucharest, Bucharest, Romania

I. Kempf
Unité Mycoplasmologie-Bactériologie, Ploufragan, France

B. Klumperman
Stellenbosch University, Stellenbosch, South Africa

B. Mordorski
Albert Einstein College of Medicine, Bronx, NY, United States

K. Naghmouchi
Université de Tunis El Manar, Tunis, Tunisia

L. Ruiz
Complutense University of Madrid, Madrid, Spain

S.K. Samal
Ghent University, Ghent, Belgium; Centre for Nano- and Biophotonics, Ghent, Belgium

N. Škalko-Basnet
Drug Transport and Delivery Research Group, Department of Pharmacy, Faculty of Health Sciences, University of Tromsø The Arctic University of Norway, Tromsø, Norway

S. Szunerits
Lille University of Science and Technology, Villeneuve d'Ascq, France

E. Teirlinck
Ghent University, Ghent, Belgium; Centre for Nano- and Biophotonics, Ghent, Belgium

N. Tison
Lille University of Science and Technology, Villeneuve d'Ascq, France; Haute *École* Louvain en Hainaut, HELHa, Mons, Belgium

A.D.P. van Staden
Stellenbosch University, Stellenbosch, South Africa

Ž. Vanić
Department of Pharmaceutical Technology, Faculty of Pharmacy and Biochemistry, University of Zagreb, Zagreb, A Kovačića 1, Zagreb, Croatia

Preface

Bacteria usually must adhere to a cellular surface in order to cause disease. In some settings, the adherent microbe physically invades the cellular layer and gains access to inner milieu, as *Salmonella* or *Shigella* would in the intestine. In other scenarios, such as the airway of an individual with cystic fibrosis, adherent *Pseudomonas aeruginosa* establishes a colony (a "biofilm") on the surface of the bronchial epithelium that resists antimicrobial therapy. Its presence provokes a chronic inflammatory process that gradually destroys the airway. Similarly, infections of the urinary tract usually occur when a strain of *Escherichia coli* attaches to the surface of the bladder, increases in number, and migrates upwards toward the kidneys. Furthermore, the surfaces of medical devices that come into contact with bacteria generally become colonized, be they catheters that are placed in a vein or artery, the urinary tract, or the airway. Bacteria seem to love to grow on surfaces.

To treat bacterial infections we almost universally administer antibiotics, which are physically quite small, relative to the size of the microbe. The antibiotic we introduce swarms around the bacterial cell. If the concentration of the antibiotic in the blood, or in the urinary tract, or on the surface of the airway is high enough, sufficient numbers of antibiotic molecules will penetrate the bacterial cell and disable it, the sooner the better.

In this volume, *Functionalized Nanomaterials for the Management of Microbial Infection*, Drs. Boukherroub, Szunerits, and Drider suggest that we ought to consider exploiting the predilection that microbes have for surfaces in our design of antibiotic therapeutics. The basic idea is to use nanoparticles, which present a small but discrete surface that can physically contact the bacterial cell. If the surface properties of the nanoparticle are optimal for the purpose intended, the particle and bacterial cell will adhere to one another. If the nanoparticle is decorated with antibiotics that can act despite being tethered to a surface, then the bacterial cell will find itself facing a wall of deadly agents. (Antimicrobial peptides, including many of the bacteriocins described in this volume, can inflict damage on the bacterial membrane and/or proteoglycan wall, even when covalently bound to a surface). We learn in this volume that lipids, various polymers, and even graphene, can be exploited as the nanoparticle carrier.

The development of nanoparticle antibiotics is very exciting. We all await the outcome of their safety and efficacy in appropriate clinical trials.

M. Zasloff
Georgetown University Hospital, Washington, DC, United States

Resistance to Antibiotics and Antimicrobial Peptides: A Need of Novel Technology to Tackle This Phenomenon

Y. Belguesmia[1], I. Kempf[2], N. Tison[1,3], K. Naghmouchi[4] and D. Drider[1]

[1]*Lille University of Science and Technology, Villeneuve d'Ascq, France* [2]*Unité Mycoplasmologie-Bactériologie, Ploufragan, France* [3]*Haute École Louvain en Hainaut, HELHa, Mons, Belgium* [4]*Université de Tunis El Manar, Tunis, Tunisia*

CHAPTER OUTLINE

1.1 RESISTANCE TO ANTIBIOTICS

Antimicrobials used for human or animal health include many different families, such as β-lactams, aminoglycosides, macrolides, tetracyclines, and polypeptides, which differ according to their chemical structures, mechanisms of action, and spectra. Antibiotics are drugs of natural, synthetic, or semi-synthetic origin that have, at low concentrations, the capacity to kill or to inhibit the growth of bacteria whilst causing little or no damage to the host. They can target the bacterial cell wall (β-lactam antibiotics, vancomycin, and bacitracin), the protein synthesis (macrolides, aminoglycosides, tetracyclines, and chloramphenicol), affect nucleic acid function (quinolones, rifampicin, and nitrofurans), or acid folic synthesis (trimethoprim and sulfonamides). Bacteria can be inherently resistant to a specific family of antimicrobials for various reasons: for example, mycoplasmas are naturally resistant to β-lactam antibiotics because they have no cell wall, and Gram-negative bacteria are intrinsically resistant to macrolides as these molecules are too large to enter the bacteria through their cell wall. Antimicrobial resistance can be acquired in previously susceptible species, either by mutations, such as modifications of the target DNA-gyrase which renders *Enterobacteriaceae* resistant to quinolones or fluoroquinolones. Mutations can also result in the increased expression of efflux pumps able to export one or several classes of antimicrobials. Bacteria can also acquire resistance genes from other bacteria, by conjugation, transformation, or transduction. The acquired genes can encode antibiotic-inactivating enzymes (e.g., β-lactamases and aminoglycosides-hydrolyzing enzymes) or modified molecules which are no longer the target of the antibiotic (e.g., altered penicillin-binding-protein PBP2a encoded by the *mec*A gene in *Staphylococcus aureus* resistant to methicillin), protect the target (e.g., *tet*(O) gene in *Campylobacter* spp. encoding a protein protecting the ribosome against tetracycline) or a new efflux pump (e.g., *tet*(A) gene in *Escherichia coli* encoding a tetracycline efflux pump).

The polypeptide antibiotics belonging to the polymyxin group were discovered in the 1940s and were first used in both human and animal medicines. Polymyxins are non-ribosomal cyclic lipopeptides mainly used from the late 1950s as a topical treatment in human medicine. The polymyxin E (colistin) has been used particularly for the treatment of Gram-negative bacterial infections. However, in the 1970s, clinical use of these molecules was limited due to their serious nephrotoxicity and neurotoxicity after parenteral administration under systematic use [1]. Because of this toxicity, the use of polymyxins soon decreased, but colistin remained in frequent use in animals, mainly for the control of enteric infections [2]. Recently the emergence of bacteria resistant to most antibiotic families, including carbapenems, renewed the interest for colistin as a last-resort treatment option for infections caused by multidrug-resistant *Enterobacteriaceae*, *Pseudomonas aeruginosa*, and *Acinetobacter baumannii* [3].

The basic structure of polymyxins is composed of a cyclic decapeptide bound to a fatty acid chain at amino terminus. Polymyxin B and polymyxin E have almost identical primary sequence with difference at position 6 where D-Phe (D-phenylalanine) in polymyxin B is replaced by D-Leu (D-leucine) in polymyxin E (Fig. 1.1) [4]. The biosynthetic gene cluster of polymyxin is called *pmx* cluster, including five open reading frames, namely, *pmxA, pmxB, pmxC, pmxD*, and *pmxE*. Synthesis of polymyxin implies three polymyxin synthetases, PmxA, PmxB, and PmxE, and transported by two membrane transport proteins, PmxC and PmxD. The polymyxin structure determines its activity, the assembly order of modules for amino acid during biosynthesis of polymyxin is PmxE–PmxA–PmxB, the fatty acid: 6-methyloctanoic acid or isooctanoic acid; Thr, threonine;

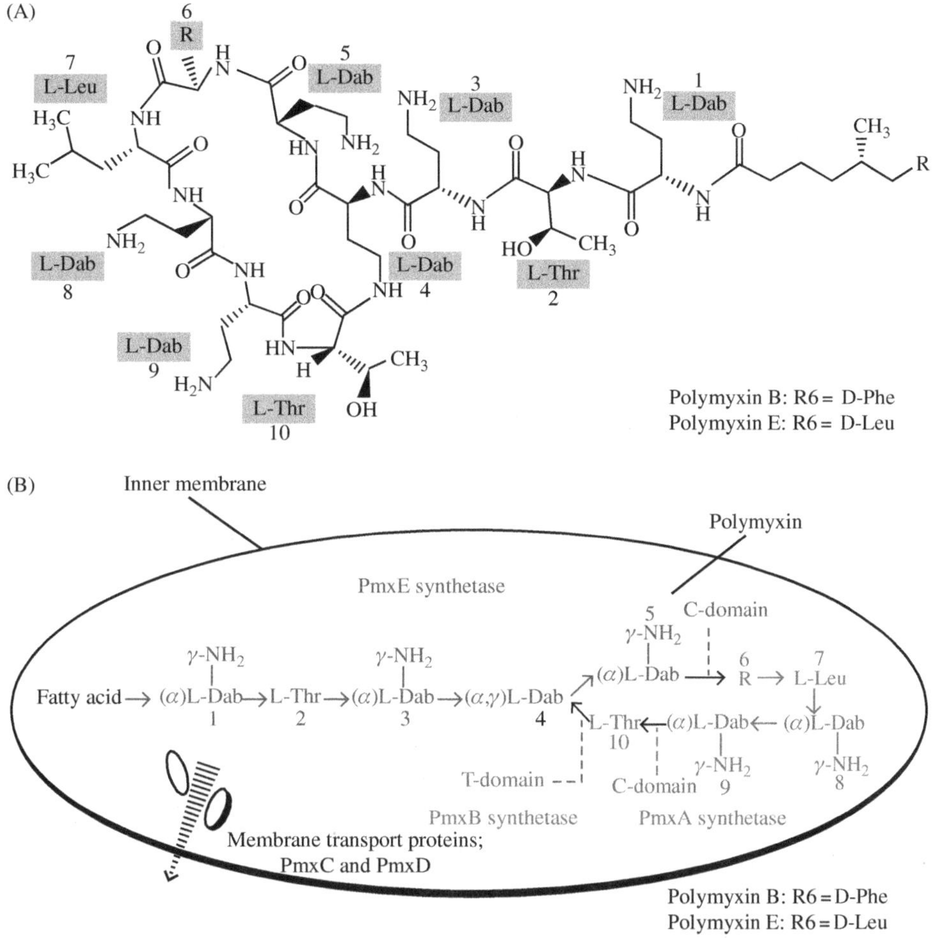

■ **FIGURE 1.1** Chemical structure of polymyxin (A); polymyxin biosynthesis in *Paenibacillus polymyxa* (B). *Adapted from Yu Z, Qin W, Lin J, Fang S, Qiu J. Antibacterial mechanisms of polymyxin and bacterial resistance. BioMed Res Int 2015;679109. doi:10.1155/2015/679109.*

Phe, phenylalanine; Leu, leucine; Dab, α,γ-diaminobutyric acid, α and γ refer to the respective—NH₂ involved in peptide linkage (Fig. 1.1).

Colistin is a bactericidal drug that binds to phospholipids in the outer cell membrane of Gram-negative bacteria and to lipopolysaccharide (LPS), by a hydrophobic interaction between the nine-carbon fatty acid side chain of colistin and the fatty acid portion of lipid A. It competitively displaces divalent cations from the phosphate groups of membrane lipids, which leads to disruption of the outer cell membrane, leakage of intracellular contents, and bacterial death (Fig. 1.2).

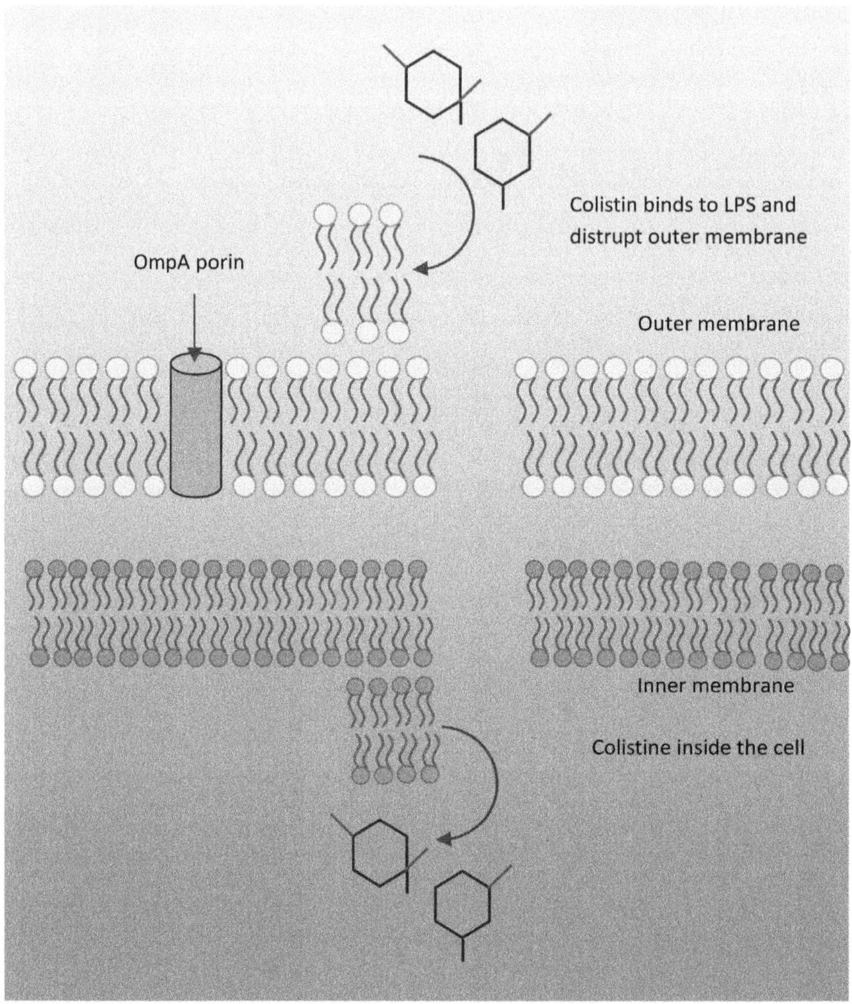

■ **FIGURE 1.2** Antimicrobial mode of action of polymyxin against Gram-negative bacterial membranes. LPS, Lipopolysaccharide. *Adapted from [5] Biswas S, Brunel JM, Dubus JC, Reynaud-Gaubert M, Rolain JM. Colistin: an update on the antibiotic of the 21st century. Expert Rev Anti-infect Ther 2012;10(8):917–34.*

Resistance to polymyxins has been investigated in intrinsically resistant bacterial species, in vitro mutants and in clinical isolates. The electrostatic interaction between positively charged Dab residues, on polymyxin, and negatively charged phosphate groups, on lipid A of LPS, represents the critical step of bactericidal activity of polymyxins. Reduction of this initial electrostatic attraction can occur when net negative charge of the bacterial outer membrane (OM) is reduced via lipid A modification, implying increasing resistance to polymyxin [1]. In bacteria the most common polymyxin-resistance mechanism is due to the shielding of phosphates on lipid A with positively charged groups, such as phosphoethanolamine (pEtN) and L-4-aminoarabinose (L-Ara4N), which is mediated by PhoP-PhoQ regulatory system encoded by *phoP* locus [6,7]. The two-component systems (TCS) PhoP/PhoQ and PmrA/PmrB (also called BasR/BasS), which are the major regulatory systems involved in the LPS modification in Gram-negative bacteria, can be activated by various environmental stimuli and, furthermore, modifications in their sequences result in their constitutive activation that leads to an overexpression of the LPS modification systems. In many bacteria, the Pho/PhoQ TCS activates the PmrA/PmrB TCS via PmrD. Phosphorylation of PmrA leads to the upregulation of the *pmrCAB* and *arnBCADTEF-pmrE* (or *pmrHFIJKLM-ugd*) operons conducting respectively to the synthesis and covalent additions of phosphoethanolamine (pEtN) and 4-deoxyaminoarabinose (L-Aara4N) to the lipid A ([8,9]). PhoP is phosphorylated by PhoP-PhoQ system under low Mg^{2+}, low pH, and polymyxin, promoting the *pmrD* gene to express PmrD protein. With the help of PmrD and in the presence of Fe^{3+}, transcription of the PmrA-activated gene is induced by PmrA-PmrB system. After phosphorylation, PmrA-P activates transcription of LPS modification loci (i.e., *wzz, pmrG, cptA, ugd, pbgP,* and *pmrC*) [1]. The O-antigen synthesis is controlled by products of *wzz* gene. The PmrG and CptA proteins are responsible for the phosphorylation modification of heptose (I) and heptose (II), respectively. Lipid A can be phosphorylated with phosphoethanolamine (pEtN), encoded by *pmrC*, or L-4-aminoarabinose (L-Ara4N), encoded by *ugd* and *pbgP* (Fig. 1.3).

These modifications of the surface charges of the LPS reduce the affinity of the positively charged colistin and thus increase the minimum inhibitory concentrations of this antimicrobial. Intrinsic resistance to polymyxins is encountered in Gram-negative bacteria such as *Proteus* spp., *Serratia* spp., *Edwardsiella tarda*, and *Burkholderia cepacia*. In most of these bacteria, the L-Ara4N or pEtN modifications of the LPS are constitutive and contribute to the resistance. Examples of the resistance mechanisms described in the main bacterial targets of polymyxins are now given below.

Modifications of the LPS are observed in polymyxin-resistant *Salmonella enterica* isolates. Mutations in PmrA/PmrB and PhoP/PhoQ TCS result in the activation of (1) e *arnBCADTEF* and *pmrCAB* operons, leading to the previously described modifications of the lipid A, and (2) *cptA* gene, regulated by *pmrA* gene, allowing the addition of pEtN to the LPS core. Additionally, the deacylation of lipid A by *pagL* activated by PhoP can increase the polymyxin-resistance [9].

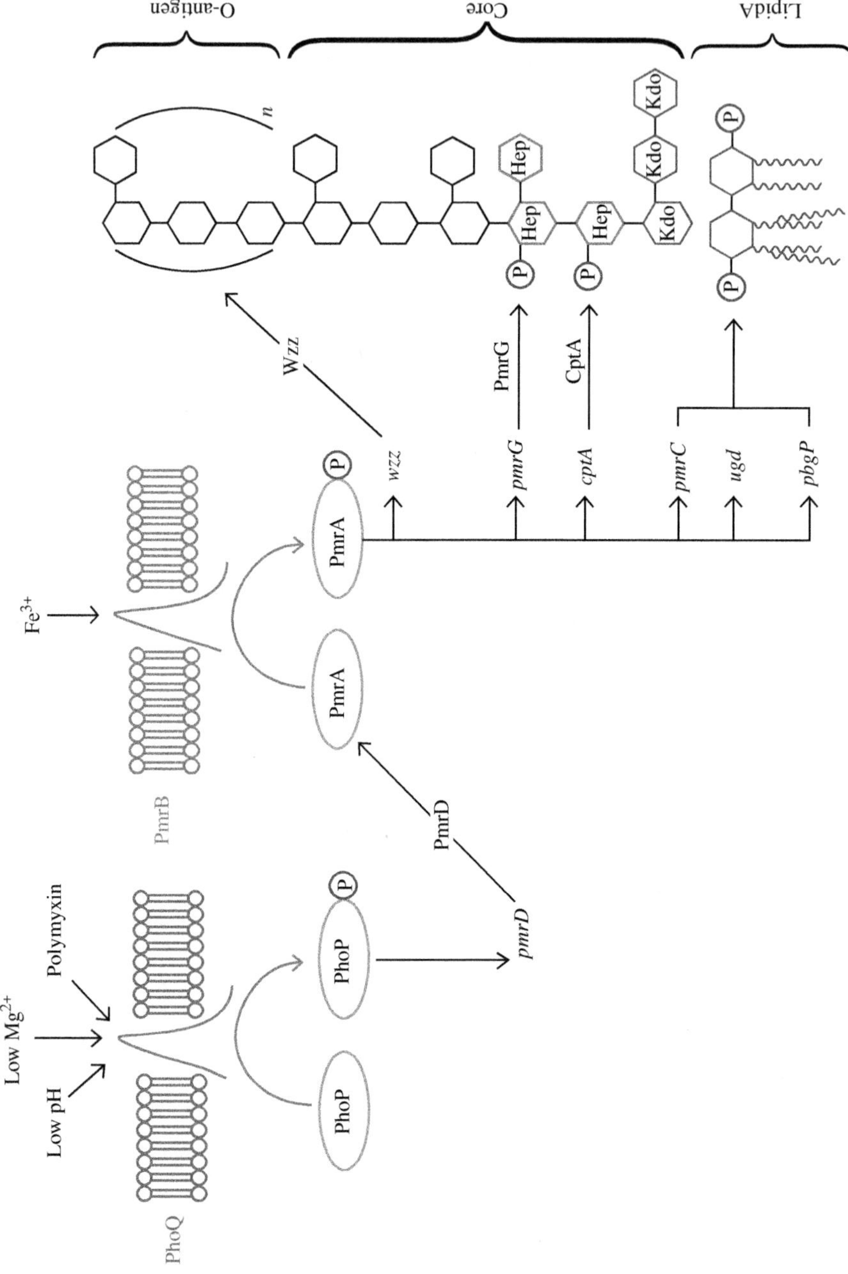

FIGURE 1.3 PhoP-PhoQ two-component system of bacterial resistance to polymyxin [1].

In *Klebsiella pneumoniae* (*K. pneumoniae*) and *Klebsiella oxytoca*, the *mgrB* gene codes for a transmembrane protein, which exerts a negative feedback on the PhoP/PhoQ system. In the resistant isolates, the genetic alterations of this gene or of its promoter increased the transcription of *pmrA* and *arnBCADTEF*, and thus the addition of Lara4N to the lipid A resulted in resistance to colistin [9,10]. Further mechanisms included mutations in the *pmrB* gene ([8]). Capsular polysaccharides of *K. pneumoniae* can be implicated in the resistance to polymyxins. Indeed, Formosa et al. [11] observed that colistin removed the capsule from colistin-susceptible strains but not from the resistant strains. Finally, the efflux pump AcrAB of *K. pneumoniae* was shown to play a major role in the resistance to polymyxin B and also to the host antimicrobial peptides [12].

Clinical isolates of *A. baumannii* resistant to colistin were studied and the resistance possibly was due to the complete inactivation of the lipid A synthesis pathway, because of mutations, deletions or insertions in the *lpxA*, *lpxC* or *lpxD* genes causing the total loss of the LPS surface [13]. Mutations in the *pmrA* or *pmrB* increased expression of these genes [14] leading thereof to the modification of the lipid A by addition of pEtN, through a drop of the negative charges of LPS [15]. These modifications appeared responsible for the colistin-resistance but were shown to induce a biological cost as evidenced in vitro retarded growth or attenuated virulence [16].

Five two-component systems, PhoP/PhoQ, PmrA/PmrB, ParR/ParS, ColR/ColS, and CprR/CprS, are involved, in addition to L-Ara4N, to the LPS modifications, and thus in polymyxin resistance described in *P. aeruginosa*. The trapping of polymyxins by the capsule and overexpression of the OM protein OprH were reported to increase resistance options adopted by *P. aeruginosa* [6].

To date, the resistance mechanisms are limited to alterations of chromosomal genes. Nevertheless, Liu et al. [17] reported the emergence of the plasmid-mediated colistin resistance mechanism *mcr-1* in *Enterobacteriaceae* isolates of both animal and human origin in China, arguing a possibility of horizontal gene transmission. The *mcr*-1 gene codes for a pEtN transferase enzyme resulting in the addition of pEtN to lipid A. The deduced amino acid sequence of the *mcr*-1 gene product was found to be close to pEtN–transferase of polymyxin-producing bacteria, *Paenibacillus* spp., suggesting a potential transfer of the gene from such bacteria to *E. coli*. Afterwards, the *mcr*-1 gene was reported in bacteria isolated in Asia, Africa, and Europe [18–21]. The dissemination of this plasmid-mediated colistin resistance constitutes a real concern and strategies to limit the use of colistin are expected shortly.

1.2 RESISTANCE TO BACTERIOCINS PRODUCED BY GRAM-POSITIVE BACTERIA (GPB)

Bacteria are frequently exposed to cationic antimicrobial peptides produced by the eukaryotic hosts, which are known as host defense peptides, or to those produced by the prokaryotic competitors known as bacteriocins [22]. Bacteriocins were first described as pore-forming AMP, causing rapid efflux of cytoplasmic compounds [23]. Resistance to bacteriocins produced by GPB involves various mechanisms, including changes in the cell wall thickness, cell-membrane composition, or genetic response, which lead to the loss of the bacteriocin-envelope affinity or the degradation of the bacteriocins [22].

According to Collins et al. [24], the mechanisms involved in the resistance to bacteriocin can be divided into two groups: the acquired resistance group, developed by a formerly susceptible strain, and the innate resistance group, which is found intrinsically in a particular genera or species. Remarkably, bacteriocins belonging to lantibiotics, at least to some of them, are active against the target strains through two modes of action. At micromolar concentration, the nisin-like lantibiotics, having a positively charged C-terminal part, could bind to the negatively charged phospholipids of the cellular membrane, causing unstable pores [25,26]. At nanomolar concentration, the nisin-like lantibiotics could bind specifically to pyrophosphate lipid II (undecaprenyl-pyrophosphoryl-MurNAc- (pentapeptide) -GlcNAc), precursor of peptidoglycan, located at the membrane of the target cell, causing disturbance of the cell wall synthesis [27,28]. Further well-characterized lantibiotics are those belonging to the lacticin 481 group, which consists of at least 16 members, including lacticin 481, streptococcin A-FF22, mutacin II, nukacin ISK-1, and salivaricins, acting via a pore-forming mechanism that causes membrane potential dissipation and intracellular compounds leakage [29,30]. Lacticin 481 group lantibiotics are known to act on lipid II, but with a mechanism that is different from that of nisin. This observation was based on in vitro studies which showed that the interaction of nisin with zwitterionic lipids is very low, whereas this interaction is very high with lacticin 481 group [30]. Remarkably, mersacidin, which has a high sequence homology with the lantibiotics of lacticin 481 group, does not interfere with the action of nisin. These two lantibiotics do not antagonize each other to bind to lipid II, proving that the regions of lipid II involved in the binding of the molecules to the cell membrane are different [30,31]. Other lantibiotics seem to be endowed of particular mode of action. Indeed, mutacin II, another lacticin 481-lantibiotic group, inhibits the enzymes involved in the energy metabolism and do not cause any pore-forming action [30]. AMP of the cinnamycin group, including duramycin B and C, act on phosphatidyle-thanolamins and phospholipase A2 [32]. Bacteria are known to develop resistance to these AMP by expressing MprF, a unique enzyme that modifies anionic phospholipids with L-lysine or L-alanine, reducing the affinity for bacteriocins by introducing positive charges into the membrane surface. The lysyl or alanyl groups are derived from aminoacyl tRNAs and are usually transferred to phosphatidylglycerol (PG). *S. aureus* produce Lys-PG, which leads to AMP resistance when an additional flippase (transmembrane lipid transporter proteins) domain of MprF is present [33]. This allows translocation of Lys-PG at the outer leaflet of the membrane and exposes it to the extra-cellular environment [33] (Fig. 1.4).

To be noted that GPB, such as *Lactobacillus casei* or *Listeria monocytogenes*, can resist nisin by producing or modifying their membrane phospholipids composition [35]. Indeed, Breuer and Radler [36] reported that production of polyanionic, phosphate-containing polysaccharides with rhamnose and galactose subunits conferred protection to *L. casei* against the bactericidal action of nisin. Mantovani and Russell [37] indicated that a strain of *Streptococcus bovis*, which was initially sensitive to nisin, can

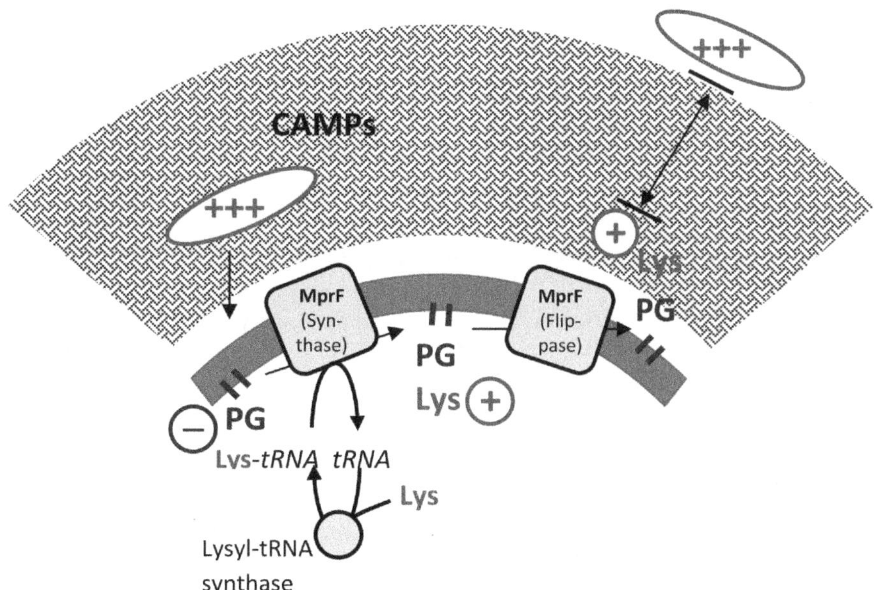

■ FIGURE 1.4   Mode of action of *Staphylococcus aureus* MprF. Lys-PG is synthesized from Lys-tRNA and PG by the synthase domain of MprF. Lys-PG can only neutralize the outer surface of the membrane upon translocation to the outer cytoplasmic membrane leaflet, which is facilitated by the flippase domain of MprF, leading to CAMPs (cationic antimicrobial peptides) repulsion. *Adapted from [34] Ernst CM, Staubitz P, Mishra NN, Yang SJ, Hornig G, Kalbacher H, et al. The bacterial defensin resistance protein MprF consists of separable domains for lipid lysinylation and antimicrobial peptide repulsion. PLoS Pathog 2009;5:e1000660.*

modify its lipoteichoic acid content to increase its resistance level to this lantibiotic. It is also known that growth conditions could play a determinant role in the emergence of resistant phenotypes, as was observed for innate bacteriocin resistance of *L. monocytogenes* 412, which exhibited higher resistance to nisin and pediocin in stationary-phase than log-phase [38]. Conversely, other studies found that the bacteriocin resistance of cells exposed to acid and osmotic stresses was not affected [39,40]. Mechanisms of molecular resistance to nisin were studied in depth for *S. aureus*, a GPB responsible for many infections [41]. Remarkably, this bacterium has two inducible ABC transporters, BraDE is involved in the bacitracin sensing and signaling through BraSR, whereas VraDE acts specifically as a detoxification module and in conferring resistance to bacitracin and nisin [42]. The ABC systems are already known to be involved in the immunity mechanism in the lantibiotics producer strains, whereas for lantibiotics, e.g., pep5 and lactocin S, the immunity is provided only by a lipoprotein called NiSi and located at the membrane, which acts by blocking the action of the lantibiotic and preventing thereof the formation of pores [43]. The expression of NiSi immunity system of *Lactococcus lactis*—a nisin producing strain—in a sensitive *Bacillus subtilis* strain conferred resistance to nisin but not to subtilin, a lantibiotic belonging to the same family of bacteriocins [43]. The expression of the gene *nuk*, conferring immunity for nukacin, in *L. lactis* strain, provides efficient resistance to both nukacin and lacticin 481, but not against nisin A [44]. The immune mimicry is a mechanism contributing to protection specifically toward bacteriocins. However, the modes of action of lantibiotics are different from one to another, which is a limiting cross-resistance phenomenon [45]. Importantly, nonbacteriocin-producing strains could carry genes coding for functional homologs of bacteriocin immunity systems, known as the orphan immunity genes [46]. This phenomenon was described for lantibiotics [47] and, previously, for class II bacteriocins [48,49]. Heterologous expression of these orphan immunity genes conferred protection against the cognate bacteriocin [46]. Another mechanism of resistance to lantibiotics involved the enzymatic degradation of the bacteriocin of this class by specific enzymes called the Dha reductase, found in different *Bacillus* strains. The principal role of the aforementioned enzymes consists in reducing dehydroalanine (Dha) to alanine (Ala) [50]. Sun et al. [51] described such mechanisms of lantibiotic resistance in nonnisin-producing *L. lactis* strains as harboring specific nisin resistance gene (nsr). This gene encodes a 35-kDa nisin resistance protein (*nsr*), conferring nisin resistance by inactivating the nisin by cutting its C-terminal "tail" [51].

Class II bacteriocins are known to cause leakage of inorganic phosphate (Pi), partial dissipation of the membrane potential, and K^+ and amino acids efflux upon pore-forming [52–55]. Of particular interest are the class IIa bacteriocins, which are small anti-listerial peptides that act directly on the membrane after their bind by electrostatic interactions, and then cause a pore-forming action through their hydrophobic C-terminal part or by recognition of a docking molecule [53]. Other factors may play a role in the activity of class IIa bacteriocins, such as the lipid composition, the pH and the charge to the membrane [35]. Class IIa bacteriocins are heat-stable peptides with low molecular weight <10 kDa and have common secondary structure organization, with hydrophobic/amphiphilic α-helix in the C-terminal part, and β-sheets structure in the N-terminal part. This part contains a conserved consensus amino-acids sequence YGNGV or YGNGL, followed by (X) $C(X)_4C(X)V(X)_4A$ (X denotes any amino acid) [53]. Positively charged residues are located mostly in the hydrophilic N-terminal region and appear to play an important role in the reconnaissance of the target cell. This allows for the binding of the bacteriocin by anchoring to a specific membrane receptor, such as the mannose permease, which acts as a docking molecule [56].

Immunity against class IIa bacteriocin was ascribed to a cytoplasmic protein which captures and sequesters the bacteriocins [57,58]. Furthermore, Johnsen et al. [59] demonstrated that the C-terminal portions of these immunity proteins have a region which interacts with the C-terminal α helical region of class IIa bacteriocins. This immunity appeared to be strong and sometimes conferred cross-resistance toward other bacteriocins, as shown for sakacin P immunity protein that conferred immunity against the action of sakacin P and pediocin PA-1/AcH [49]. Resistance to class II bacteriocins can be achieved by different strategies, leading to diverse levels of resistance. Studies of the mechanisms revealed the high stability of the resistance levels in the microorganisms developing them [60,61]. Thus, the low level of resistance in *L. monocytogenes* and *Enterococcus faecalis*, with a minimal inhibition concentration (MIC) of 2- to 4-fold compared to the sensitive strain, was attributed to alteration occurring in the membrane lipid composition [35,62]. It is to be noted that *L. monocytogenes* are able to easily develop a nonsensitive variant to class IIa bacteriocins [35]. This resistance was conferred to the cell membrane plasticity resulting from the modification in the fatty acid composition and resistant strains displayed higher C16:0 content, contributing therefore to the membrane rigidity, and thereby impeding the AMP insertion [35]. Studies performed on *E. faecalis* have established that the composition of the unsaturated fatty acids increase significantly in the presence of mundticin KS [63]. Resistant strains have cell membranes with higher amounts of amino-containing phospholipid and lower amounts of phosphatidylglycerol and cardiolipin than cell membranes of sensitive strains [63]. Another level of resistance to class IIa bacteriocins, designed as an intermediary level of resistance, was reported in *L. monocytogenes* and *E. faecalis*. Indeed, these strains displayed in this case MIC values ranging from 100- to 500-fold increase comparative to the wild-type (sensitive strain). Calvez et al. [64] reported three genes associated with the intermediate resistance of *E. faecalis* JH2-2 to divercinV41 that is a class IIa bacteriocin. These genes were: the *rpoN* gene coding for the σ^{54} factor, an alternative subunit of RNA polymerase; a *glpQ* gene coding for a putative glycerophosphoryl diester phosphoesterase (GlpQ) and a *pde* gene coding for a putative phosphoesterase (PDE), which belongs to the DHH proteins family. However the resistance conferred by mutations on these three genes appeared graduated. The *rpoN* mutant appeared more resistant than the *glpQ* mutant, which was more resistant than the *pde* mutant. Interestingly, all these mutations induced resistance to class IIa bacteriocins only [64]. The high level of resistance to class IIa bacteriocins resulted from the inactivation of the *mptACD* operon, which encodes the EIIMant mannose permease of the phosphotransferase system (PTS) [65,66]. Molecular studies unveiled that inactivation of *rpoN* gene in *L. monocytogenes* and *E. faecalis* conferred resistance to mesentericin Y105; a class IIa bacteriocin produced by *Leuconostoc mesenteroides* Y105 [67,68]. The σ^{54} factor controls the transcription of various genes, among which the operon encoding an *mpt* mannose permease involved in the transport of sugars type PTS (phosphotransferase system) and which was described as a docking molecule for class IIa bacteriocins [53,65]. This conferred a high-level resistance to class IIa bacteriocin in *L. monocytogenes* and *E. faecalis*, inducing a very important increase of MIC value, reaching more than 1000-fold more than that of the sensitive strains [69]. Moreover, Fimland et al. [48] reported the existence of an interaction between the class IIa bacteriocin and the target membrane mediated by tryptophan residues "Trp18, Trp33 and Trp41" and the C-terminal part of the peptide. The mannose permease is composed of four subunits, IIAMan, IIBMan, IICMan, and IIDMan, and sometimes by only three subunits, when the subunits A and B are combined. The σ^{54} factor positively regulates the expression of the *mpt* operon in association with an activator protein named ManR in *E. faecalis* and MptR in *L. monocytogenes*. Inactivation of the genes encoding these activators, as well as those encoding IIAMan subunits IIBMan and mannose transport protein IIDMan, induced the emergence of a resistant phenotype toward the mesentericin Y105 [66]. The subunit IIDMan was identified as the docking molecule for class IIa bacteriocins [65,66]. The heterologous expression of the membrane subunit IICMan of *L. monocytogenes* in *L. lactis* rendered this bacterium sensitive to class IIa bacteriocins. Class IIa bacteriocins were found to bind to this membrane subunit, allowing the pore formation, modifying the membrane permeability and leading to the cell death [53].

Similarly, class IIb bacteriocins, called two-peptide bacteriocins, of which inhibitory activity requires the combined action of two individual peptides, usually act by pore-forming leading to the dissipation of the proton motive force and leak of K^+ ions [23,70,71]. The cationic charges of the peptides might interact with the anionic charges of the membrane through electrostatic interactions, and form pores by the penetration of the hydrophobic α helices into the membrane [53]. Formerly, it was established that a docking molecule was required for the activity of these bacteriocins [72]. The immunity of the class IIb bacteriocins producing strains is conferred by a single immunity protein located at the cell membrane which acts as ABC transport pump system permitting the detoxification by efflux of the bacteriocin [73,74].

Recently, Kjos et al. [75] showed that mutations in *uppP/bacA* gene in sensitive *L. lactis* strains led to a high level of resistance to lactococcin G. The *uppP* (*bacA*) gene encodes an undecaprenyl pyrophosphate phosphatase, a membrane protein involved in peptidoglycan synthesis, which seem to be a target for lactococcin G. This was confirmed by the expression of the Lactococcal *uppP* gene in the non-sensitive *Streptococcus pneumoniae*, which converted this host to a sensitive phenotype thereby inhibiting by lactococcin G [75]. Studies on class IIc bacteriocins, known as circular bacteriocins, were performed mainly on enterocin AS-48, which is extremely active against GPB [76]. Class IIc bacteriocins act on the cell membrane by causing the leakage of the intracellular compounds [77]. The study of the mode of action realized on gassericin A [78], enterocin AS-48 [79], and A carnocyclin [80] showed that these circular cationic bacteriocins bind to the cell membrane and lead to pore-forming spontaneously upon insertion into the bacterial membranes. Further, the immunity to circular bacteriocins is conferred by a combination of an immunity protein and the action of an efflux ABC transporter system [81]. Enterocin AS-48 is highly active on all *Listeria* species; however, *Listeria* AS-48-adapted cells can appear spontaneously at sub-inhibitory concentrations [82]. These adapted *Listeria* cells exhibited a cross-resistance to nisin and showed an increased resistance to muramidases. Similarly to class IIa bacteriocins, the adapted cells could carry modifications on their membrane fatty acid composition, mainly harboring a high amount of branched fatty acids and C15:0 An-to-C17:0 [82].

Finally, the class IId bacteriocins have multiple modes of action. Lactococcin A induced variations and dissipation of the membrane potential causing thereof a loss of the intracellular amino acids upon the pore formation [83]. This phenomenon suggested a membrane receptor, as this peptide was not active on the liposomes [83]. Another bacteriocin belonging to this group called the lactococcin 972, which is secreted by *L. lactis* IPLA 972, acts on the biosynthesis of the cell wall itself [84], by inhibiting a precursor of the wall, the *N*-acetylglucosamin, causing defect to the wall integrity. This same bacteriocin seems to possess a second mode of action that consists in disturbing the septum formation during the cell division [84]. Recently, a study on *L. lactis* mutants resistant to lactococcin 972 characterized the mechanism involved in the resistance to this bacteriocin. The gene *llmg2447*, which may encode a putative extracytoplasmic function (ECF) anti-sigma factor, is located downstream of the non-functional ECF gene sigX(pseudo) and could confer high lactococcin 972 resistance in *L. lactis* [85]. The putative membrane protein, encoded by *llmg2447*, was identified as a main factor in resistance to the cell wall-active lactococcin 972, although the way in which it protects *L. lactis* remains unknown. Further attention should be given to the presence of an homologous system in *Streptococcus agalactiae*, a bacterium often involved in mastitis and other infectious diseases [85].

Class III bacteriocins, so-called bacteriolysins, are larger than 10 kDa and heat-sensitive proteins. The structure and the mode of action of these bacteriocins are completely different from those of the other classes of bacteriocins. Helveticin J produced by *Lactobacillus helveticus* [86,87], the zoocin A produced by *Streptococcus zooepidermicus* [88], the enterolysin A produced by *E. faecalis* [89], the millericin B produced by *Streptococcus milleri* [90], and the M18 linocine produced by *Brevibacterium linens* [91] are the most known bacteriolysins. The target of these bacteriocins is assumed to be the cell wall, and the action is the hydrolysis of peptidoglycan [88,89]. Interestingly, the zoocin A has a narrow spectrum of action, whereas enterolysin A and B millericin have a wider spectrum of activity [88–90]. Study on sensitivity/resistance to enterolysin A showed relation with binding ability/disability of its cell wall binding (CWB) domain. Absence of the specific receptor on the cell surface, which is a lipoteichoic acid necessary for binding of the enzyme molecule, is sufficient to obtain a resistant phenotype to such bacteriocins [92].

1.3   CONCLUSION AND PROSPECTS

Although resistance developed against polymyxins and bacteriocins remains to be deeply studied, recent studies show the potential of their combination as promising drug therapies [93–96]. In addition to their synergism, the combination of colistin and bacteriocins was reported to diminish colistin cytotoxicity toward mammalian cells [96]. Previously, Naghmouchi et al. [95] unveiled the benefits allocated to the combination of class I and IIa bacteriocins, nisin A and pediocin PA-1/AcH, respectively, with polymyxin-E on resistant strains of Gram-positive *L. monocytogenes* and Gram-negative *E. coli* strains. More recently, Al Atya et al. [97] demonstrated that colistin and carbapenems combinations could hamper the growth, at least in vitro, of the extended spectrum β-lactamase (ESBL) producing *E. coli* strains isolated from children with urinary tract infections [97].

Another promising concept for overcoming bacterial resistance is the use of nanotechnology. Indeed, the nanoparticles have emerged as an efficient means of fight drug-resistant bacteria grown in planktonic and biofilm cultures [98]. The mode of action of nanoparticles is dependent on their physico-chemical properties, particularly their surface properties which are highly versatile and can be easily modulated through ligand engineering to generate particles with new and emergent properties [98]. Size, charge, and electrostatic properties seem to be critical factors in the antimicrobial activity of nanoparticles [99–101]. Early studies of nanoparticle-microbe interactions show that cationic gold nanoparticles (AuNPs) were endowed with toxicity against bacteria, due to the hydrophobic and cationic nature of AuNPs which develop spatiotemporal aggregate patterns on the bacterial surface [102]. Then, Mahmoudi and Serpooshan [103] successfully functionalized silver nanoparticles ring-coated in superparamagnetic iron oxide nanoparticles (SPIONS) associated with gaps, demonstrating high antimicrobial activity and good compatibility with healthy cells [103]. Currently, nanoparticles appear to be versatile materials in antimicrobial treatment process design, with modifiable surface properties enabling suitable actions for targeting specific sites of microorganisms, and opening up a wide area of application as a promising therapeutic approach in the near future.

ACKNOWLEDGMENTS

Research on colistin and alternatives to colistin was supported by ANR research grant through a collaborative project to DD (ANR Sincolistin).

REFERENCES

[1] Yu Z, Qin W, Lin J, Fang S, Qiu J. Antibacterial mechanisms of polymyxin and bacterial resistance. BioMed Res Int 2015:679109. http://dx.doi.org/10.1155/2015/679109.

[2] Kempf I, Fleury MA, Drider D, Bruneau M, Sanders P, Chauvin C, et al. What do we know about resistance to colistin in *Enterobacteriaceae* in avian and pig production in Europe? Int J Antimicrob Agents 2013;42:379–83.

[3] Catry B, Cavaleri M, Baptiste K, Grave K, Grein K, Holm A, et al. Use of colistin-containing products within the European Union and European Economic Area (EU/EEA): development of resistance in animals and possible impact on human and animal health. Int J Antimicrob Agents 2015;46:297–306.

[4] Velkov T, Thompson PE, Nation RL, Li J. Structure-activity relationships of poly-myxin antibiotics. J Med Chem 2010;53(5):1898–916.

[5] Biswas S, Brunel JM, Dubus JC, Reynaud-Gaubert M, Rolain JM. Colistin: an update on the antibiotic of the 21st century. Expert Rev Anti-infect Ther 2012;10(8):917–34.

[6] Moskowitz SM, Ernst RK, Miller SI. PmrAB, a two-component regulatory system of *Pseudomonas aeruginosa* that modulates resistance to cationic antimicrobial peptides and addition of aminoarabinose to lipid A. J Bacteriol 2004;186(2):575–9.

[7] Breazeale SD, Ribeiro AA, McClerren AL, Raetz CRH. A formyltransferase required for polymyxin resistance in *Escherichia coli* and the modification of lipid A with 4-amino-4-deoxy-L-arabinose: identification and function of UDP-4-deoxy-4-formamido-L-arabinose. J Biol Chem 2005;280(14):14154–67.

[8] Jayol A, Poirel L, Brink A, Villegas MV, Yilmaz M, Nordmann P. Resistance to colistin associated with a single amino acid change in protein PmrB among *Klebsiella pneumoniae* isolates of worldwide origin. Antimicrob Agents Chemother 2014;58:4762–6.

[9] Olaitan AO, Morand S, Rolain JM. Mechanisms of polymyxin resistance: acquired and intrinsic resistance in bacteria. Front Microbiol 2014;5:643. http://dx.doi.org/10.3389/fmicb.2014.00643.

[10] Poirel L, Jayol A, Bontron S, Villegas MV, Ozdamar M, Turkoglu S, et al. The mgrB gene as a key target for acquired resistance to colistin in *Klebsiella pneumoniae*. J Antimicrob Chemother 2015;70:75–80.

[11] Formosa C, Herold M, Vidaillac C, Duval RE, Dague E. Unravelling of a mechanism of resistance to colistin in *Klebsiella pneumoniae* using atomic force microscopy. J Antimicrob Chemother 2015;70:2261–70.

[12] Padilla E, Llobet E, Domenech-Sanchez A, Martinez-Martinez L, Bengoechea JA, Alberti S. *Klebsiella pneumoniae* AcrAB efflux pump contributes to antimicrobial resistance and virulence. Antimicrob Agents Chemother 2010;54:177–83.

[13] Moffatt JH, Harper M, Adler B, Nation RL, Li J, Boyce JD. Insertion sequence ISAba11 is involved in colistin resistance and loss of lipopolysaccharide in *Acinetobacter baumannii*. Antimicrob Agents Chemother 2011;55:3022–4.

[14] Lesho E, Yoon EJ, McGann P, Snesrud E, Kwak Y, Milillo M, et al. Emergence of colistin-resistance in extremely drug-resistant *Acinetobacter baumannii* containing a novel *pmrCAB* operon during colistin therapy of wound infections. J Infect Dis 2013;208:1142–51.

[15] Pelletier MR, Casella LG, Jones JW, Adams MD, Zurawski DV, Hazlett KR, et al. Unique structural modifications are present in the lipopolysaccharide from colistin-resistant strains of *Acinetobacter baumannii*. Antimicrob Agents Chemother 2013;57:4831–40.

[16] Beceiro A, Moreno A, Fernandez N, Vallejo JA, Aranda J, Adler B, et al. Biological cost of different mechanisms of colistin resistance and their impact on virulence in *Acinetobacter baumannii*. Antimicrob Agents Chemother 2014;58:518–26.

[17] Liu YY, Wang Y, Walsh TR, Yi LX, Zhang R, Spencer J, et al. Emergence of plasmid-mediated colistin resistance mechanism MCR-1 in animals and human beings in China: a microbiological and molecular biological study. Lancet Infect Dis 2015;16(2):161–8. doi:10.1016/S1473-3099(15)00424-7.

[18] Hasman H, Hammerum AM, Hansen F, Hendriksen RS, Olesen B, Agerso Y, et al. Detection of mcr-1 encoding plasmid-mediated colistin-resistant *Escherichia*

coli isolates from human bloodstream infection and imported chicken meat, Denmark 2015. Euro Surveill 2015;20(49). http://dx.doi.org/10.2807/1560-7917. ES.2015.20.49.30085.

[19] Malhotra-Kumar S, Xavier BB, Das AJ, Lammens C, Hoang HT, Pham NT, et al. Colistin-resistant *Escherichia coli* harbouring mcr-1 isolated from food animals in Hanoi, Vietnam. Lancet Infect Dis 2016;16(3):286–7. doi:10.1016/S1473-3099(16)00014-1.

[20] Olaitan AO, Chabou S, Okdah L, Morand S, Rolain JM. Dissemination of the mcr-1 colistin resistance gene. Lancet Infect Dis 2015;16(2):147. doi:10.1016/S1473-3099(15)00540-X.

[21] Webb HE, Granier SA, Marault M, Millemann Y, den Bakker HC, Nightingale KK, et al. Dissemination of the mcr-1 colistin resistance gene. Lancet Infect Dis 2015;16(2):144–5. doi:10.1016/S1473-3099(15)00538-1.

[22] Bastos MC, Coelho ML, Santos OC. Resistance to bacteriocins produced by Gram-positive bacteria. Microbiology 2015;161:683–700.

[23] Moll GN, Clark J, Chan WC, Bycroft BW, Roberts GC, Konings WN, et al. Role of transmembrane pH gradient and membrane binding in nisin pore formation. J Bacteriol 1997;179:135–40.

[24] Collins B, Guinane CM, Cotter PD, Hill C, Ross RP. Assessing the contributions of the LiaS histidine kinase to the innate resistance of *Listeria monocytogenes* to nisin, cephalosporins, and disinfectants. Appl Environ Microbiol 2012;78:2923–9.

[25] Driessen AJ, van den Hooven HW, Kuiper W, van de Kamp M, Sahl HG, Konings RN, et al. Mechanistic studies of lantibiotic-induced permeabilization of phospholipid vesicles. Biochemistry 1995;34:1606–14.

[26] Moll GN, Konings WN, Driessen AJ. The lantibiotic nisin induces transmembrane movement of a fluorescent phospholipid. J Bacteriol 1998;180(24):6565–70.

[27] Wiedemann I, Breukink E, van Kraaij C, Kuipers OP, Bierbaum G, de Kruijff B, et al. Specific binding of nisin to the peptidoglycan precursor lipid II combines pore formation and inhibition of cell wall biosynthesis for potent antibiotic activity. J Biol Chem 2001;276:1772–9.

[28] Breukink E, de Kruijff B. The lantibiotic nisin, a special case or not? Biochim Biophys Acta 1999;1462:223–4.

[29] Wiedemann I, Böttiger T, Bonelli RR, Schneider T, Sahl HG, Martínez B. Lipid II-based antimicrobial activity of the lantibiotic plantaricin C. Appl Environ Microbiol 2006;72(4):2809–14.

[30] Dufour A, Hindré T, Haras D, Le Pennec J-P. The biology of lantibiotics from the lacticin 481group is coming of age. FEMS Microbiol Rev 2007;31:134–67.

[31] Diep DB, Nes IF. Ribosomally synthesized antibacterial peptides in Gram-positive bacteria. Curr Drugs Targets 2002;3:107–22.

[32] Märki F, Hänni E, Fredenhagen A, Oostrum J. Mode of action of the lanthionine-containing peptide antibiotics duramycin, duramycin B and C, and cinnamycin as indirect inhibitors of phospholipase A2. Biochem Pharmacol 1991;42(10):2027–35.

[33] Ernst CM, Peschel A. Broad-spectrum antimicrobial peptide resistance by MprF-mediated aminoacylation and flipping of phospholipids. Mol Microbiol 2011;80:290–9.

[34] Ernst CM, Staubitz P, Mishra NN, Yang SJ, Hornig G, Kalbacher H, et al. The bacterial defensin resistance protein MprF consists of separable domains for lipid lysinylation and antimicrobial peptide repulsion. PLoS Pathog 2009;5:e1000660.

[35] Naghmouchi K, Drider D, Kheadr E, Lacroix C, Prevost H, Fliss I. Multiple characterizations of *Listeria monocytogenes* sensitive and insensitive variants to divergicin M35, a new pediocin-like bacteriocin. J Appl Microbiol 2006;100:29–39.

[36] Breuer B, Radler F. Inducible resistance against nisin in *Lactobacillus casei*. Arch Microbiol 1995;65:114–8.

[37] Mantovani HC, Russell JB. Nisin resistance of *Streptococcus bovis*. Appl Environ Microbiol 2001;67(2):808–13.

[38] Jydegaard AM, Gravesen A, Knøchel S. Growth condition-related response of *Listeria monocytogenes* 412 to bacteriocin inactivation. Lett Appl Microbiol 2000;31:68–72.

[39] Begley M, Cotter PD, Hill C, Ross RP. Glutamate decarboxylase-mediated nisin resistance in *Listeria monocytogenes*. Appl Environ Microbiol 2010;76:6541–6.

[40] van Schaik W, Gahan CG, Hill C. Acid-adapted *Listeria monocytogenes* displays enhanced tolerance against the lantibiotics nisin and lacticin 3147. J Food Protect 1999;62:536–9.

[41] Chipolombwe J, Török ME, Mbelle N, Nyasulu P. Methicillin-resistant *Staphylococcus aureus* multiple sites surveillance: a systemic review of the literature. Infect Drug Resist 2016;12:35–42.

[42] Hiron A, Falord M, Valle J, Débarbouillé M, Msadek T. Bacitracin and nisin resistance in *Staphylococcus aureus*: a novel pathway involving the BraS/BraR two-component system (SA2417/Bacteriocin resistance mechanisms SA2418) and both the BraD/BraE and VraD/VraE ABC transporters. Mol Microbiol 2011;81:602–22.

[43] Stein T, Heinzmann S, Solovieva I, Entian KD. Function of *Lactococcus lactis* nisin immunity genes nisI and nisFEG after coordinated expression in the surrogate host *Bacillus subtilis*. J Biol Chem 2003;278(1):89–94.

[44] Aso Y, Sashihara T, Nagao J, Kanemasa Y, Koga H, Hashimoto T, et al. Characterization of a gene cluster of *Staphylococcus warneri* ISK-1 encoding the biosynthesis of and immunity to the lantibiotic, nukacin ISK-1. Biosci Biotechnol Biochem 2004;68:1663–71.

[45] Draper LA, Cotter PD, Hill C, Ross RP. Lantibiotic resistance. Microbiol Mol Biol Rev 2015;79:171–91.

[46] Draper LA, Tagg JR, Hill C, Cotter PD, Ross RP. The *spiFEG* locus in *Streptococcus infantarius* subsp. *infantarius* BAA-102 confers protection against nisin U. Antimicrob Agents Chemother 2012;56:573–8.

[47] Draper LA, Ross RP, Hill C, Cotter PD. Lantibiotic immunity. Curr Protein Pept Sci 2008;9:39–49.

[48] Fimland G, Eijsink VG, Nissen-Meyer J. Mutational analysis of the role of tryptophan residues in an antimicrobial peptide. Biochemistry 2002;41:9508–15.

[49] Fimland G, Eijsink VG, Nissen-Meyer J. Comparative studies of immunity proteins of pediocin-like bacteriocins. Microbiology 2002;148:3661–70.

[50] Sonomoto K, Yokota A. Lactic acid bacteria and bifidobacteria: current progress in advanced research. Norfolk, UK: Horizon Scientific Press; 2011.2011.286 pages.

[51] Sun Z, Zhong J, Liang X, Liu J, Chen X, Huan L. Novel mechanism for nisin resistance via proteolytic degradation of nisin by the nisin resistance protein NSR. Antimicrob Agents Chemother 2009;53:1964–73.

[52] Chen Y, Montville TJ. Efflux of ions and ATP depletion induced by pediocin PA-1 are concomitant with cell death in *Listeria monocytogenes* Scott A. J Appl Bacteriol 1995;79:684–90.

[53] Drider D, Fimland G, Héchard Y, McMullen LM, Prévost H. The Continuing story of class IIa bacteriocins. Microbiol Mol Biol Rev 2006;70:564–82.

[54] Herranz C, Chen Y, Chung HJ, Cintas LM, Hernandez PE, Montville TJ, et al. Enterocin P selectively dissipates the membrane potential of *Enterococcus faecium* T136. Appl Environ Microbiol 2001;67:1689–92.

[55] Herranz C, Cintas LM, Hernandez PE, Moll GN, Driessen AJ. Enterocin P causes potassium ion efflux from *Enterococcus faecium* T136 cells. Antimicrob Agents Chemother 2001;45:901–4.

[56] Yan LZ, Gibbs AC, Stiles ME, Wishart DS, Vederas JC. Analogues of bacteriocins: antimicrobial specificity and interactions of leucocine A with its enantiomer, carnobacteriocin B2, and truncated derivatives. J Med Chem 2000;43:4579–8451.

[57] Diep DB, Skaugen M, Salehian Z, Holo H, Nes IF. Common mechanisms of target cell recognition and immunity for class II bacteriocins. Proc Natl Acad Sci USA 2007;104:2384–9.

[58] Eijsink VGH, Axelsson L, Diep DB, Håvarstein LS, Holo H, Nes IF. Production of bacteriocins by lactic acid bacteria; an example of biological warfare and communication. Antonie Leeuwenhoek 2002;81:639–54.

[59] Johnsen L, Dalhus B, Leiros I, Nissen-Meyer J. A bacterial Immunity protein conferring immunity to the antimicrobial activity of the pediocin-like bacteriocin enterocin A. J Biol Chem 2005;280:19045–50.

[60] Dykes GA, Hastings JW. Fitness costs associated with class IIa bacteriocin resistance in *Listeria monocytogenes* B73. Lett Appl Microbiol 1998;26:5–8.

[61] Rekhif N, Atrih A, Lefebvre G. Selection and properties of spontaneous mutants of *Listeria monocytogenes* ATCC 15313 resistant to different bacteriocins produced by lactic acid bacteria strains. Curr Microbiol 1994;28:237–41.

[62] Vadyvaloo V, Arous S, Gravesen A, Héchard Y, Chauhan-Haubrock R, Hastings JW, et al. Cell-surfacealterations in class IIa bacteriocin-resistant *Listeria monocytogenes* strains. Microbiology 2004;150:3025–33.

[63] Sakayori Y, Muramatsu M, Hanada S, Kamagata Y, Kawamoto S, Shima J. Characterization of *Enterococcus faecium* mutants resistant to mundticin KS, a class IIa bacteriocin. Microbiology 2003;149(10):2901–8.

[64] Calvez S, Rincé A, Auffray Y, Prévost H, Drider D. Identification of new genes associated with intermediate resistance of *Enterococcus faecalis* to divercin V41, a pediocin-like bacteriocin. Microbiology UK 2007;153:1609–18.

[65] Dalet K, Cenatiempo Y, Cossart P, Héchard Y. A s54-dependent PTS permease of the mannose family is responsible for sensitivity of *Listeria monocytogenes* to mesentericin Y105. Microbiology UK 2001;147:3263–9.

[66] Héchard Y, Pelletier C, Cenatiempo Y, Frère J. Analysis of θ54-dependent genes in *Enterococcus faecalis*: a mannose PTS permease (EIIMan) is involved in sensitivity to a bacteriocin, mesentericin Y105. Microbiology UK 2001;147:1575–80.

[67] Héchard Y, Derijard B, Letellier F, Cenatiempo Y. Characterization and purification of mesentericin Y105, an anti-*Listeria* bacteriocin from *Leuconostoc mesenteroides*. J Gen Microbiol 1992;138:2725–31.

[68] Robichon D, Gouin E, Debarbouille M, Cossart P, Cenatiempo Y, Hechard Y. The rpoN (σ^{54}) gene from *Listeria monocytogenes* is involved in resistance to mesentericin Y105, an antibacterial peptide from *Leuconostoc mesenteroides*. J Bacteriol 1997;179(23):7591–4.

[69] Calvez S, Prévost H, Drider D. Relative expression of genes involved in the resistance/sensitivity of *Enterococcus faecalis* JH2-2 to recombinant divercin RV41. Biotechnol Lett 2008;30:1795–800.

[70] Castellano P, Raya R, Vignolo G. Mode of action of lactocin 705, a two-component bacteriocin from *Lactobacillus casei* CRL705. Int J Food Microbiol 2003;85:35–43.

[71] Oppegard C, Rogne P, Emanuelsen L, Kristiansen PE, Fimland G, Nissen-Meyer J. The two-peptide class II bacteriocins: structure, production, and mode of action. J Mol Microbiol Biotechnol 2007;13:210–9.

[72] Hauge HH, Nissen-Meyer J, Nes IF, Eijsink VG. Amphiphilic alpha-helices are important structural motifs in the alpha and beta peptides that constitute the bacteriocin lactococcin G-enhancement of helix formation upon alpha-beta interaction. Eur J Biochem 1998;251:565–72.

[73] Cintas LM, Casaus P, Herranz PE, Håvarstein LS, Holo H, Hernández PE, et al. Biochemical and genetic evidence that *Enterococcus faecium* L50 produces enterocins L50A and L50B, the sec-dependent enterocin P, and a novel bacteriocin secreted without an N-terminal extension termed enterocin Q. J Bacteriol 2000;182:6806–14.

[74] Nissen-Meyer J, Rogne P, Oppegard C, Haugen HS, Kristiansen PE. Structure-function relationships of the non-lanthionine-containing peptide (class II) bacteriocins produced by Gram-positive bacteria. Curr Pharm Biotechnol 2009;10:19–37.

[75] Kjos M, Oppegård C, Diep DB, Nes IF, Veening JW, Nissen-Meyer J, et al. Sensitivity to the two-peptide bacteriocin lactococcin G is dependent on UppP, an enzyme involved in cell-wall synthesis. Mol Microbiol 2014;92(6):1177–87.

[76] Abriouel H, Valdivia E, Gálvez A, Maqueda M. Response of *Salmonella choleraesuis* LT2 spheroplasts and permeabilized cells to the bacteriocin AS-48. Appl Environ Microbiol 1998;64:4623–6.

[77] Martin-Visscher LA, van Belkum JM, Vederas JC. Genetics, biosynthesis structure and mode of action of AMP from Gram-positive bacteria. Class IIc or circular bacteriocins Drider D, Rebuffat S, editors. prokaryotic and antimicrobial peptides: from genes to biotechnologies. New-York, NY: Springer; 2010.

[78] Kawai Y, Kemperman R, Kok J, Saito T. The circular bacteriocins gassericin A and circularin A. Curr Protein Pept Sci 2004;5:393–8.

[79] Gálvez A, Maqueda M, Martinez-Bueno M, Valdivia E. Permeation of bacterial cells, permeation of cytoplasmic and artificial membrane vesicles, and channel formation on lipid bilayers by peptide antibiotic AS-48. J Bacteriol 1991;173:886–92.

[80] Gong X, Martin-Visscher LA, Nahirney D, Vederas JC, Duszyk M. The circular bacteriocin, carnocyclin A, forms anion-selective channels in lipid bilayers. Biochim Biophys Acta 2009;1788:1797–803.

[81] Franz CMAP, van Belkum MJ, Holzapfel WH, Abriouel H, Gálvez A. Diversity of enterococcal bacteriocins and their grouping in a new classification scheme. FEMS Microbiol Rev 2007;31:293–310.

[82] Mendoza F, Maqueda M, Gálvez A, Martínez-Bueno M, Valdivia E. Antilisterial activity of peptide AS-48 and study of changes induced in the cell envelope properties of an AS-48-adapted strain of *Listeria monocytogenes*. Appl Environ Microbiol 1999;65(2):618–25.

[83] Van Belkum MJ, Kok J, Venema G, Holo H, Nes IF, Konings WN, et al. The bacteriocin lactococcin A specifically increases permeability of lactococcal cytoplasmic

membranes in a voltage-independent, protein-mediated manner. J Bacteriol 1991;173(24):7934–41.

[84] Martinez B, Fernandez M, Suarez JE, Rodriguez A. Synthesis of lactococcin 972, a bacteriocin produced by *Lactococcus lactis* IPLA 972, depend on the expression of a plasmid-encoded bicistronic operon. Microbiology 1999;145:3155–61.

[85] Roces C, Pérez V, Campelo AB, Blanco D, Kok J, Kuipers OP, et al. The putative lactococcal extracytoplasmic function anti-sigma factor *llmg2447* determines resistance to the cell wall-active bacteriocin lcn972. Antimicrob Agents Chemother 2012;56(11):5520–7.

[86] Joerger MC, Klaenhammer TR. Characterization and purification of helveticin J and evidence for a chromosomally determined bacteriocin produced by *Lactobacillus helveticus* 481. J Bacteriol 1986;167:439–46.

[87] Joerger MC, Klaenhammer TR. Cloning, expression, and nucleotide sequence of the *Lactobacillus helveticus* 481 gene encoding the bacteriocin helveticin. J Bacteriol 1990;172:6339–47.

[88] Simmonds RS, Simpson WJ, Tagg JR. Cloning and sequence analysis of zooA, a *Streptococcus zooepidemicus* gene encoding a bacteriocin-like inhibitory substance having a domain structure similar to that of lysostaphin. Gene 1997;189:255–61.

[89] Hickey RM, Twomey DP, Ross RP, Hill C. Production of enterolysin A by a raw milk enterococcal isolate exhibiting multiple virulence factors. Microbiology 2003;149:655–64.

[90] Beukes M, Hastings JW. Self-protection against cell wall hydrolysis in *Streptococcus milleri* NMSCC 061 and analysis of the millericin B operon. Appl Environl Microbiol 2001;67:3888–96.

[91] Valdes-Stauber N, Scherer S. Isolation and characterization of Linocin M18, a bacteriocin produced by *Brevibacterium linens*. Appl Environ Microbiol 1994;60:3809–14.

[92] Malinicova L, Dubikova K, Piknova M, Pristas P, Javorsky P. Peptidoglycan hydrolase enterolysin a recognizes lipoteichoic acid chains in the cell walls of sensitive bacteria. Protein Pept Lett 2012;19(9):924–9.

[93] Draper LA, Cotter PD, Hill C, Ross RP. The two peptide lantibiotic lacticin 3147 acts synergistically with polymyxin to inhibit Gram-negative bacteria. BMC Microbiol 2013;26(13):212.

[94] Lebel G, Piché F, Frenette M, Gottschalk M, Grenier D. Antimicrobial activity of nisin against the swine pathogen *Streptococcus suis* and its synergistic interaction with antibiotics. Peptides 2013;50:19–23. http://dx.doi.org/10.1016/j.peptides.2013.09.014.

[95] Naghmouchi K, Belguesmia Y, Baah J, Teather R, Drider D. Antibacterial activity of class I and IIa bacteriocins combined with polymyxine E against resistant variants of *Listeria monocytogenes* and *Escherichia coli*. Res Microbiol 2011;162:99–107.

[96] Naghmouchi K, Baah J, Hober D, Jouy E, Rubrecht C, Sané F, et al. Synergistic effect between colistin and bacteriocins in controlling Gram-negative pathogens and their potential to reduce antibiotic toxicity in mammalian epithelial cells. Antimicrob Agents Chemother 2013;57:2719–25.

[97] Al Atya AK, Drider-Hadiouche K, Vachee A, Drider D. Potentialization of β-lactams with colistin: in case of extended spectrum β-lactamase producing *Escherichia coli* strains isolated from children with urinary infections. Res Microbiol 2016;167(3):215–21.

[98] Gupta A, Landis RF, Rotello VM. Nanoparticle-Based Antimicrobials: Surface Functionality is Critical. F1000Res 2016;16:5. pii: F1000 Faculty Rev-364. http://dx.doi.org/10.12688/f1000research.7595.1. eCollection 2016.

[99] Jiang Z, Le ND, Gupta A, Rotello VM. Cell surface-based sensing with metallic nanoparticles. Chem Soc Rev 2015;44:4264–74.

[100] Li X, Yeh YC, Giri K, Mout R, Landis RF, Prakash YS, et al. Control of nanoparticle penetration into biofilms through surface design. Chem Commun (Cambridge England) 2015;51(2):282–5.

[101] Peulen TO, Wilkinson KJ. Diffusion of nanoparticles in a biofilm. Environ Sci Technol 2011;45(8):3367–73.

[102] Goodman CM, McCusker CD, Yilmaz T, Rotello VM. Toxicity of gold nanoparticles functionalized with cationic and anionic side chains. Bioconjugate Chem 2004;15:897–900.

[103] Mahmoudi M, Serpooshan V. Silver-coated engineered magnetic nanoparticles are promising for the success in the fight against antibacterial resistance threat. ACS Nano 2012;6(3):2656–64.

The Role of the Food Chain in the Spread of Antimicrobial Resistance (AMR)

L. Ruiz[1] and A. Alvarez-Ordóñez[2]

[1]Complutense University of Madrid, Madrid, Spain [2]University of León, León, Spain

CHAPTER OUTLINE

2.1 INTRODUCTION

Antimicrobials have been used in human and veterinary medicine for more than 70 years and have greatly contributed to both the fight against pathogenic bacteria and improvements in human health. However, in recent decades, the emergence and spread of antimicrobial-resistant bacteria, with reduced sensitivity or complete insensitivity to one or more antimicrobial agents, has started to challenge clinicians and researchers alike. Antimicrobial resistance (AMR) has been defined as a global pandemic [1], one of the major challenges of the 21st century [2], and a potential worldwide catastrophe [3]. In the European Union, an estimated 25,000 patients die yearly and about €1.5 billion are spent on additional healthcare costs related to antibiotic-resistant bacteria [4], while in the United States it is estimated that 2 million people are infected yearly with antimicrobial-resistant bacteria and that there are 23,000 associated deaths per year [5]. Estimates for the number of infections and deaths associated with AMR in other countries are not available but are expected to be of similar magnitude. Acquired resistance to some antimicrobials is already widespread to such an extent that their value for the treatment of life-threatening infections is already compromised, e.g., tetracycline and sulfonamide [6]. Moreover, the reliability of some other newer antimicrobials that play a major role in the treatment of serious infections nowadays and for which good alternative treatments may be limited (listed by [7]) is already reduced or under serious threat because of increasing levels of acquired resistance. Indeed, for any new antimicrobial class it is almost inevitable that at some point a spontaneous random genetic mutation will result in acquisition of resistance, leading to a new variant of a previously very sensitive microorganism that is less vulnerable to the corresponding antimicrobial. Furthermore, the presence of the antibiotic itself represents a strong selective pressure, favoring selective enrichment of resistant derivative strains that outcompete the original sensitive ones. Unlike intrinsic resistance typically encountered in all strains of a particular taxonomic group and presumed not to be transferable, acquired resistance, which is not present in all strains of a particular group but is acquired through horizontal gene transfer (HGT) of AMR genes located in mobile genetic elements, may be easily transferred within a population or even between different species. Understanding the genetic determinants of AMR is therefore crucial in evaluating the risk of AMR spread, since certain genetic elements are more prone to HGT and therefore can potentially spread to phylogenetically distant bacteria encountered in the same habitat [8] (Fig. 2.1).

The high priority given to the AMR challenge is reflected in high level policy documents such as the European Commission Action Plan against the rising threat of AMR [9] and the WHO Global Strategy for the Containment of AMR [10]. Emergence and spread of AMR has been usually attributed to the misuse or indiscriminate use of antibiotics as therapeutic drugs in human and animal health care or as growth promoters in veterinary husbandry. In addition, there is a growing concern over the possibility of AMR transmission via the food chain [11]. Indeed, the European strategic action plan on antibiotic resistance highlights that AMR is a food safety issue, because resistant bacteria and resistance genes can spread from food animals to humans through the food chain. In addition, it proposes the following strategic objectives: (1) address the interconnections between bacterial resistance and antibiotic use in human and animal health, including the food chain; and (2) prevent and control the

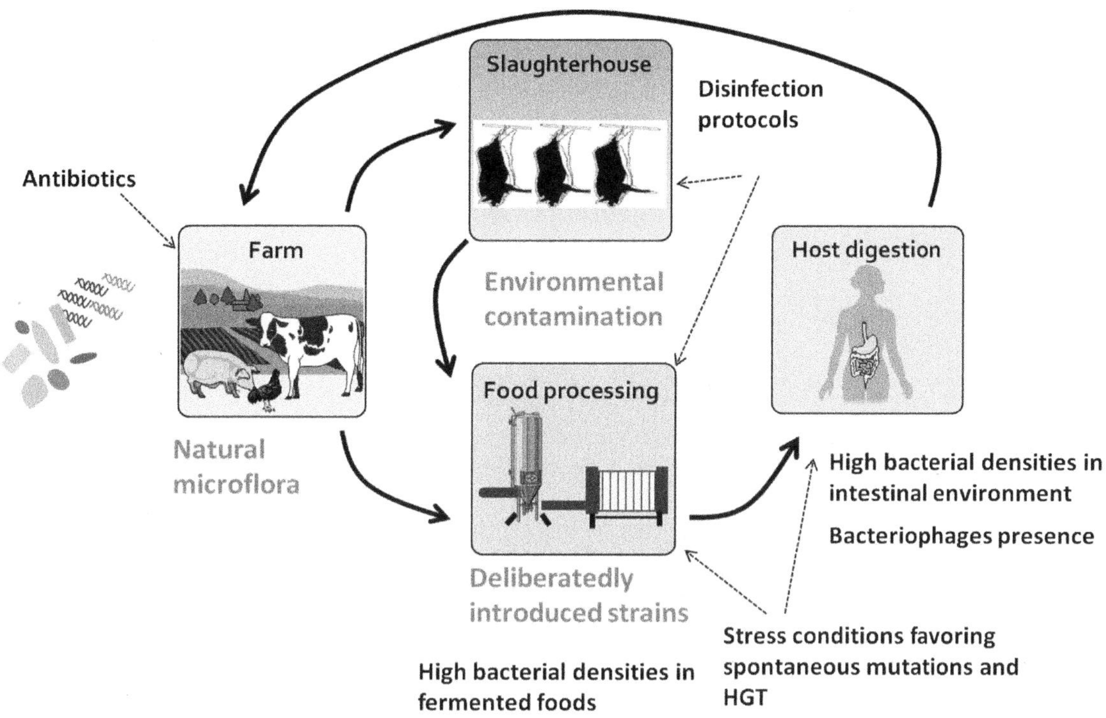

■ **FIGURE 2.1** Origin of microorganisms harboring transmissible antibiotic resistance genes (light) along the food chain and factors that may favor their horizontal spread (dark).

development and spread of antibiotic resistance in the veterinary and agricultural sectors [12]. Although the relative contribution of the food chain to the global burden of infections caused by antimicrobial-resistant microorganisms remains still unknown, it is generally recognized that controlling the emergence and spread of antimicrobial-resistant bacteria and AMR genes in primary food production and food processing must be a priority in order to reduce the occurrence of untreatable infections.

2.2 AMR IN FOODBORNE PATHOGENIC BACTERIA

Antibiotic resistance acquisition in food associated pathogens is particularly worrying since it may result in easily spread and difficult to treat infections that can be deadly in some cases. Therefore, continuous surveillance of AMR along the food chain is essential to understand its role in the emergence and spread of antimicrobial-resistant pathogenic microorganisms. The most important sources of surveillance at the EU level are the joint European Centre for Disease Prevention and Control (ECDC)/European Food Safety Authority (EFSA) annual reports on AMR in zoonotic bacteria. Since 2011, data on isolates from food and animals have been combined with human infection data [13]. The latest ECDC/EFSA annual report on AMR provides data from 28 Member States (MS) and three other European countries during 2013. The report gives information on AMR regarding *Salmonella* and *Campylobacter* isolates from humans, foods of animal origin and animals, and methicillin-resistant *Staphylococcus aureus* (MRSA) and indicator *Escherichia coli* and enterococci isolates from animals and foods of animal origin[14].

The foodborne pathogens of primary concern with regard to AMR in Europe at present are *Salmonella* and *Campylobacter*. Based on the European surveillance data, contaminated poultry meat, eggs, pork, and beef are considered to be important in relation to transmission of antibiotic-resistant *Salmonella*. A recent EU outbreak of *Salmonella* Stanley was traced to contamination in the turkey production chain [15]. *Salmonella* isolates normally exhibit multidrug resistance to various agents, including tetracyclines, sulfonamides, streptomycin, kanamycin, chloramphenicol, and some of the β-lactam antibiotics (penicillins and cephalosporins) [16,17]. It should be noted that the percentage of isolates resistant to these antibiotics has decreased or remained stable since 1996. Conversely, resistance to drugs such as amoxicillin/clavulanic acid, ceftriaxone, ceftiofur, and nalidixic acid follows an increasing trend [18]. According to the last ECDC/EFSA annual report, high proportions of human *Salmonella* isolates obtained from cases of infection during 2013 were resistant to ampicillin (36.1%), sulfonamides (35.7%), and tetracyclines (34.5%). Multiresistance was high (31.8%) in the EU, with very high occurrence in some countries. Some of the investigated serovars exhibited very or extremely high multiresistance, such as monophasic *S.* Typhimurium 1,4,[5],12:i: (83.8%) and *S.* Kentucky (67.3%). Interestingly, multiresistance was generally high in *Salmonella* isolates from broilers (56%), pigs (37.9%), and turkeys (73.0%), and low levels of "microbiological" co-resistance to ciprofloxacin and cefotaxime occurred in *Salmonella* from broilers, laying hens, and/or pigs. Resistance levels observed were generally higher in isolates from pigs and turkeys than from broilers, laying hens, breeding hens, and cattle [14].

In the case of *Campylobacter*, isolates are frequently resistant to quinolones, macrolides, lincosamides, chloramphenicol, aminoglycosides, tetracycline, ampicillin, and other β-lactams, cotrimoxazole, and tylosin [19,20]. Resistance to quinolones has emerged during the last 20 years, fitting the use of fluoroquinolones (mainly enrofloxacin) in the veterinary medicine [19]. Poultry meat is considered to be the main vehicle of resistant *Campylobacter* isolates [6,14]. The Food and Drug Administration (FDA) highlighted the significance of this in their assessment of the human health impact of fluoroquinolone-resistant *Campylobacter* attributed to the consumption of chicken [21]. The use of the fluoroquinolone enrofloxacine for the purpose of treating bacterial infections in poultry was discontinued in 2005 in the United States as a result of that assessment. According to the latest ECDC/EFSA annual report, the percentage of clinical isolates resistant to fluoroquinolones in EU in 2013 was extremely high in some MSs. Indeed, very high occurrence of resistance to the clinically important antimicrobial ciprofloxacin was observed in human *Campylobacter* isolates with more than half (54.6%) of Campylobacter *jejuni* and two-thirds (66.6%) of *Campylobacter coli* isolates being resistant [14]. Levels of resistance observed among poultry isolates to ciprofloxacin, nalidixic acid, and tetracyclines were considered as high, even extremely high, at 54.6–68.8%, 52.3–63.9%, and 41.4–70.4%, respectively. It is worth noting that statistically increasing trends in resistance to these antimicrobials were observed for several MSs, for both *Ca. jejuni* and *Ca. coli* poultry isolates.

MRSA has been a major cause of hospital-acquired infections for more than three decades, and is frequently detected in meat and products derived from animals. In particular, MRSA clonal complex 398 (MRSA CC398) is quite common among pigs, although it has been also isolated from cattle, dogs, horses, and chickens [22]. Occupational transmission of MRSA from food animals to humans is well known and transfer of MRSA through the food chain has also been well documented [23,24]. Interestingly, nasal carriage of livestock-associated MRSA CC398 among farmers and other persons in contact with animals has been reported [25]. At the EU level, only three MSs reported data on occurrence of MRSA during 2013 and a large degree of variation was observed among them in occurrence in pigs (from 20.8% to 97.8% of animals/herd slaughter batches were positive in slaughterhouse monitoring) [14].

In the United States, four major surveillance networks track foodborne disease and antibiotic resistance: FoodNet (Foodborne Diseases Active Surveillance Network), the National Animal Health Monitoring System (NAHMS), the National Antimicrobial Resistance Monitoring System (NARMS), and PulseNet [26]. In Canada, two major networks track antimicrobial resistance: the Canadian Integrated Program for Antimicrobial Resistance Surveillance (CIPARS) and FoodNet Canada. However, while a number of developed countries have national surveillance programs, in low- and middle-income countries, antibiotic resistance of foodborne pathogens remains largely unaddressed [26].

2.3 AMR IN FOOD-RELATED BENEFICIAL MICROBES

Bacteria deliberately introduced in food products, including fermentation starters and health-promoting probiotic microorganisms, hereinafter jointly referred to as "beneficial food-associated bacteria," can also acquire AMR and contribute to its spread throughout the food chain [11]. Indeed, from a microbiological point of view, fermented foods harbor high densities of a nondiverse microflora dominated by the starter microorganisms, which are particularly well adapted to the food matrix and can outcompete other microorganisms present in the raw food material. Lactic acid bacteria (LAB), including the genera *Lactobacillus, Lactococcus, Pediococcus, Leuconostoc,* or *Enterococcus,* and fungi including *Saccharomyces, Kluyveromyces, Pichia, Kloeckera, Candida, Penicillium,* or *Aspergillus,* are among the microorganisms most commonly used to ferment dairy, vegetable and meat food products. Moreover, health-promoting probiotic bacteria such as some strains of *Lactobacillus* or *Bifidobacterium,* among others, are incorporated at high bacterial densities into functional foods, so that a sufficiently high number of viable cells reach the intestine to exert their health-promoting beneficial effects. Beneficial food associated bacteria represent an important reservoir of transmittable antibiotic resistance genes which could further spread following ingestion by the host. During gastrointestinal passage, food associated bacteria face a number of stress factors (e.g., acid pH or bile salts) that can favor rescue of adaptive resistance to stress factors and antibiotics [27] and promote gene transfer [28]. In the intestine, the ingested bacteria encounter a densely populated environment where microorganisms are in close association, which may favor transmission of genetic determinants of AMR, even to phylogenetically distant bacteria and potential pathogens [28]. These will then be released into the environment initiating a spread chain that can go way beyond the original food product [29].

It is difficult to estimate the likelihood of horizontal gene transfer events of beneficial microbes associated AMR genes since it is highly conditioned by a number of environmental factors. For instance, although transmission of AMR genes from a probiotic *Lactobacillus* strain could not be evidenced in the human intestine [30], vancomycin resistance was demonstrated to be transferred in a mice model [31]. Moreover, simultaneous antibiotic and probiotic administration were shown to favor antibiotic resistance spread from ingested probiotic strains [32], underlining the need to establish surveillance systems and control measures to prevent AMR spread also for beneficial food-associated bacteria, which are consumed at high doses in certain fermented foods.

In Europe, starter microorganisms are traditionally recognized as safe via a positive QPS (Qualified Presumption of Safety) assessment by EFSA [33], and determination of intrinsic and acquired AMR, including characterization of genetic determinants of AMR, is essential before any strain can receive the QPS status. Moreover, EFSA includes antibiotic resistance determination as criteria check for probiotic characterization and provides recommendations on resistance cut-offs for most clinically relevant antibiotics and bacterial species, which can be used to determine whether particular strains are sensitive or resistant to a certain antibiotic [34,35]. Reduced sensitivity to antibiotics is not uncommon in bacteria isolated from food, including both starters and probiotic microorganisms [36–39]. However, the present concern of HGT and AMR spread depends on the characteristics of AMR genetic determinants. Intrinsic resistances are considered not to be transferable and could even be useful to restore the gut microbiota following an antibiotic course [40]. Conversely, resistances conferred by single genes or small gene clusters, and those included in mobile genetic elements (plasmids, transposons, etc.) are the most prone to spread. Indeed, HGT is a demonstrated route of AMR genes transmission between LAB under laboratory conditions [41–43], but also in vitro in food and intestinal models [44–47] and in vivo [31,48,49]. Hereinafter we will focus the discussion exclusively on those acquired antibiotic resistances susceptible to HGT spread.

There is increasing evidence that supports the role of LAB, which comprise most of the food starters and probiotics utilized to date, as an AMR reservoir and a source for AMR transmission [50–53]. Indeed, several LAB are also highly represented in the microbiota of the human and animal gut, which is a significant point for genetic exchange between food-associated LAB and resident colonizers [54,55]. The fact that the intestinal microbiota is a frequent environmental source to screen for new potential probiotic candidate strains justifies the need to check new probiotic candidates for AMR [34]. *Lactobacillus, Streptococcus, Lactococcus, Pediococcus*, and *Leuconostoc* generally display intrinsic resistance to ciprofloxacin, gentamicin, and streptomycin, but are sensitive to erythromycin, chloramphenicol, tetracycline, and β-lactams. Several lactobacilli isolated from Italian fermented dairy products [56], vegetables [57], and meat products [42,58], a probiotic *Lactobacillus reuteri* [38], and various *Leuconostoc* strains isolated from fermented dairy products were found to harbor tetracycline resistance genes [45]. Regarding *Bifidobacteria*, they generally display intrinsic resistances to mupirocin and aminoglycosides and are usually sensitive to macrolides, vancomycin, chloramphenicol, and β-lactams [40]. *Bifidobacteria* resistant to streptomycin and erythromycin were detected, although resistance was due to chromosomally acquired point mutations and therefore were not transmissible [59]. However, acquired tetracycline resistance is also frequent in several bifidobacterial probiotic strains belonging to *Bifidobacterium longum, Bifidobacterium animalis, Bifidobacterium bifidum, Bifidobacterium breve, Bifidobacterium pseudocatenulatum*, and *Bifidobacterium thermpophilum* [40,60]. It is worth highlighting that genes conferring resistance to tetracycline as *tetM* or *tetW*, are frequently associated to plasmid determinants and a family of transposons [52,61]. They are widely spread in both beneficial and pathogenic bacteria, and their presence is currently considered as ubiquitous [62]. *Bifidobacteria* strains resistant to tetracycline usually harbor the *tetW* genes within their chromosome, frequently surrounded by transposase elements, supporting thereof their acquisition through HGT events [63]. Recent reports on *Lactobacillus* and *Lactococcus* strains isolated from food products evidenced that a high number of isolates harboring AMR genes, among which *tetM*, conferring resistance to tetracycline as above-mentioned and *ermB*, conferring resistance to erythromycin, were among the most highly represented [50,64]. Erythromycin resistance genes were found in numerous *Lactobacillus* strains of human and food origins [56,58,65], and were frequently associated with plasmids [30,66,67]. In opposition to that, in some *Bifidobacteria* strains intended for food supplements, erythromycin resistance was attributed to structural point mutations within the rRNA-encoding genes, and, therefore, it was considered a nontransmissible resistance [59]. In other cases, resistance to erythromycin in *Bifidobacteria* was associated to mobile genetic elements. Indeed, this is the case of *ermS* gene, which is associated with transposon Tn5432 in *B. thermophilum* and *B. animalis* [68]. Remarkably, this family of erythromycin resistant transposons is widely distributed, even in somepathogenic bacteria [69]. On the other hand, aminoglycoside resistance genes were also reported in *Lactobacillus casei* [47], *Lactobacillus delbrueckii* [70], *Lactobacillus acidophilus, Lactobacillus rhamnosus, Lactobacillus paracasei, Lactobacillus plantarum, Lactobacillus paraplantarum*, and *Pediococcus* isolated from food [47,71–73]. Interestingly, *aadE*, a gene involved in resistance to streptomycin, identical in sequence to a *C. jejuni* gene was described to be present within a plasmid in *Pediococcus acidilactici* [72]. Resistance to aminoglycosides was also detected in *Leuconostoc* strains [45].

Remarkably, the simultaneous presence of resistance genes to multiple antibiotics was reported in several LAB, leading to a major concern. For instance, a *Lactobacillus johnosonii* strain was found to be resistant to erythromycin, tetracycline and clindamycin and harbored both *tetW* and *ermB* [60]. Three out of 75 tested strains of *Lactobacillus rhamnosus* were resistant to clindamycin, erythromycin, and streptomycin, and one strain was resistant to streptomycin and tetracycline [74]. In a recent study, 29 out of 101 tested *La. acidophilus* strains were found to be resistant to one antibiotic and 15 of them displayed simultaneous resistance to more than one tested antibiotic [75]. Mobile genetic elements conferring multiresistance deserve special attention. The plasmid pRE25, originally found in an *Enterococcus faecalis* strain isolated from raw fermented sausages, was capable of transferring resistance to kanamycin, neomycin, streptomycin, clindamycin, lincomycin, azithromycin, clarithromycin, erythromycin, roxithromycin, tylosin, chloramphenicol, and nourseothricin sulfate by in vitro conjugation to Lactococcus *lactis* and *Listeria innocua* strains [76]. In addition, pRE25 also transferred these AMR phenotypes in an intestinal model to *Listeria monocytogenes* and other commensal members of the gut microbiota [77]. Similarly, the *aadE-sat4-aphA-3* cluster of a *Leuconostoc* strain resistant to streptomycin, streptothricin, kanamycin, and neomycin is almost identical to a gene cluster previously detected in *Enterococci*, *Staphylococci*, and *Campylobacter* [45]. The flanking sequences of this gene cluster displayed strong similarity to plasmid replication proteins and transposase elements and identical clusters were previously associated to mobile genetic elements, arguing that it may be horizontally transferred [78].

2.4 ROUTES OF TRANSMISSION OF AMR THROUGHOUT THE FOOD CHAIN

2.4.1 Foods of Animal Origin

Foods of animal origin are considered an important source of antimicrobial-resistant bacteria entering the food chain [6]. The main focus of concern is the acquisition of antibiotic-resistant bacteria and AMR genes by the animal on the farm and their transmission to humans through the food chain. Once antibiotic-resistant bacteria have become established in an animal, they may persist for long periods of time after the antimicrobial has become undetectable in the animal's body. Foodborne transmission of antimicrobial-resistant bacteria from food animals to people is well documented [6,79].

Acquisition of AMR by animals on farm is mainly driven by the selective pressure caused by antibiotic consumption in veterinary husbandry. The use of antibiotics in food-producing animals would select for resistance to these agents in zoonotic microorganisms that would be then foodborne transmitted to humans. Antibiotics have been used in veterinary husbandry for more than 60 years to maintain a consistent supply of healthy animals entering the food chain. An estimated 50% of all antibiotics consumed in North America and Europe are used in food-producing animals [34]. Antibiotics are used in veterinary husbandry as therapeutic agents, for the treatment and control of many types of infections, or, in some countries, as growth promoters, to improve the efficiency of food utilization and weight gain. As several antibiotics are used as therapeutic agents in both human and animal medicine, the emergence of AMR to those agents in animal husbandry could compromise their therapeutic value for the treatment of human infection. The antibiotic classes of highest priority in this regard are quinolones, third and fourth generation of cephalosporins and macrolides [80]. The use of antibiotics as AGP is still common in several countries but has been banned at EU level. Avoparcin was the first antibiotic removed from the EU market as AGP as a precautionary measure aimed at preserving the clinical utility of vancomycin [81]. Later on, the growth promoters bacitracin zinc, spiramycin and tylosin phosphate, virginiamycin, olaquindox, carbadox, avilamycin, flavophospholipol, monensin, and salinomycin were also banned [82–84]. Since 2006, only coccidiostats and histomonostats are allowed as feed additives at EU level [85]. Nevertheless, the actual impact of the antibiotic ban on the evolution of the prevalence of antibiotic-resistant human infections is controversial. Indeed, although the complete removal of growth-promoting antibiotics has resulted in a decrease in the total consumption of antibiotics in animal husbandry [86] and has been associated with a reduction of vancomycin-resistant enterococci (VRE) in animal and human feces and food samples [86,87], several authors point out that the antibiotic ban has led to an increase in the use of therapeutic antibiotics, with a subsequent impact on animal health and welfare and financial consequences for farmers [88].

2.4.2 **Foods of Nonanimal Origin**

The magnitude of antibiotic use in agriculture is small in comparison to the extent of use in humans and animal husbandry. Most antimicrobials used in plant agriculture are fungicides. Indeed, there are no antibiotics authorized for use in plant agriculture in the EU [89], while in the United States some antibiotics (streptomycin and oxytetracycline) have been used for decades in preventive treatments to control microorganisms affecting fruit and vegetables [85].

Genetically modified plants or transgenic crops have received great attention regarding agriculture-associated AMR. In order to produce genetically modified plants the gene that will confer the new trait is normally coupled to an AMR marker gene. Treating the cells after gene transfer with an antibiotic allows only the successfully transformed cells to survive. Although the marker gene serves no purpose after this procedure, it remains part of the transgenic plant. Some concerns exist about the risk of horizontal transfer through natural transformation of AMR marker genes to soil or gut microorganisms. Indeed, some researchers have demonstrated the successful transfer of transgenic-borne AMR genes to bacteria [90]. Nevertheless, successful transfer of AMR genes has been observed only when facilitated by the existence of sequence homology between the transgene and the genome of the recipient microorganism and recovery of plant DNA by naturally occurring bacteria has not been demonstrated yet. Overall, genetically modified plants are not considered at the moment an important route for transmission of AMR since the potential for AMR genes to be transferred to bacteria is thought to be low as it would require a series of biologically complex and unlikely events to occur [91].

Foods of nonanimal origin are now recognized as relevant sources of foodborne infection worldwide. Indeed, foods of nonanimal origin have been linked to about 10% of all foodborne outbreaks occurring from 2007 to 2011 in EU [92]. The following pathogen/food combinations were identified by EFSA as the most relevant in this regard: *Salmonella* and leafy greens eaten raw followed by *Salmonella* and bulb and stem vegetables, *Salmonella* and tomatoes, *Salmonella* and melons, and pathogenic *E. coli* and fresh pods, legumes, or grain [92]. All foods acting as vehicles for pathogenic bacteria have the potential to transmit AMR, although surveillance data on AMR in bacteria from foods of nonanimal origin are very limited. Thus, it is reasonable to assume that they can be a potential reservoir of AMR. Foods of nonanimal origin may become contaminated with antimicrobial-resistant bacteria during primary production or at a later stage. Of particular interest in this regard is both the water used for irrigation and the use of manure and compost as fertilizers. Surface waters used for irrigation have been implicated in the transmission of water-borne pathogens to fresh produce [93]. Antimicrobial-resistant bacteria were found in many types of environmental waters, including coastal waters, creeks, rivers, lakes, wells, agricultural runoff, and groundwater [94]. In addition, irrigation waters were suggested as a potential pathway for human exposure to wildlife strains of antimicrobial-resistant bacteria [95]. Wastewater and wastewater treatment plants are considered as important reservoirs of antimicrobial resistant microorganisms, AMR genes, and mobile genetic elements encoding AMR determinants [96]. Sewage and wastewater treatment plants may serve as interfaces between different environmental niches, such as hospitals and surface waters, and provide mixing opportunities between environmental bacteria and human pathogens in the presence of selective agents or pollutants such as drugs, heavy metals, or biocides, therefore facilitating AMR spread between these habitats. The use of manure (excrement from farm animals) or compost to fertilize crops can also serve as a route for the emergence and spread of AMR through foods of nonanimal origin. Microorganisms (including antimicrobial-resistant bacteria) present in animal feces could therefore contaminate food when manure or compost is used as a soil amendment or when manure contaminates irrigation waters or comes into contact with food, with leafy vegetables eaten raw being of particular concern [85].

2.4.3 Primary Production and Food Processing Environments and Food Processing Technologies

2.4.3.1 Role of Biocides in AMR Acquisition and Spread

Biocides, which can be defined as active substances and preparations containing one or more active compounds intended to inactivate or exert a controlling effect on harmful microorganisms, are widely used for the maintenance of the required levels of hygiene at farms, slaughterhouses, and food-processing premises [97]. In farms, they are used for the cleaning and disinfection of areas associated with livestock animals including farm buildings, beddings, equipment, boot baths, and transportation vehicles, among others. In addition, a range of biocides are habitually used in slaughterhouses and food production and processing areas for disinfection of equipment and surfaces in order to control colonization by potentially hazardous microorganisms. Some of the most widely used biocides are alcohols, aldehydes, chlorine, and chlorine- releasing agents (sodium hypochlorite, chlorhexidine), iodine, peroxygen compounds (hydrogen peroxide, peracetic acid), phenolic type compounds, quaternary ammonium compounds (benzalkonium chloride), bases (sodium hydroxide, potassium hydroxide, sodium carbonate), and acids (mineral and organic acids).

Biocides are generally employed at farms, slaughterhouses, and food industries at concentrations well above the minimal inhibitory concentrations (MIC) of all major target microorganisms and should therefore be able to guarantee microbial inactivation and avoiding biocide resistance development. However, suboptimal biocide concentrations do occur in selected niches (e.g., under objects or in cracks and crevices and other harborage sites) or as a consequence of improper use. Presence of organic matter (known to inactivate some biocides such as chlorinated compounds) may reduce their efficacy, while erroneous formulation, inappropriate storage of formulations, and inadequate distribution of the compound on surfaces and equipment can result in a decrease in active biocide concentration at some locations in premises [98]. Niches with low biocide concentrations may also occur through rinsing of frequently cleaned and disinfected areas or through biocide application to wet surfaces, with a consequent dilution of the compound to concentrations that may be sublethal for microorganisms. In addition, wastewater disposal lines can provide permanent contact of microorganisms with low concentrations of biocides. For instance, sub-lethal concentrations of representative quaternary ammonium compounds such as dialkyldimethylammonium compounds and benzalkonium chlorides have been identified in river sediment (26 and 1.5 ppm, respectively) and wastewater from hospitals (1.5–4 ppm) [99,100]. The same is likely to be true for wastewater from farms, slaughterhouses and food processing facilities. Therefore, microbial communities colonizing environments within farms, slaughterhouses, and food production facilities (including food-contact environments) are recurrently exposed to subinhibitory biocide concentrations, and this may have an impact on microbial ecology and food safety.

Selection of biocide-resistant microorganisms by exposure to suboptimal biocide concentrations was described to occur for some biocide compounds and microbial species (e.g., *Li. monocytogenes*—quaternary ammonium compounds, sodium hypochlorite; *Salmonella* Typhimurium—triclosan; *E. coli*—trisodium phosphate, sodium nitrite, sodium hypochloride) [101–103]. Resistance to biocides represents a potentially major public health issue as it can contribute to the increased persistence of pathogenic and spoilage microorganisms in the food chain. Bacterial persistence, which can be defined as the survival for extended periods of time in a particular location is a great concern for food industries since it can lead to the repeated contamination of food with spoilage or pathogenic microorganisms, thus seriously impacting on the health of consumers and causing great economic losses to food businesses. A good example of association between persistence and biocide resistance is the case of *Li. monocytogenes* 6179 strain. *Listeria monocytogenes* 6179 is a serotype 1/2a strain, isolated at Teagasc Food Research Centre (Ireland) from particular environments at a cheese processing facility repeatedly over a period of 12 years [104,105]. The sequencing of *Li. monocytogenes* 6179 genome allowed identification of a novel transposon, Tn*6188*, which comprised three consecutive transposase genes (*tnp*ABC), a *qacH* gene encoding a small multidrug resistance protein family (SMR) transporter responsible for the export of quaternary ammonium compounds, and a putative *tetR* family transcriptional regulator upstream of the transporter [106]. Investigations also showed that exposure to benzalkonium chloride (BAC) caused an increase in *qacH* expression and that a *qacH* deletion mutant strain had lower BAC tolerance than the wild type strain. Recent studies carried out also in Ireland have identified various persistent strains undistinguishable from *Li. monocytogenes* 6179 by pulsed field gel electrophoresis analysis that were recurrently isolated from environments and food of five different seafood industries during 2013 and 2014. All these strains harbored the Tn*6188* transposon and had significantly higher MIC against BAC than other strains isolated during the same time frame and lacking the Tn*6188* transposon [107]. Bacterial isolates such as these, which carry biocide resistance determinants, persist in the industrial settings for long periods of time and are a likely source of food contamination events.

The suspicion that exposure to suboptimal concentrations of biocides may select for increased resistance against clinically relevant antibiotics has also recently arisen. Some studies carried out in the last decade comparing biocide and antibiotic resistance of collections of strains from major foodborne pathogens revealed a correlation between resistance to both agents [108–110]. Unfortunately, the small data sets used and the lack of appropriate statistical analyses limited the findings of these studies. More recently, Coelho et al. [111] carried out an elegant study using machine learning methodologies to analyze antibiotic and biocide susceptibility of the largest collection of isolates so far tested for this purpose (1632 isolates), in this case from *St. aureus*. They described that reduced susceptibility to two biocides, clorhexidine and BAC, which belong to different structural families, was associated with resistance to several antibiotics (amoxicillin/clavulanate, cefuroxime, cefaclor, cefpodoxime, clindamycin, erythromycin, clarithromycin, azithromycin, telithromycin, ciprofloxacin, levofloxacin, gatifloxacin, and moxifloxacin). Other authors isolated stable mutant strains with increased resistance to one or several antibiotics after exposure to sublethal biocide concentrations. For instance, Langsrud et al. [112] reported that serial cultivation of two *E. coli* strains in the presence of subinhibitory concentrations of BAC resulted in an increased resistance to various antibiotics (ampicillin, penicillin G, norfloxacin, nalidixic acid, kanamycin, gentamicin, chloramphenicol, tetracycline, and erythromycin), with MIC values 1.5–20-fold higher than those observed for control cultures. Randall et al. [113] described that exposure to an aldehyde-based disinfectant gave rise to *S.* Typhimurium mutants with a decreased susceptibility to ciprofloxacin in various bacterial strains. Karatzas et al. [102], following extended treatment of *S.* Typhimurium with three widely used farm disinfectants (a blend of oxidizing compounds; a quaternary ammonium disinfectant containing formaldehyde and glutaraldehyde;

and a biocide composed of organic acids and surfactants), obtained one stable individual variant from each treatment that exhibited reduced susceptibility to a range of antibiotics (ciprofloxacin, chloramphenicol, tetracycline, and ampicillin). Whitehead et al. [114] isolated two mutants of *S.* Typhimurium after a single exposure to in-use concentrations of two biocides (a mixture of aldehydes and quaternary ammonium compounds, and a halogenated tertiary amine compound) that had a broad reduced susceptibility to nalidixic acid, chloramphenicol, tetracycline, and ciprofloxacin. Webber et al. [115] showed that four different biocides (a mixture of aldehydes and quaternary ammonium compounds, a quaternary ammonium compound, an oxidative compound and a halogenated tertiary amine compound) selected *S.* Typhimurium multidrug-resistant mutants with decreased antibiotic susceptibility.

It is not fully known how exposure to biocide compounds selects for increased resistance against antibiotics, but two mechanisms are postulated, cross-resistance and coresistance. Since certain biocides and antibiotics share the same cellular targets, it is likely that some biocide resistance determinants and advantageous mutations leading to increased biocide resistance may be also responsible for the acquisition of antibiotic resistance in microbial populations. For instance, some multidrug efflux pumps, which have more than one substrate that might be chemically unrelated to each other, can confer simultaneous resistance to antibiotics and biocides when they are overexpressed [116]. This is known as cross-resistance. In addition, even for those biocides which do not share a target with antibiotics, reduced susceptibility to antibiotics can be due to the horizontal transfer of various different resistance determinants (antibiotic resistance determinants and biocide resistance determinants) associated together on common genetic elements like plasmids, phages, integrons, or transposons, which can spread to other strains, species or genera [117]. This is known as coresistance. An example of this phenomenon are class 1 integrons, which are known to contain a wide range of gene cassettes including some encoding resistance to different antibiotics and quaternary ammonium compounds. Indeed, the existence of environmental reservoirs of class 1 integrons has been known for some time and the class 1 integron-integrase gene *intI1* has been recently proposed as a biomarker of selective pressures imposed by anthropogenic pollution [118,119]. Microorganisms (including nonpathogenic and unculturable) present at food and food-related environments can play an important role in the coresistance events. Indeed, they are good reservoirs of antimicrobial resistance (AMR) genes and can be facilitators for AMR gene dissemination in various environmental ecosystems, including food ecosystems [120].

2.4.3.2 Food Processing Technologies and AMR

Foods are complex environments in which microorganisms may face natural stress conditions such as limited nutrient availability, adverse pH, osmolarity, oxidation, and extreme temperatures, among others. In addition, industrial food preservation regimes commonly rely upon imposing extreme physical and chemical stresses with the aim to inactivate or limit the growth of pathogenic bacteria. Thus, a variety of preservation technologies such as thermal processing, high hydrostatic pressure, pulsed electric fields, radiation, refrigeration, and drying impose a challenge to bacterial cells and can determine the fate of foodborne pathogens along the food chain [121]. Bacteria have evolved a range of adaptive strategies (stress responses) to overcome such challenges. The stress responses are global, complex systems of defense. They comprise networks to adapt to changing environments and to survive under adverse conditions. Adaptations sometimes derive from the acquisition of stochastic genomic mutations which are positively selected and fixed in the microbial population due to the beneficial phenotype they confer under selective pressure environments. In other occasions, a transient stress response generally consisting of a characteristic change in the pattern of gene expression occurs. Bacteria are able to do so by upregulating certain genes and downregulating others to maintain viability [121]. Stress-adapted microorganisms are particularly challenging to the food industry as they can survive food processing conditions. In addition, the adaptation to stress is sometimes associated with the cross-protection against a wide variety of apparently unrelated challenges, including antibiotics. On one side antimicrobials themselves are growth-inhibiting stressors that often elicit protective responses in bacteria, thus provoking their own resistance-promoting responses. In addition, cellular permeability to antibiotics is known to change in response to factors such as shifts in temperature, presence of specific inducers, reactive oxygen species, or specific metabolic signals [122]. Reduced susceptibility to antibiotics in response to environmental stresses may be mediated through modification of the lipopolysaccharide (LPS) or other components of the cellular envelopes, or through enhanced expression of multiple efflux pumps that extrude the antibiotic once it has reached the cytoplasm decreasing the toxic activity [122,123]. Furthermore, it has been observed that processing stresses can increase HGT of antibiotic resistance genes by conjugation or transformation [122,124]. Nevertheless, it is still contentiously disputed whether such induction of AMR through food processing technologies actually occurs in the food industry.

2.5 **CONCLUSIONS**

In the last decades, it has become apparent that AMR is a food safety issue, since antibiotic-resistant bacteria and AMR genes can spread from food animals to humans through the food chain. Nevertheless, the relative contribution of the food chain to the global burden of infections caused by antimicrobial-resistant microorganisms still remains unknown. Research efforts must be devoted in the coming years to characterizing the actual impact of the food chain on the total AMR figures worldwide. Implementation of more complete and coordinated surveillance systems, involving analysis of not only food animal samples, but also foods of non-animal origin and environmental samples from primary production and food processing facilities, will facilitate this complex task and will allow the elucidation of the role of thus far neglected routes of acquisition and spread of AMR. The integration of novel technologies, such as genomic and metagenomic tools, into those surveillance systems will undoubtedly provide new clues for the control of AMR throughout the food chain.

REFERENCES

[1] EASAC. Tackling antibacterial resistance in Europe. European Academies Science Advisory Council, June 2007. Available from: <http://www.leopoldina-halle.de/easac-report07.pdf>; 2007.

[2] WHO. World Health Organization. Related WHO publications and links on antimicrobial resistance. Available from: <http://www.who.int/foodborne_disease/resistance/publications/en/index.html>; 2009a.

[3] European Parliament. Antibiotic resistance. IP/A/STOA/ST/2006-4. Available from: <http://www.europarl.europa.eu/stoa/publications/studies/stoa173_en.pdf>; 2006.

[4] ECDC/EMEA. The bacterial challenge: time to react. EMEA doc. ref. EMEA/576176/2009. Available from: <http://ecdc.europa.eu/en/publications/Publications/0909_TER_The_Bacterial_Challenge_Time_to_React.pdf>; 2009.

[5] CDC. Antibiotic resistance threats in the United States, 2013. Available from: <http://www.cdc.gov/drugresistance/threat-report-2013/>; 2013.

[6] EFSA Foodborne antimicrobial resistance as a biological hazard. Scientific opinion of the panel on biological hazards. EFSA J 2008;765:1–87.

[7] WHO. Report of the 1st Meeting of the WHO AGISAR. Available from: <http://apps.who.int/medicinedocs/index/assoc/s16735e/s16735e.pdf>; 2009b.

[8] Rossi F, Rizzotti L, Felis GE, Torriani S. Horizontal gene transfer among microorganisms in food: current knowledge and future perspectives. Food Microbiol 2014;42:232–43.

[9] European Commission. Communication from the Commission to the European Parliament and the Council, Action plan against the rising threats from Antimicrobial Resistance. Available from: <http://ec.europa.eu/dgs/health_food-safety/docs/communication_amr_2011_748_en.pdf>; 2011.

[10] WHO. Global action plan on antimicrobial resistance. Available from: <http://apps.who.int/iris/bitstream/10665/193736/1/9789241509763_eng.pdf?ua=1>; 2015.

[11] Verraes C, Van Boxstael S, Van Meervenne E, Van Coillie E, Butaye P, Catry B, et al. Antimicrobial resistance in the food chain: a review. Int J Environ Res Public Health 2013;10:2643–69.

[12] WHO. European strategic action plan on antibiotic resistance. Available from: <http://www.euro.who.int/__data/assets/pdf_file/0008/147734/wd14E_AntibioticResistance_111380.pdf?ua=1>; 2011a.

[13] WHO. Antimicrobial resistance: global report on surveillance. Geneva: World Health Organization. Available from: <http://apps.who.int/iris/bitstream/10665/112642/1/9789241564748_eng.pdf>; 2014.

[14] EFSA EU summary report on antimicrobial resistance in zoonotic and indicator bacteria from humans, animals and food in 2013. EFSA J 2015;13:4036.

[15] ECDC/EFSA. Multi-country outbreak of *Salmonella* Stanley infections – Third update, 8 May 2014. Stockholm and Parma. Available from: <http://ecdc.europa.eu/en/publications/Publications/salmonella-stanley-multi-country-outbreak-assessment-8-May-2014.pdf>; 2014.

[16] Alcaine SD, Warnick LD, Wiedmann M. Antimicrobial resistance in nontyphoidal *Salmonella*. J Food Protect 2007;70:780–90.

[17] Olsen SJ, Ying M, Davis MF, Deasy MP, Holland B, Iampietro L, et al. Multidrug-resistant *Salmonella* Typhimurium infection from milk contaminated after pasteurization. Emerg Infect Dis 2004;10:932–5.

[18] Foley SL, Lynne AM. Food animal-associated *Salmonella* challenges: pathogenicity and antimicrobial resistance. J Animal Sci 2008;86:E173–87.

[19] Engberg J, Keelan M, Gerner-Smidt P, Taylor D. Antimicrobial resistance in *Campylobacter* Aaerstrup FM, editor. Antimicrobial Resistance in Bacteria of Animal Origin. Washington, DC: ASM Press; 2006. p. 269–91.

[20] Koluman A, Dikici A. Antimicrobial resistance of emerging foodborne pathogens: status quo and global trends. Crit Rev Microbiol 2013;39:57–69.

[21] FDA. Withdrawal of enrofloxacin for poultry. Available from: <http://www.fda.gov/AnimalVeterinary/SafetyHealth/RecallsWithdrawals/ucm042004.htm>; 2000.

[22] Cuny C, Friedrich A, Kozytska S, Layer F, Nubel U, Ohlsen K, et al. Emergence of methicillin-resistant *Staphylococcus aureus* (MRSA) in different animal species. Int J Med Microbiol 2010;300:109–17.

[23] Hanselman BA, Kruth SA, Rousseau J, Low DE, Willey BM, McGeer A, et al. *Staphylococcus aureus* colonization in veterinary personnel. Emerg Infect Dis 2006;12:1933–8.

[24] Lewis HC, Molbak K, Reese C, Aarestrup FM, Selchau M, Sorum M, et al. Pigs as source of methicillin-resistant *Staphylococcus aureus* CC398 infections in humans, Denmark. Emerg Infect Dis 2008;14(9):1383–9.

[25] Oppliger A, Moreillon P, Charrière N, Giddey M, Morisset D, Sakwinska O. Antimicrobial resistance of *Staphylococcus aureus* strains acquired by pig farmers from pigs. Appl Environ Microbiol 2012;78:8010–4.

[26] Lammie SL, Hughes JM. Antimicrobial resistance, food safety, and one health: the need for convergence. Annu Rev Food Sci Technol 2016;7:13.1–13.26.

[27] Kheadr E, Dabour N, Le Lay C, Lacroix C, Fliss I. Antibiotic susceptibility profile of bifidobacteria as affected by oxgall, acid, and hydrogen peroxide stress. Antimicrob Agents Chemother 2007;51:169–74.

[28] Feld L, Schjørring S, Hammer K, Licht TR, Danielsen M, Krogfelt K, et al. Selective pressure affects transfer and establishment of a *Lactobacillus plantarum* resistance plasmid in the gastrointestinal environment. J Antimicrob Chemother 2008;61:845–52.

[29] Hart WS, Heuzenroeder MW, Barton MD. A study of the transfer of tetracycline resistance genes between *Escherichia coli* in the intestinal tract of a mouse and a chicken model. J Vet Med B Infect Dis Vet Public Health 2006;53:333–40.

[30] Egervärn M, Lindmark H, Olsson J, Roos S. Transferablity of a tetracycline resistance gene from probiotic *Lactobacillus reuteri* to bacteria in the gastrointestinal tract of humans. Antonie Van Leeuwenhoek 2010;97:189–200.

[31] Mater DD, Langella P, Corthier G, Flores MJ. A probiotic *Lactobacillus* strain can acquire vancomycin resistance during digestive transit in mice. J Mol Microbiol Biotechnol 2008;14:123–7.

[32] Saarela M, Maukonen J, von Wright A, Vilpponen-Salmela T, Patterson AJ, Scott KP, et al. Tetracycline susceptibility of the ingested *Lactobacillus acidophilus* LaCH-5 and *Bifidobacterium animalis* Subsp. *lactis* Bb-12 strains during antibiotic/probiotic intervention. Int J Antimicrob Agents 2007;29:271–80.

[33] EFSA Introduction of a Qualified Presumption of Safety (QPS) approach for assessment of selected microorganisms referred to EFSA. EFSA J 2007;587:1–16.

[34] WHO. World Health Organization. Antimicrobial resistance. Fact sheet No194. Available from: <http://www.who.int/mediacentre/factsheets/fs194/en/>; 2002.

[35] EFSA Guidance on the assessment of bacterial susceptibility to antimicrobials of human and veterinary importance. EFSA J 2012;10:2740.

[36] Wong A, Ngu DYS, Dan LA, Ooi A, Lim RLH. Detection of antibiotic resistance in probiotics of dietary supplements. Nutr J 2015;14:95.

[37] Fraqueza MJ. Antibiotic resistance of lactic acid bacteria isolated from dry-fermented sausages. Int J Food Microbiol 2015;212:76–88.

[38] Kastner S, Perreten V, Bleuler H, Hugenschmidt G, Lacroix C, Meile L. Antibiotic susceptibility patterns and resistance genes of starter cultures and probiotic bacteria used in food. Syst Appl Microbiol 2006;29:145–55.

[39] Liu C, Zhang ZY, Dong K, Yuan JP, Guo XK. Antibiotic resistance of probiotic strains of lactic acid bacteria isolated from marketed foods and drugs. Biomed Environ Sci 2009;22:401–12.

[40] Gueimonde M, Sánchez B, de los Reyes-Gavilán CG, Margolles A. Antibiotic resistance in probiotic bacteria. Front Microbiol 2013;4:202.

[41] Thumu SC, Halami PM. Phenotypic expression, molecular characterization and transferability of erythromycin resistance genes in *Enterococcus* spp. isolated from naturally fermented food. J Appl Microbiol 2014;116:689–99.

[42] Gevers D, Huys G, Swings J. In vitro conjugal transfer of tetracycline resistance from *Lactobacillus* isolates to other gram-positive bacteria. FEMS Microbiol Lett 2003;225:125–30.

[43] Matter DD, Langella P, Corthier G, Flores MJ. Evidence of vancomycin resistance gene transfer between enterococci of human origin in the gut of mice harbouring human microbiota. J Antimicrob Chemother 2005;56(5):975–8.

[44] Toomey N, Monaghan A, Fanning S, Bolton DJ. Assessment of antimicrobial resistance transfer between lactic acid bacteria and potential foodborne pathogens using *in vitro* methods and mating in a food matrix. Foodborne Pathog Dis 2009;6:925–33.

[45] Flórez AB, Campedelli I, Delgado S, Alegría A, Salvetti E, Felis GE, et al. Antibiotic susceptibility profiles of dairy *Leuconostoc*, analysis of the genetic basis of atypical resistances and transfer of genes *in vitro* and in a food matrix. PLoS One 2016;11:e0145203.

[46] Gazzola S, Fontana C, Bassi D, Cocconcelli PS. Assessment of tetracycline and erythromycin resistance transfer during sausage fermentation by culture-dependent and -independent methods. Food Microbiol 2012;30:348–54.

[47] Ouoba LI, Lei V, Jensen LB. Resistance of potential probiotic lactic acid bacteria and bifidobacteria of African and European origin to antimicrobials: determination and transferability of the resistance genes to other bacteria. Int J Food Microbiol 2008;121:217–24.

[48] Jacobsen L, Wilcks A, Hammer K, Huys G, Gevers D, Andersen SR. Horizontal transfer of tet(M) and erm(B) resistance plasmids from food strains of *Lactobacillus plantarum* to *Enterococcus faecalis* JH2-2 in the gastrointestinal tract of gnotobiotic rats. FEMS Microbiol Ecol 2007;59:158–66.

[49] Boguslawska J, Zycka-Krzesinska J, Wilcks A, Bardowski J. Intra- and interspecies conjugal transfer of Tn916-like elements from *Lactococcus lactis in vitro* and *in vivo*. Appl Environ Microbiol 2009;75:6352–60.

[50] Devirgilis C, Zinno P, Perozzi G. Update on antibiotic resistance in foodborne *Lactobacillus* and *Lactococcus* species. Front Microbiol 2013;4:301.

[51] Kyselková M, Jirout J, Vrchotová N, Schmitt H, Elhottová D. Spread of tetracycline resistance genes at a convenctional dairy farm. Front Microbiol 2015;6:536.

[52] Mendonça AA, de Lucena BT, de Morais MM, de Morais Jr. MA. First identification of Tn916-like element in industrial strains of *Lactobacillus vini* that spread the *tet-M* resistance gene. FEMS Microbiol Lett 2016;363(3).

[53] Nawaz M, Wang J, Zhou A, Ma C, Wu X, Moore JE, et al. Characterization and transfer of antibiotic resistance in lactic acid bacteria from fermented food products. Curr Microbiol 2011;62:1081–9.

[54] Sommer MO, Dantas G, Church GM. Functional characterization of the antibiotic resistance reservoir in the human microflora. Science 2009;325:1128–31.

[55] Marshall BM, Ochieng DJ, Levy SB. Commensals: underappreciated reservoir of antibiotic resistance. Microbe 2009;4:231–8.

[56] Comuniam R, Daga E, Dupré I, Paba A, Devirgilis C, Piccioni V, et al. Susceptibility to tetracycline and erythromycin *of Lactobacillus paracasei* strains isolated from traditional Italian fermented foods. Int J Food Microbiol 2010;138:151–6.

[57] Casado Muñoz MC, Benomar N, Lerma LL, Gálvez A, Abriouel H. Antibiotic resistance of *Lactobacillus pentosus* and *Leuconostoc pseudomesenteroides* isolated from naturally-fermented Aloreña table olives throughout fermentation process. Int J Food Microbiol 2014;172:110–8.

[58] Zonenschain D, Rebecchi A, Morelli L. Erythromycin- and tetracycline-resistant lactobacilli in Italian fermented dry sausages. J Appl Microbiol 2009;107:1559–68.

[59] Sato T, Lino T. Genetic analysis of the antibiotic resistance of *Bifidobacterium bifidum* strain Yakult YIT 4007. Int J Food Microbiol 2010;137:254–8.

[60] Ammor MS, Flórez AB, van Hoek AH, de los Reyes-Gavilán CG, Aarts HJ, Margolles A, et al. Molecular characterization of intrinsic and acquired antibiotic resistance in lactic acid bacteria and bifidobacteria. J Mol Microbiol Biotechnol 2008;14:6–15.

[61] Raftis EJ, Forde BM, Claesson MJ, O'Toole PW. Unusual genome complexity in *Lactobacillus salivarius* JCM1046. BMC Genom 2014;15:771.

[62] Rizzotti L, La Fioia F, Dellaglio F, Torriani S. Molecular diversity and transferability of the tetracycline resistance gene tet(M), carried on Tn916-1545 family transposons, in enterococci from a total food chain. Antonie Van Leeuwenhoek 2009;96:43–52.

[63] Kazimierczak KA, Flint HJ, Scott KP. Comparative analysis of sequences flanking tet(W) resistance genes in multiple species of gut bacteria. Antimicrob Agents Chemother 2006;50:2632–9.

[64] Aquilanti L, Garofalo C, Osimani A, Silvestri G, Vignaroli C, Clementi F. Isolation and molecular characterization of antibiotic-resistant lactic acid bacteria from poultry and swine meat products. J Food Protect 2007;70:557–65.

[65] Klare I, Konstabel C, Werner G, Huys G, Vankerckhoven V, Kahlmeter G, et al. Antimicrobial susceptibilities of *Lactobacillus*, *Pediococcus* and *Lactococcus* human isolates and cultures intended for probiotic or nutritional use. J Antimicrob Chemother 2007;59:900–12.

[66] Feld L, Bielak E, Hammer K, Wilcks A. Characterization of a small resistance plasmid pLFE1 from the food-isolate *Lactobacillus plantarum* M345. Plasmid 2009;61:159–70.

[67] Gfeller KY, Roth M, Meile L, Teuber M. Sequence and genetic organization of the 19.3-kb erythromycin- and dalfopristin- resistance plasmid pLME300 from *Lactobacillus fermentum* ROT1. Plasmid 2003;50:190–201.

[68] Van Hoek AH, Mayrhofer S, Domig KJ, Aarts HJ. Resistance determinant erm(X) is borne by transposon Tn5432 in *Bifidobacterium thermophilum* and *Bifidobacterium animalis* subsp. *lactis*. Int J Antimicrob Agents 2008;31:544–8.

[69] Rosato AE, Lee BS, Nash KA. Inducible macrolide resistance in *Corynebacterium jeikeium*. Antimicrob Agents Chemother 2001;45(7):1982–9.

[70] Zhou N, Zhang JX, Fan MT, Wang J, Guo G, Wei XY. Antibiotic resistance of lactic acid bacteria isolated from Chinese yogurts. J Dairy Sci 2012;95:4775–83.

[71] Jaimee G, Halami PM. Emerging resistance to aminoglycosides in lactic acid bacteria of food origin - an impending menace. Appl Microbiol Biotechnol 2016;100:1137–51.

[72] O'Connor EB, O'Sullivan O, Stanton C, Danielsen M, Simpson PJ, Callanan MJ, et al. pEOC01: a plasmid from *Pediococcus acidilactici* which encodes an identical streptomycin resistance (*aadE*) gene to that found in *Campylobacter jejuni*. Plasmid 2007;58:115–26.

[73] Rojo-Bezares B, Sáenz Y, Poeta P, Zarazaga M, Ruiz-Larrea F, Torres C. Assessment of antibiotic susceptibility within lactic acid bacteria strains isolated from wine. Int J Food Microbiol 2006;111:234–40.

[74] Korhonen JM, Van Hoek AH, Saarela M, Huys G, Tosi L, Mayrhofer S, et al. Antimicrobial susceptibility of *Lactobacillus rhamnosus*. Benefic Microb 2010;1:75–80.

[75] Maryhofer S, van Hoek AH, Mair C, Huys G, Aarts HJ, Kneifel W, et al. Antibiotic susceptibility of members of the *Lactobacillus acidophilus* group using broth

microdilution and molecular identification of their resistance determinants. Int J Food Microbiol 2010;144(1):81–7.

[76] Teuber M, Schwarz F, Perreten V. Molecular structure and evolution of the conjugative multiresistance plasmid pRE25 of *Enterococcus faecalis* isolated from a raw-fermented sausage. Int J Food Microbiol 2003;88:325–9.

[77] Haugh MC, Tanner SA, Lacroix C, Stevens MJ, Meile L. Monitoring horizontal antibiotic resistance gene transfer in a colonic fermentation model. FEMS Microbiol Ecol 2011;78:210–9.

[78] Qin S, Wang Y, Zhang Q, Chen X, Shen Z, Deng F, et al. Identification of a novel genomic island conferring resistance to multiple aminoglycoside antibiotics in *Campylobacteri coli*. Antimicrob Agents Chemother 2012;56:5332–9.

[79] WHO. Tackling antibiotic resistance from a food safety perspective in Europe. Available from: <http://www.euro.who.int/__data/assets/pdf_file/0005/136454/e94889.pdf>; 2011b.

[80] WHO Critically important antimicrobials for human medicine: categorization for the development of risk management strategies to contain antimicrobial resistance due to non-human antimicrobial use. Copenhagen: Report of the second WHO expert meeting; 2007.29.31 May 2007. Available from: <http://www.who.int/foodborne_disease/resistance/antimicrobials_human.pdf>.

[81] OJEC Commission Directive 97/6/EC of 30 January 1997 amending Council Directive 70/524/EEC concerning additives in feedingstuffs. Off J Eur Commun 1997;L35:11–13.

[82] OJEC Council Regulation (EC) No 2821/98 of 17 December 1998 amending, as regards withdrawal of the authorization of certain antibiotics, Directive 70/524/EEC concerning additives in feedingstuffs. Off J Eur Commun 1998;L351:4–8.

[83] OJEC Commission Regulation (EC) No 2788/98 of 22 December 1998 amending Council Directive 70/524/EEC concerning additives in feedingstuffs as regards the withdrawal of authorization for certain growth promoters. Off J Eur Commun 1998;L347:31–2.

[84] OJEU Regulation (EC) No 1831/2003 of the European Parliament and of the Council of 22 September 2003 on additives for use in animal nutrition. Off J Eur Union 2003;L268:29–43.

[85] Capita R, Alonso-Calleja C. Antibiotic-resistant bacteria: a challenge for the food industry. Crit Rev Food Sci Nutr 2013;53:11–48.

[86] Casewell M, Friis C, Marco E, McMullin P, Philips I. The European ban on growth-promoting antibiotics and emerging consequences for human and animal health. J Antimicrob Chemother 2003;52:159–61.

[87] Singer RS, Finch R, Wegener HC, Bywater R, Walters J, Lipsitch M. Antibiotic resistance – the interplay between antibiotic use in animals and human beings. Lancet Infect Dis 2003;3:47–51.

[88] Phillips I. Withdrawal of growth-promoting antibiotics in Europe and its effects in relation to human health. Int J Antimicrob Agents 2007;30:101–7.

[89] European Commission, 2009. EU pesticides database. Available from: <http://ec.europa.eu/food/plant/pesticides/eu-pesticides-database/public/?event=homepage&language=EN>.

[90] Demanèche S, Sanguin H, Poté J, Navarro E, Bernillon D, Mavingui P, et al. Antibiotic-resistant soil bacteria in transgenic plant fields. Proc Natl Acad Sci USA 2008;105:3957–62.

[91] EFSA Consolidated presentation of the joint Scientific Opinion of the GMO and BIOHAZ Panels on the "Use of antibiotic resistance genes as marker genes in genetically modified plants" and the Scientific Opinion of the GMO Panel on "Consequences of the Opinion on the use of antibiotic resistance genes as marker genes in genetically modified plants on previous EFSA assessments of individual GM plants". EFSA J 2009;1108:1–8.

[92] EFSA Scientific Opinion on the risk posed by pathogens in food of non-animal origin. Part 1 (outbreak data analysis and risk ranking of food/pathogen combinations). EFSA J 2013;11:3025.

[93] Gerba CP. The role of water and water testing in produce safety Fan X, Niemira BA, Doona CJ, Feeherty FE, Gravani RB, editors. Microbial safety of fresh produce. New York, NY: Wiley and Sons; 2009. p. 129–42.

[94] Baquero F, Martínez J, Cantón R. Antibiotics and antibiotic resistance in water environments. Curr Opin Biotechnol 2008;19:260–5.

[95] Allen HK, Donato J, Wang HH, Cloud-Hansen KA, Davies J, Handelsman J. Call of the wild: antibiotic resistance genes in natural environments. Nat Rev Microbiol 2010;8:251–9.

[96] Schlüter A, Szczepanowski R, Pühler A, Top EM. Genomics of IncP-1 antibiotic resistance plasmids isolated from wastewater treatment plants provides evidence for a widely accessible drug resistance gene pool. FEMS Microbiol Rev 2007;31:449–77.

[97] EU European Union Regulation (EU) No 528/2012. Off J Eur Union 2012;27:1–123.

[98] SCENIHR Assessment of the antibiotic resistance effects of biocides Scientific committee on emerging and newly identified health risks. Brussels, Belgium: European Commission; 2009. Available from: <http://ec.europa.eu/health/ph_risk/committees/04_scenihr/docs/scenihr_o_021.pdf>.

[99] Li XL, Brownawell BJ. Quaternary ammonium compounds in urban estuarine sediment environments – a class of contaminants in need of increased attention? Environ Sci Technol 2010;44:7561–8.

[100] Martínez-Carballo E, Sitka A, González-Barreiro C, Kreuzinger N, Fürhacker M, Scharf S, et al. Determination of selected quaternary ammonium compounds by liquid chromatography with mass spectrometry. Part I. Application to surface, waste and indirect discharge water samples in Austria. Environ Pollut 2007;145:489–96.

[101] Capita R, Riesco-Peláez F, Alonso-Hernando A, Alonso-Calleja C. Exposure of *Escherichia coli* ATCC 12806 to sublethal concentrations of food-grade biocides influences its ability to form biofilm, resistance to antimicrobials, and ultrastructure. Appl Environ Microbiol 2014;80:1268–80.

[102] Karatzas KAG, Webber MA, Jorgensen F, Woodward MJ, Piddok LJV, Humphrey TJ. Prolonged treatment of *Salmonella enterica* serovar Typhimurium with commercial disinfectants selects for multiple antibiotic resistance, increased efflux and reduced invasiveness. J Antimicrob Chemother 2007;60:947–55.

[103] Lundén J, Autio T, Markkula A, Hellström S, Korkeala H. Adaptive and cross-adaptive responses of persistent and non-persistent *Listeria monocytogenes* strains to disinfectants. Int J Food Microbiol 2003;82:265–72.

[104] Fox E, Hunt K, O'Brien M, Jordan K. *Listeria monocytogenes* in Irish Farmhouse cheese processing environments. Int J Food Microbiol 2011;145:S39–45.

[105] Fox EM, Leonard N, Jordan K. Physiological and transcriptional characterization of persistent and nonpersistent *Listeria monocytogenes* isolates. Appl Environ Microbiol 2011;77:6559–69.

[106] Müller A, Rychli K, Muhterem-Uyar M, Zaiser A, Stessl B, Guinane CM, et al. Tn6188 – A novel transposon in *Listeria monocytogenes* responsible for tolerance to benzalkonium chloride. PLoS One 2013;8:e76835.

[107] Leong D, Alvarez-Ordóñez A, Zaouali S, Jordan K. Examination of *Listeria monocytogenes* in seafood processing facilities and smoked salmon in the Republic of Ireland. J Food Protect 2015;78:2184–90.

[108] Copitch JL, Whitehead RN, Webber MA. Prevalence of decreased susceptibility to triclosan in *Salmonella enterica* isolates from animals and humans and association with multiple drug resistance. Int J Antimicrob Agents 2010;36:247–51.

[109] Lambert RJW. Comparative analysis of antibiotic and antimicrobial biocide susceptibility data in clinical isolates of methicillin-sensitive *Staphylococcus aureus*, methicillin-resistant *Staphylococcus aureus* and *Pseudomonas aeruginosa* between 1989 and 2000. J Appl Microbiol 2004;97:699–711.

[110] Beier RC, Anderson PN, Hume ME, Poole TL, Duke SE, Crippen TL, et al. Characterization of *Salmonella entérica* isolates from turkeys in comercial processing plants for resistance to antibiotics, disinfectants, and a growth promoter. Foodborne Pathog Dis 2011;8:593–600.

[111] Coelho JR, Carriço JA, Knight D, Martínez JL, Morrissey I, Oggioni MR, et al. The use of machine learning methodologies to analyse antibiotic and biocide susceptibility in *Staphylococcus aureus*. PLoS One 2013;8:e55582.

[112] Langsrud S, Sundheim G, Holck AL. Cross-resistance to antibiotics of *Escherichia coli* adapted to benzalkonium chloride or exposed to stress-inducers. J Appl Microbiol 2004;96:201–8.

[113] Randall LP, Cooles SW, Coldham NG, Penuela EG, Mott AC, Woodward MJ, et al. Commonly used farm disinfectants can select for mutant *Salmonella enterica* serovar Typhimurium with decreased susceptibility to biocides and antibiotics without compromising virulence. J Antimicrob Chemother 2007;60:1273–80.

[114] Whitehead RN, Overton TW, Kemp CL, Webber MA. Exposure of *Salmonella enterica* serovar Typhimurium to high level biocide challenge can select multidrug resistant mutants in a single step. PLoS One 2011;6:e22833.

[115] Webber MA, Whitehead RN, Mount M, Loman NJ, Pallen MJ, Piddock LJV. Parallel evolutionary pathways to antibiotic resistance selected by biocide exposure. J Antimicrob Chemother 2015;70:2241–8.

[116] Hernández A, Ruiz FM, Romero A, Martínez JL. The binding of triclosan to SmeT, the repressor of the multidrug efflux pump SmeDEF, induces antibiotic resistance in *Stenotrophomonas maltophilia*. PLoS Pathog 2011;7:e1002103.

[117] Bjorland J, Steinum T, Kvitle B, Sunde M, Heir E. Widespread distribution of disinfectant resistance genes among *Staphylococci* of bovine and caprine origin in Norway. J Clin Microbiol 2005;43:4363–8.

[118] Gaze WH, Zhang L, Abdouslam NA, Hawkey PM, Calvo-Bado L, Royle J, et al. Impacts of anthropogenic activity on the ecology of class 1 integrons and integrin-associated genes in the environment. ISME J 2011;5:1253–61.

[119] Gillings MR, Gaze WH, Pruden A, Smalla K, Tiedje JM, Zhu YG. Using the class 1 integron-integrase gene as a proxy for anthropogenic pollution. ISME J 2015;9:1269–79.

[120] Wang HH, Manuzon M, Lehman M, Wan K, Luo H, Wittum TE, et al. Food commensal microbes as a potentially important avenue in transmitting antibiotic resistance genes. FEMS Microbiol Lett 2006;254:226–31.

[121] Alvarez-Ordóñez A, Broussolle V, Colin P, Nguyen-The C, Prieto M. The adaptive response of bacterial food-borne pathogens in the environment, host and food: implications for food safety. Int J Food Microbiol 2015;213:99–109.

[122] McMahon MA, Xu J, Moore JE, Blair IS, McDowell DA. Environmental stress and antibiotic resistance in food-related pathogens. Appl Environ Microbiol 2007;73:211–7.

[123] Martínez JL, Rojo F. Metabolic regulation of antibiotic resistance. FEMS Microbiol Rev 2011;35:768–89.

[124] Rodrigo D, Sampedro F, Silva A, Palop A, Martínez A. New food processing technologies as a paradigm of safety and quality. Brit Food J 2010;112:467–75.

Chapter 3

Penetrating the Bacterial Biofilm: Challenges for Antimicrobial Treatment

E. Teirlinck[1,2], S.K. Samal[1,2], T. Coenye[1] and K. Braeckmans[1,2]

[1]Ghent University, Ghent, Belgium [2]Centre for Nano- and Biophotonics, Ghent, Belgium

CHAPTER OUTLINE

3.1 INTRODUCTION

3.1.1 Antimicrobial Treatment: In Desperate Need of a Wind of Change

Without urgent, coordinated action by many stakeholders, the world is headed for a post-antibioticera, in which common infections and minor injuries which have been treatable for decades can once again kill [1]

Dr. Keiji Fukuda, WHO's Assistant Director-General, Health Security

Functionalized Nanomaterials for the Management of Microbial Infection.

The increasing antibiotic resistance against commonly used antimicrobial agents is a worldwide threat, and has reached alarming levels. Multidrug resistant bacteria cause the death of 25,000 Europeans each year, with an estimated burden on the healthcare system of €1.5 billion annually [2]. According to a recent study published by the Centres for Disease Control (CDC) nearly half a million people in the United States are infected annually with the life-threatening bacterium *Clostridium difficile*. Among these, at least 29,000 people died as a consequence of the infection [3]. Despite the obvious urgent need for new antibiotics, the number of approvals is low [4]. As the impact on public health and economics is tremendous, the World Health Organization (WHO) has made bacterial resistance one of its central focuses. Also, the WHO warned in its last global report that the 21st century can truly be the beginning of the "post-antibiotic era," a period in which people worldwide can die from common infections, because the available antibiotic treatments become ineffective, including the so-called "last resort" antibiotics [1]. Obtaining a better understanding of the resistance mechanisms and the concomitant development of improved treatments is key for the future, as acknowledged in the May 2015 World Health Assembly action plan that attempts to manage the crisis globally (Box 1.1) [5].

BOX 1.1 THE WHO'S ACTION PLAN TO COMBAT ANTIMICROBIAL RESISTANCE [5]

The goal of the WHO's global action plan is on the one hand to prolong antibiotic treatments that are still successful in eradicating resistant bacteria, and on the other hand to decrease the number of emerging/existing resistant infections worldwide. Their strategy is essentially based on 5 cornerstones:

1. More communication with different sectors (including healthcare, animal science, political authorities…) in order to increase awareness of the resistance problem
2. Increase academic, industrial and economic research to maximize the knowledge about the incidence of antimicrobial resistance, related microorganisms, mechanisms of resistance development, effectiveness of existing and newly approaching treatments, cost of antimicrobial resistance
3. Reduce the amount of infections
4. Limit misuse of antimicrobial agents
5. Maximize the economic support to drug development, health care institutions, vaccine supply.

3.1.2 **The Biofilm Lifestyle**

Intensive research has been conducted to obtain a better understanding of how bacteria become multiresistant to a range of antibiotics, and how this results in chronic, difficult to eradicate infections. In vitro and in vivo observations of persistent infections showed that bacteria are often organized as bacterial aggregates or biofilms, which contributes to a great extent to the problem of persistent infections [6]. In these biofilms, the bacterial cells are packed together into dense large clusters, surrounded by a self-produced matrix of extracellular polymeric substances (EPS), including proteins, DNA, and polysaccharides [7]. These so-called sessile bacteria can communicate with each other in a highly organized fashion, which enables them to respond and adapt instantly to changing environmental conditions [8]. Furthermore, the organized architecture of biofilms also acts as a protective barrier that protects biofilm bacteria from the host immune response [9].

One of the typical properties of biofilms is their increased tolerance to antibiotic agents. They can withstand 10–1000 times higher antibiotic concentrations compared to their free-floating planktonic counterparts [10]. This raises the question of how to combat these bacterial biofilm-related infections. In order to find a solution to this pressing problem, one must unravel the precise mechanisms behind the increased biofilm tolerance. Distinctions can be made between four distinct, but intertwined processes that lead to increased tolerance [10–12]:

1. The existence of different microenvironments within a biofilm, such as pockets of anaerobic regions or low pH due to local accumulation of acidic waste products, that may antagonize the action of antibiotics [13].
2. The presence of a subpopulation of difficult to eradicate persister cells.
3. The ability to counteract oxidative stress.
4. The slow or incomplete penetration of antibiotics through biofilms.

In this book chapter we will focus on the last tolerance mechanism. Limited mass transfer through biofilms is a recognized mechanism of tolerance and is caused for two main reasons. First, due to the fact that the bacterial cells are packed together in large dense cell clusters composed of hundreds or thousands of bacteria, they are more difficult to reach by antibiotics as compared to free-floating planktonic cells [14,15]. Second, the cells and cell clusters are surrounded by a sticky matrix, sometimes referred to as the "house of biofilm cells" [16]. The biofilm matrix is a conglomeration of polysaccharides, proteins and extracellular DNA, providing strength and protection to the biofilm [17]. Improving antibiotics penetration through biofilms is an important strategy to achieve a better treatment efficacy of both existing and potential new antibiotics (Box 1.1).

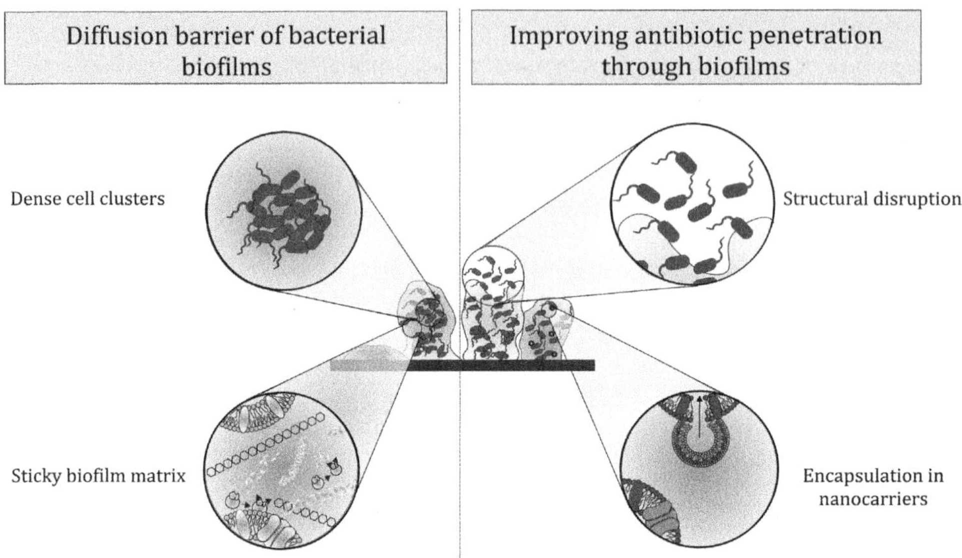

■ **FIGURE 3.1** Schematic representation of the architecture of bacterial biofilms and different approaches to improve antibiotic penetration. *Left*: Two important features of biofilms contribute to their limited diffusivity to antimicrobial agents. First, cells are packed into dense cell clusters, due to which antibiotics cannot easily reach deep cell layers. Second, sessile cells are surrounded by a biofilm matrix that can hinder antibiotic diffusion, due to physicochemical interactions of antibiotics with matrix constituents. *Right*: Improved antibiotic penetration can essentially be obtained by two therapeutic approaches. One the one hand, the biofilm structure can be degraded in order to release cells from their protective environment. On the other hand, antibiotics can be shielded from interactions with biofilm matrix constituents by encapsulation in nanocarriers.

In this chapter, we will first look more closely into the structural organization of the bacterial biofilm in relation to hindered antibiotics penetration (Fig. 3.1). Special attention will be given to the role of dense cell clusters, the sticky biofilm matrix and the presence of degrading enzymes [11,12,18]. Next, we will explain different therapeutic approaches to improving antibiotics penetration through bacterial biofilms. Both physical methods to destruct the biofilm's organization as the use of antibiotic-enclosing nanoparticles will be discussed. At the end of the chapter, the reader will have gained more insight into the different challenges that antibiotics encounter while diffusing through bacterial biofilms, and will have become more acquainted with different strategies that are currently explored to overcome this problem.

3.2 HOW BIOFILMS POSE A CHALLENGE TO THE DIFFUSION OF ANTIBIOTICS

3.2.1 Mass Transport Phenomena in Biofilms

At first, biofilms were considered as simple systems of four planar layers lying on top of each other. These four layers were, from the bottom to the top, the substratum, the biofilm, the bulk liquid, and a possible gas space [19]. It was accepted that passive diffusion was the predominant mode of transport within biofilms. Only in the liquid bulk layer residing on top of the biofilm, convective flow could happen [20,21]. With the introduction of confocal laser scanning microscopy (CLSM), this view on the structural organization of biofilms changed radically [22,23]. Confocal microscopy enabled 3-D high-resolution imaging of biofilms [24], revealing their complex heterogeneous architectures [25]. The bacterial cells were found to be organized in dense clusters of hundreds to even thousands of cells surrounded by a matrix of polymeric material that holds the cell aggregates together and contributes to the strength of the entire biofilm [9]. CLSM also revealed that biofilms contain voids—open channels that run through the biofilm and are connected to the bulk liquid [26]. Microscopic evaluation confirmed the possibility of liquid flow through these channels [27]. In cell clusters, on the other hand, liquid flow was hindered, and became stagnant [27]. Hence, two modes of mass transport are active inside biofilms. Inside the voids, both convection and diffusion are possible, whereas in the cell clusters diffusion is the principal mass transport mechanism [28]. This was an important discovery, as the voids can transfer solutes quickly throughout the biofilm, but do not give direct access to the inner parts of the cell clusters [29].

Because of the heterogeneity of biofilms and different penetration pathways (via voids and dense cell clusters), contradicting findings were published in literature regarding the existence of a diffusion barrier. Indeed, many studies state that antibiotics move fairly rapidly through biofilms [30–32], however, at the same time, hindered penetration of antibiotics has also been frequently observed, as described above [33–39]. Confusion about the existence of a diffusion barrier is confounded by the way it is experimentally investigated. If the arrival of antibiotics at the base of the biofilm is taken as a measure for diffusion [29], often one will conclude that there is no diffusion barrier since antibiotics can reach those parts fairly easily through the large voids. However, this does not necessarily mean that the antibiotics have efficiently penetrated deep into the cell clusters [10]. Another important issue is that it may not be sufficient to measure the arrival of antibiotics at a certain location in the biofilm. Also the rate of diffusion is an important aspect. It is possible that an antibiotic molecule is indeed transported through the biofilm, but that its rate of penetration is significantly reduced. When the sessile cells experience gradually increasing antibiotic concentrations instead of an immediate full antibiotic dose, they can have the time to mount a defensive response [40,41]. Increased production of alginate has been reported by treating *Pseudomonas aeruginosa* biofilms with sub-inhibitory imipenem concentrations, hence resulting in a more viscous biofilm matrix [42]. Sub-inhibitory concentrations of β-lactams increased colonic acid production in *Escherichia coli* biofilms, resulting in a more matured biofilm matrix [43].

In conclusion, although limitation of mass transfer remains controversial, the controversy is inspired by differences in measurements, the complex structure of biofilms and the fact that it depends on the specific combination of biofilms and antibiotics [44]. One can safely conclude that a diffusion barrier can exist as evidenced by various studies for many antibiotics.

3.2.2 **Dense Cell Clusters**

To reach the cells in the deeper layers of the clusters, the antibiotics have to gradually diffuse through the narrow spaces in between the cells. A first question is whether there is a size limit for diffusion into the dense cell clusters.

De Beer et al. examined the diffusion of three fluorescent dyes with different molecular weights in mixed biofilms of *Klebsiella pneumoniae, P. aeruginosa,* and *Pseudomonas fluorescens* [39,45]. After microinjecting the dyes in the interstitial voids and cell clusters separately, fluorophore diffusion was analyzed. By evaluating how fast fluorophores can diffuse out of the initial spot, the local diffusion coefficient of the molecules could be measured in selected areas—rather than reporting an average diffusion coefficient for the entire biofilm. Fluorescein (MW 332) exhibited similar diffusion in the voids and in the cell clusters, whereas dyes of higher molecular weight such as TRITC-IgG (MW 150,000) and phycoerythrin (MW 240,000) did not diffuse or were slowed down by 40% in the cell clusters, respectively. However, likely these findings are not caused by steric hindrance, but rather by binding interactions with biofilm constituents as penetration of TRITC-IgG was improved by adding bovine serum albumin [39,45]. Indeed, in a recent study it was confirmed that nanoparticles up to around 100nm can penetrate into dense biofilm clusters of *Burkholderia multivorans* and *P. aeruginosa* virtually as efficiently as small dextrans of 4kDa [33]. The authors used inert PEGylated polystyrene nanoparticles and liposomes ranging from 40 to 550nm, which were visualized in their journey toward the biofilm cell clusters by confocal microscopy (Fig. 3.2). A general consensus about the optimal size of nanoparticles was obtained: particles smaller than 100–130nm are able to diffuse in the cell clusters, while penetration into the clusters was gradually less efficient for larger particles.

This study confirms that the size of antibiotics is not likely to play a significant role in the diffusion into dense cell clusters [33]. What does restrict

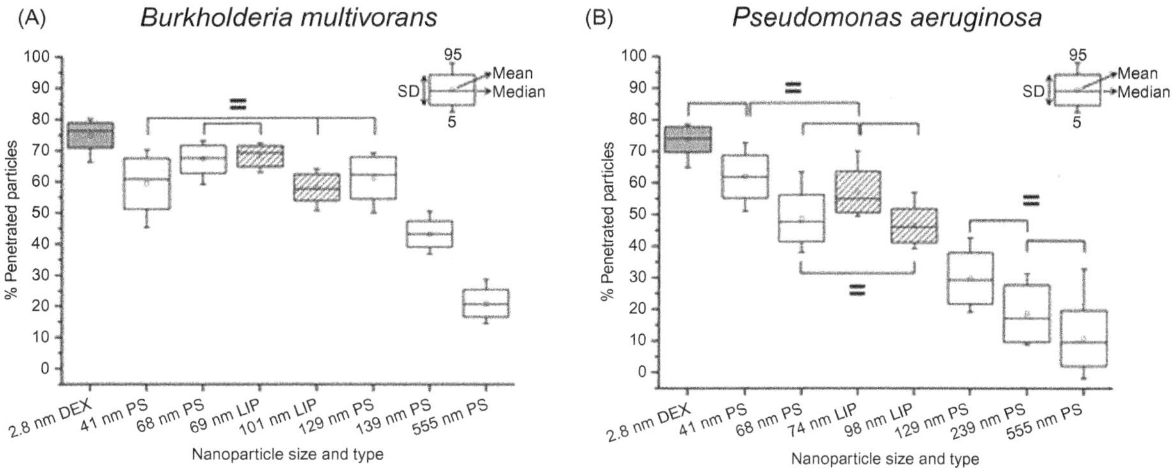

■ **FIGURE 3.2** The percentage of penetrated polystyrene nanoparticles (open), liposomes (dashed) and control molecules (FITC–dextran 4000 Da, gray) in *Burkholderia multivorans* LMG18825 (left, A) and *Pseudomonas aeruginosa* LMG27622 biofilms (right, B) [33].

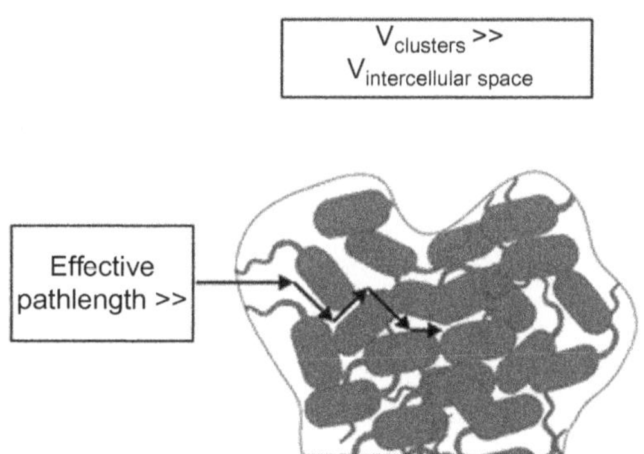

■ **FIGURE 3.3** Restricted antibiotic diffusion into dense cell clusters of bacterial biofilms.

net antibiotics influx into dense clusters is the fact that most of the cluster volume is occupied by the bacterial cells, while the space in between the cells through which the antibiotics have to diffuse constitutes only a minor volume fraction (Fig. 3.3). This evidently limits the amount of antibiotics present within the clusters at any time, potentially leading to a suboptimal dose and ineffective treatment. In addition, as the antibiotics have to diffuse in between densely packed cells, the effective path length toward the cluster center is increased [29], further contributing to a delayed exposure of the deeper cell layers, which have more time to mount a defensive response.

3.2.3 **Binding Interactions Between Antibiotics and Matrix Constituents**

The second important characteristic of biofilms that contributes to their decreased diffusivity is their heterogeneous matrix that surrounds the cells and cell clusters. Antibiotic penetration in the biofilm can be substantially delayed or even inhibited by (transient) binding to matrix constituents [46]. In addition, antibiotics may become inactivated by enzymatic degradation or modification.

3.2.3.1 Physicochemical Binding to Matrix Polymers

It has been demonstrated in many studies that hindered diffusion can be the result of electrostatic interaction between antibiotic molecules and matrix polymers of opposite surface charge [46]. For example, aminoglycosides such as tobramycin and gentamicin penetrate biofilms of mucoid *P. aeruginosa* slower than β-lactam antibiotics (Fig. 3.3). This is caused by the fact that the aminoglycosides, unlike β-lactams, can bind to extracellular polymers, such as alginates [34–37]. Alginate is produced by mucoid *P. aeruginosa* strains and is implicated in bacterial adhesion to substrates such as the respiratory epithelium [47]. It is a copolymer consisting of [1,4]-linked-β-D-mannuronate and α-L-guluronate monomers which confers a negative charge to the polysaccharides (Fig. 3.4) [48]. Positively charged antibiotics, such as aminoglycosides, can bind to negatively charged alginate polymers by means of an electrostatic interaction.

Recently, it has been found that not only mucoid strains offer protection against antibiotic penetration, but also the matrix in non-mucoid strains contains elements that retard antibiotic diffusion. Tseng et al. clearly demonstrated that, unlike the neutral fluoroquinolone ciprofloxacin, positively charged tobramycin was sequestered at the outer surface of nonmucoid *P. aeruginosa* biofilms. Therefore, it could not fully penetrate the biofilm anymore. To acquire information about which component of the matrix accounts for the tobramycin sequestration, different mutants of known EPS matrix components were examined. However, no polymers could be assigned as sequestering agents for tobramycin, because the penetration of tobramycin was similar in all biofilms. Further experiments showed that the tobramycin binding sites in the biofilm matrix could be saturated by adding high concentrations of either tobramycin or divalent cations. In both cases, the penetration of tobramycin was substantially increased, because ionic interactions with the matrix components were no longer possible [49]. Billing et al. found that Psl, a known matrix component of non-mucoid *P. aeruginosa* biofilms, was responsible for sequestering

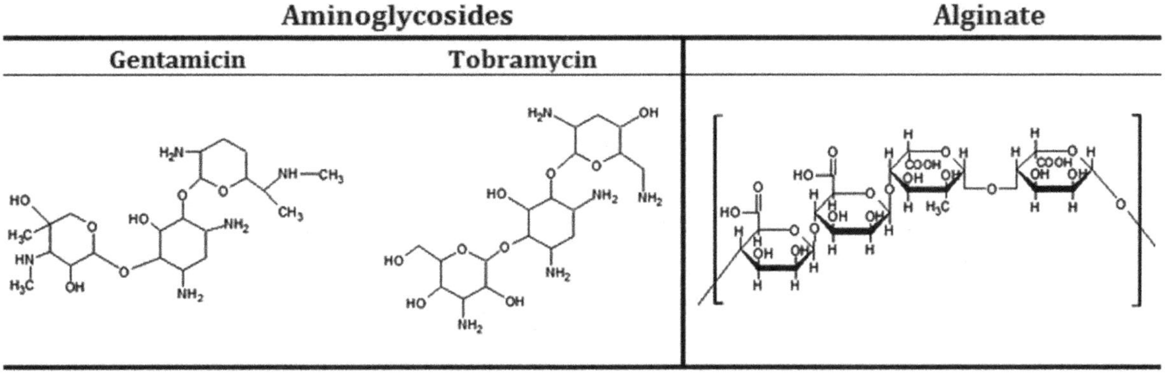

Aminoglycosides   Alginate

Gentamicin   Tobramycin

■ FIGURE 3.4 Structure of aminoglycosides gentamicin, tobramycin and the polysaccharide alginate.

tobramycin. Psl was able to interact electrostatically with tobramycin, in this way contributing to the matrix barrier function of *P. aeruginosa* biofilms [50,51]. Finally, Pel, the third known extracellular polysaccharide involved in *P. aeruginosa* biofilm development—besides alginate and Psl—has also been reported to play a role in protecting *P. aeruginosa* biofilms against aminoglycosides [52].

Hindered penetration in biofilms has also been described for biocides [53]. Chlorine, a biocide, is known to suffer from impaired penetration in biofilms [38]. This was confirmed by direct measurement of chlorine concentrations with microelectrodes in biofilms of *P. aeruginosa* and *K. pneumoniae*; the chlorine concentration inside the clusters was not more than 20% of the incubation solution [38]. A follow-up study revealed that biofilm constituents could trap chlorine in the periphery of the biofilm before it could diffuse inside the biofilm [54]. Davison et al. measured that the time chlorine needed to fully penetrate a *Staphylococcus epidermidis* cluster was about 30 minutes, while it would take ± 3s to diffuse if sorption did not occur (600 × slower) [55].

3.2.3.2 Inactivation of Antibiotics by Enzymes Present in the Biofilm Matrix

Degradative or modifying enzymes that are present in the biofilm matrix constitute yet another barrier to antibiotics in biofilms [12,46,56].

Anderl et al. showed that the inability of ampicillin to eradicate wild-type *K. pneumoniae* biofilms was caused by enzymes in the biofilm surface layer degrading ampicillin before reaching the cells (Fig. 3.5) [57]. In β-lactamase deficient strains, ampicillin readily penetrated the biofilms,

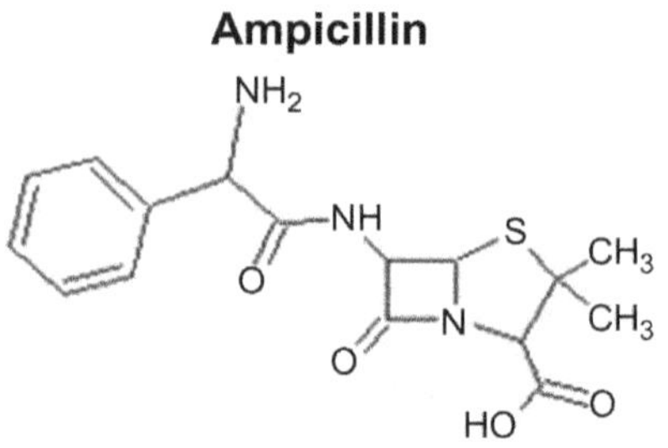

Ampicillin

■ FIGURE 3.5 Structure of ampicillin, with the characteristic β-lactam ring.

which supports the hypothesis that the penetration limitation is due to a reaction-diffusion barrier.

Besides antibiotics, biocides can also be enzymatically inactivated in biofilms [53,58,59]. Studies concerning the penetration of hydrogen peroxide into *P. aeruginosa* biofilms revealed that penetration was largely retarded, likely by catalases [60]. Catalases are enzymes that defend cells against oxidative stress by converting hydrogen peroxide into water and oxygen molecules ($2H_2O_2 \rightarrow 2\ H_2O + O_2$) [61]. To test this assumption, H_2O_2 penetration was assessed in wild-type and catalase-deficient *P. aeruginosa* biofilms. In the wild-type biofilms, H_2O_2 failed to fully penetrate the biofilm mass, as opposed to the catalase-deficient mutant biofilms, where H_2O_2 succeeded in penetrating the biofilm [60]. Hence, catalase plays a role in protecting biofilm bacteria by decomposing H_2O_2 into its degradation products H_2O and O_2 in a reaction-diffusion mechanism [62].

3.3 HOW TO IMPROVE ANTIBIOTIC DELIVERY TO BIOFILM CELLS?

In the previous section we discussed the two factors that contribute to hindered diffusion of antibiotics in biofilms, i.e., reduced penetration in dense cell clusters and binding to matrix constituents. Consequently, improved antibiotics penetration could be achieved by either releasing the cells from their protective environment [63,64], or by shielding the antibiotics from interacting with matrix constituents [65,66]. In the next sections we will discuss the various approaches that have been developed to this end.

3.3.1 Interference With the Biofilm Structure

In order for antibiotics to more easily reach all cells in biofilms, cells could be released from their protective environment by interfering with the biofilm structure [35,67–72]. Indeed, it has been observed in many studies that by degrading the biofilm structure, bacteria become more susceptible to antimicrobial treatment [71]. An additional benefit is that releasing bacteria from their protective environment can also enhance recognition by polymorphonuclear leukocytes, aiding the immune system to clear the bacterial biofilm infection [73]. Furthermore, degradation of the biofilm does not only increase antibiotic diffusion, but also the diffusion of nutrients, which activates the bacterial cells and renders them more susceptible for antibiotics [74].

Interference with the biofilm structure can be achieved by specific compounds that degrade the biofilm matrix, for instance matrix dispersants that destabilize EPS or specific biofilm components such as eDNA [63,72]. Messiaen et al. investigated the potential of NaClO and dispersin B to degrade the EPS matrix and the specific matrix component poly-β-1,6-N-acetylglucosamine (PNAG) of *Burkholderia cepacia* biofilms, respectively [75]. Destabilization of both compounds resulted in increased tobramycin susceptibility, confirming the fact that the EPS matrix plays a role in the tobramycin susceptibility of *B. cepacia* biofilms [75]. Hatch et al. demonstrated the potential of alginate lyase in reducing the diffusion barrier of mucoid *P. aeruginosa* biofilms to aminoglycosides [47]. Alginate lyase enzymatically cleaves the glycosidic bounds of alginate by β-elimination [76]. Degradation of the biofilm matrix by alginate lyase increased the diffusion of aminoglycosides into mucoid *P. aeruginosa* biofilms and improved the aminoglycoside activity [47].

Another way to interfere with biofilm formation is to target the bacterial communication system, called quorum sensing (QS). QS is the process in which bacteria release signaling molecules in a cell density-dependent manner, leading to altered gene expression. QS regulates different bacterial processes such as virulence, motility, and biofilm formation [77,78]. Although more research needs to be performed to elucidate the precise role of QS in biofilm formation [79,80], many studies reported that quorum sensing inhibitors (QSI) can be a promising new class of antimicrobial agents that affect the biofilm's structural organization [81,82]. Davies et al. demonstrated that *lasI* mutants of *P. aeruginosa* formed flat biofilms instead of the complex bacterial communities in case of wild-type biofilms. Moreover, the quorum-sensing mutants were also more susceptible to sodium dodecyl sulfate, demonstrating the potential of QSIs in increasing the susceptibility of biofilms [83]. Similar results were reported for *Burkholderia cenocepacia* [84], where QS-deficient mutants showed an altered biofilm formation and increased sensitivity to sodium dodecyl sulfate [84]. Brackman et al. investigated the potential of combining different antibiotics and QSIs to affect the susceptibility of medically relevant biofilms. Both QSIs targeting the acylhomoserine lactone-based QS-system of *P. aeruginosa* and *B. cepacia*, as the peptide-based system of *Staphylococcus aureus*, were investigated. They concluded that treating biofilms with QSIs is promising for the eradication of bacterial biofilm infections, as the combined use of QSIs and antibiotics resulted in increased killing efficiency [85].

The disadvantage of using pharmacological compounds is that their specificity restricts broad applicability [86]. Indeed, the biofilm matrix composition differs between various bacterial species and strains [87,88], and the presence of polymicrobial biofilms further increases the matrix heterogeneity [86,89]. Physical approaches, on the other hand, that interfere with the biofilm structure could be more generally applicable.

Many studies investigated the potential of ultrasonic waves to mechanically disrupt biofilms and consequently enhance antibiotic efficacy [90–94]. Williams et al. showed that ultrasonic treatment of mucoid strains of *P. aeruginosa* resulted in a significant decrease in alginate viscosity. Degradation of the mucoid polymer markedly increased the diffusion of tobramycin and piperacillin and consequently their efficacy [95]. Another example is the use of low-frequency ultrasound to disrupt biofilms of *E. coli*, which consequently enhances the efficacy of gentamycin in the eradication of the biofilm [96]. Also, Dong et al. demonstrated the advantage of applying ultrasound (and ultrasound-mediated microbubbles) in the treatment of *S. epidermidis* biofilms, as ultrasound improved the penetration of vancomycin significantly [91]. Furthermore, promising in vivo data were reported regarding the use of ultrasound for disrupting biofilm structures, such as in the treatment for chronic rhinosinusitis [97]. Also, different in vivo studies reported that ultrasound could be used to disrupt biofilms of *E. coli*, resulting in enhanced efficacy of aminoglycosides [98–100]. Carmen et al. assessed the capability of low-frequency ultrasound to enhance the activity of vancomycin on the Gram-positive bacterium *S. epidermidis*. Biofilms were formed on polyethylene disks and implanted subcutaneously in rabbits. The combination of low-frequency ultrasound and systemic vancomycin treatment reduced the number of viable bacteria, without causing detrimental tissue damage [101].

Another example is the use of laser-generated shockwaves to physically disrupt biofilms. Efficient *S. epidermidis* biofilm disruption was reported with the use of a Q-switched ND:YAG laser [70]. Nigri et al. demonstrated that the combination of laser-generated shockwaves with tobramycin could be used for efficient eradication of biofilms related to vascular prosthetic grafts. Shockwaves were generated in biofilms consisting of *S. epidermidis* and *S. aureus*, both isolated from vascular grafts. Subsequent tobramycin treatment resulted in a significant decrease in colony forming units, pointing out the potential of shockwaves to disrupt biofilm structures and to enhance the antibiotic's efficiency [102].

In conclusion, physical methods such as ultrasonic treatment have been successfully used to disrupt biofilms and release bacteria from their protective environment. Hence, the combination of physical biofilm disruption and concurrent antibiotic treatment is a promising route to explore in order to eradicate persistent biofilm infections [73,103].

3.3.2 Nanocarriers for Improved Delivery of Antibiotics in Biofilms

3.3.2.1 Nanocarriers Protect Antibiotics From Interactions With Biofilm Components

Encapsulation of antibiotics inside nanomaterials reduces physicochemical interactions between biofilm matrix components and antimicrobial molecules [18,66,104–107]. Hence, nanosized carriers can protect antibiotics from detrimental interactions with biofilm components [33]. Meers et al. investigated if encapsulation of amikacin in DPPC:Chol liposomes could be beneficial in the treatment of chronic *P. aeruginosa* biofilm infections. By CLSM it could be shown that those liposomes were able to penetrate *P. aeruginosa* biofilms, resulting in a higher antibiotics concentration near the center of the biofilm as compared to the bulk fluid. Consequently, liposomal amikacin could reduce the bacterial count of *P. aeruginosa* infections more than free amikacin [108,109]. Mugabe et al. compared the activity of liposome-encapsulated gentamicin and free gentamicin against resistant strains of *P. aeruginosa* isolates. DMPC:Chol, DPPC:Chol, and DSPC:Chol liposomes encapsulating gentamicin exhibited a higher antimicrobial activity against *P. aeruginosa* biofilms than gentamicin by itself. The authors attributed this effect to either an enhanced diffusion of the enclosed antibiotics through the biofilms, or protection provided by the liposomes against enzymatic degradation [110].

Nanocarriers also offer protection against antibiotic degradation by detrimental enzymes that can be present in the biofilm matrix [65]. Nacucchio et al. investigated if shielding piperacillin inside liposomes could offer protection against degradation by β-lactamases leading to improved bacterial efficacy. Therefore, both piperacillin containing phosphatidylcholine:cholesterol (1:1) liposomes and exogenous β-lactamase were added to *S. aureus* biofilms and staphylococcal growth was examined. Nonencapsulated piperacillin was hydrolyzed by exogenous β-lactamases and resulted in similar *S. aureus* growth as the control. Attaching piperacillin to the outside of (empty) liposomes provided protection against β-lactamases, as lower viability of *S. aureus* was observed. This is thought to be due to steric hindrance for the enzymes. However, most successful protection was provided when piperacillin was encapsulated in the interior of the liposomes, resulting in the lowest staphylococcal growth of all conditions [111].

Besides liposomal entrapment, encapsulation in polyacrylate nanoparticles has also been reported to protect antibiotics against enzymatic degradation. Turos et al. showed that nanoparticles loaded with penicillin exhibited increased antimicrobial activity against MRSA infections compared to free penicillin. Antimicrobial activity was even restored when exogenous β-lactamases were added, proving the fact that polymeric encapsulation can hide antibiotics against enzymatic attack by β-lactamases [112].

3.3.2.2 Nanocarriers can be Designed to Increase Local Antibiotic Delivery

Besides size, surface charge and hydrophobicity are important for efficient delivery of NPs inside biofilms [113,114]. For instance, Li et al. showed that anionic and neutral quantum dots could not penetrate *E. coli* biofilms, while cationic ones could [115]. Furthermore, the authors demonstrated that hydrophobic cationic quantum dots could reach sessile cells, while more hydrophilic cationic quantum dots were bound to EPS matrix components. This shows that, by rationally designing the NP outer surface through modifications of charge and hydrophobicity, nanoparticle-enclosed drugs can be targeted toward the EPS matrix (e.g., dispersing agents such as dispersin B) or bacterial aggregates (e.g., antimicrobial agents such as vancomycin) [115].

By making use of charge-based attraction of NPs to biofilm constituents, nanometer-sized delivery vehicles can reach parts in biofilms that are inaccessible to free antibiotics [114]. Duncan et al. combined encapsulated peppermint oil and cinnamaldehyde with cationic silica NPs to treat bacterial biofilm infections [116]. Self-assembly of the silica NPs at the oil-water phase resulted in capsules consisting of a core of peppermint oil and cinnamaldehyde and a shell of cationic NPs. Due to the penetration properties of the positive silica NPs, the capsules were able to diffuse into the entire biofilm, as visualized by confocal microscopy. Antimicrobial activity of the capsules was assessed against full-grown biofilms of *E. coli*, *P. aeruginosa*, methicillin-resistant *S. aureus,* and *Enterobacter cloacae*. It was found that viability significantly decreased with capsules carrying phytochemicals compared to free phytochemicals. Hence, transporting antimicrobial agents by nanometer-sized delivery vehicles directly to the biofilm-enclosed bacteria, greatly improved their therapeutic efficiency [116].

Inclusion of certain lipids such as phosphatidylinositol (PI) or stearylamine (SA) can also be used to achieve interaction between phospholipid liposomes and specific compounds in the biofilm [117]. The targeting potential of PI has been demonstrated to a variety of biofilm-enclosed bacteria such as *Streptococcus mutans, Streptococcus gordonii, Streptococcus sanguis, Proteus vulgaris,* and *S. epidermidis* [118–120]. Attractive interactions are accomplished by forming hydrogen bonds between hydroxyl groups of inositol-molecules and polymers present on the bacterial surface [121]. Stearylamine, on the other hand, can be used to achieve electrostatic interactions between positively charged SA-bearing liposomes and negative charges present in the bacterial biofilm [122]. Stearylamine is known to guide liposomes to sessile bacteria such as *S. epidermidis, S. aureus,* and to a lesser extent to *P. vulgaris* and oral bacteria such as *S. mutans* and *S. sanguis* [123,124]. When comparing the affinity of PI:DPPC liposomes and DPPC-cholesterol-SA liposomes to *S. epidermidis* biofilms, SA-liposomes possessed greater affinity for the bacteria [125]. A possible explanation could be that hydrogen bonds originating from PI-biofilm interactions are less efficient than the electrostatic interactions between SA-liposomes and negatively charged biofilm constituents [65].

The possible disadvantage of using charge-based or hydrophobic interactions is the lower specificity, as most bacteria have a negative cell wall, accessible for electrostatic interactions, and matrix components, able to interact through hydrogen bonds [126]. Indeed, to avoid detrimental side-effects, it is essential that antibiotics are directed to the pathogenic bacterial species (and/or strains), while they do not exert any bacteriostatic/bactericide effect on the commensal bacteria [127]. Precise targeting of NPs to specific bacterial species and strains can be accomplished by functionalizing nanocarriers with targeting agents such as antibodies or lectins which interact with EPS and/or sessile bacteria [114]. Highly specific targeting to *Streptococcus oralis* was accomplished by decorating DPPC:PI liposomes with antibodies raised to surface antigens of *S. oralis.* The immunoliposomes possessed high specificity to *S. oralis,* as little liposome adsorption was noticed around other oral bacteria like *S. sanguis.* For low chlorhexidine concentrations, targeted liposomal delivery resulted in enhanced growth inhibition of *S. oralis* in comparison with non-encapsulated bactericides [128,129]. Other site-directing molecules used to target NPs to bacterial biofilms include carbohydrate-binding lectins such as concanavalin A (con-A) and wheat germ agglutinin (WGA) [130,131].

Con-A selectively targets α-D-mannopyranosyl and α-D-glucopyranosyl units, while WGA has a high specificity for N-acetylneuraminic acid and *N*-acetylglucosamine, present in peptidoglycan layer of bacterial cell walls and some lipopolysaccharides [132,133]. Effective targeting to the carbohydrate receptors of *Helicobacter pylori* could be obtained by functionalizing gliadin nanoparticles with lectins, more specifically fructose-specific (UEA-I) and mannose-specific (Con A) lectins [134]. WGA functionalized DPPC:DOTAP:NHS-PEG$_{2000}$-DSPE:mPEG$_{2000}$-DSPE liposomes delivered more temoporfin to MRSA biofilms than unmodified liposomes or free temoporfin and resulted in complete eradication of the MRSA infection [133], highlighting the potential of targeted NPs to increase the antibiotic's bioavailability and hence efficacy to bacterial biofilms.

3.3.2.3 Nanocarriers Can Shuttle Antibiotics Inside Bacterial Cells

Apart from a better delivery of antibiotics deep into biofilms, nanomaterials can also aid in improved delivery of antibiotics into cells. Gram-negative (and to a lesser extent Gram-positive) bacterial cells are protected with multi-component barrier systems, restricting the permeation of antimicrobials [135,136]. As the antibiotic targets are often located inside the cell (cytosol or cytoplasmic membrane) [137], cellular impermeability seriously decreases antibiotics efficiency [135,136]. Therefore, nanocarriers—especially liposomes—that are designed to interact with cellular membranes, can be of paramount importance to increased antibiotic interaction with bacterial cells [138].

The advantage of using liposomes as carriers for antibiotics is that they are able to fuse with bacterial membranes, because the liposomal wall consists of amphiphatic phospholipids that resemble the cellular membrane. A particular class of liposomes, the so-called fluidosomes, deserve extra attention in this regard, as they have an enhanced capability to interact and fuse with phospholipid membranes [138]. Fluidosomes consist of lipids, for example, 1,2-dioleoyl-sn-glycero-3-phosphoethanolamine (DOPE) [139], that enhance the fluidity of nanocarriers. Mugabe et al. reported that liposomal entrapment of gentamicin increased the susceptibility of clinical isolates of *P. aeruginosa* toward this antibiotic [110]. Further investigation revealed that those fluidosomes guided the encapsulated aminoglycoside directly through the bacterial envelope, thereby avoiding the problem of restricted penetration through porin channels [140]. Transmission Electron Microscopy images revealed a close interaction between DPPC:Chol liposomes encapsulating aminoglycosides such as amikacin, gentamicin, or tobramycin and the cell membrane of *P. aeruginosa*. Liposome-bacterial membrane fusion was further evidenced by flow cytometry and lipid mixing studies [138,140].

Halwani et al. investigated the potential of fluidosomal encapsulation of different aminoglycosides in the treatment of *Burholderia cenocepacia* infections. Again, the same event was observed: fluidosomes were able to transfer aminoglycosides through the cell membrane of *B. cenocepacia*, thereby greatly increasing the efficiency of the antibiotics [141] Beaulac et al. investigated the suitability of fluidosomes against a wide range of pathogens including *P. aeruginosa*, *Stenotrophomonas maltophilia*, *B. cepacia*, *E. coli*, and *S. aureus* [142]. Liposomes consisting of dipalmitoylphosphatidylcholine (DPPC) and dimyristoylphosphatidylglycerol (DMPG) were loaded with subinhibitory-concentrations of tobramycin and, surprisingly, possessed strong bactericidal activity against all strains; the same but non-encapsulated amount of tobramycin (only 50% of the tobramycin MIC) could not exert this effect [142]. It was shown by negative staining that fluidosomes interacted closely with the bacterial outer membrane (Fig. 3.6). Immuno-electron microscopy images showed that more tobramycin could penetrate the bacterial cells in case of fluidosomal entrapment than in the case of free tobramycin. Integration of fluidosomal phospholipids in bacterial membranes could be confirmed by Fluorescence Activated Cell Sorting. After fusion, the fluorescent lipophilic marker PKH-2GL was detected in bacterial cells, proving that fluidosomal phospholipids were integrated in bacterial walls [143].

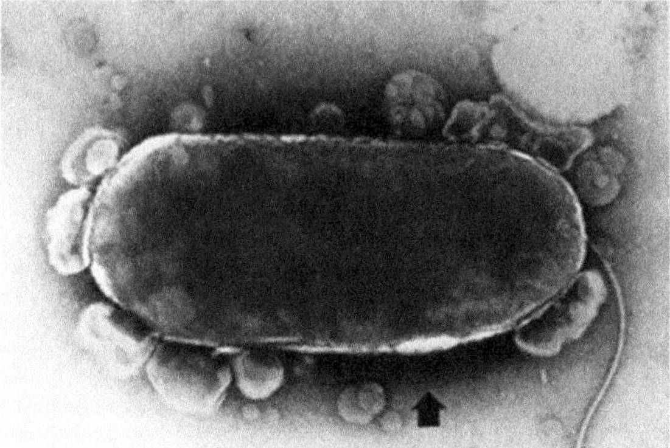

■ **FIGURE 3.6** Electron microscopy image of the interaction between fluidosomes and *P. aeruginosa* 429 by negative staining with 1% phosphotungstenic acid. Fluidosomes came in close contact with the bacterial cell membrane and eventually fusion took place (enlargement of the outer membrane) [143].

In vivo experiments confirmed the potential of fluidosomes in the treatment of bacterial infections. DPPC:DMPG liposomes containing tobramycin were tested in an in vivo rat model by infecting the rats with a mucoid strain of *P. aeruginosa*. The fluidity of the liposomal wall was important, as only fluid DPPC:DMPG liposomes (transition phase temperature around 30°C) eradicated the pulmonary mucoid *P. aeruginosa* infection. Both tobramycin loaded rigid liposomes (transition phase temperature around 40°C), as free tobramycin, were not effective in treating the *P. aeruginosa* infection, highlighting the importance of the liposomal wall fluidity for optimal fusion with bacterial membranes [144].

3.3.2.4 *Responsive Nanoparticles: Antibiotic Release on Demand*

As mentioned in the previous section, nanocarriers can be designed to shuttle antibiotics directly into the bacterial cells, thus greatly improving their cytosolic delivery. In other cases, it is more favorable to deliver agents in the proximity of the bacterial cell, rather than intracellularly, e.g., in case of DNase that degrades extracellular DNA [145]. To this end, nanocarriers can be functionalized with responsive moieties that enable triggered release of the antibiotic, once the carrier reaches the target site [146].

Many studies were conducted to investigate different triggers for antibiotic release from nanocarriers such as liposomes and polymeric NPs [147–150]. Both internal triggers such as enzymes and pH variations, and external factors such as light, heat and ultrasound can be used to release drugs out of carriers. A comprehensive discussion of this subject is beyond the scope of this chapter and therefore we want to refer the reader to recent reviews on this topic [151,152].We would like to highlight one trigger mechanism that selectively activated nanocarriers at the bacterial biofilm site, the so-called microbial responsive nanocarriers. It has been demonstrated that rhamnolipids, a certain type of virulence factor of *P. aeruginosa*, could disturb the liposomal membrane, in this way selectively releasing amikacin out of the liposomes at the microbial infection site [108,153]. Rhamnolipids were able to trigger release of amikacin from DPPC:cholesterol liposomes in a very efficient way: one rhamnolipid molecule per 100 lipids present in liposomes was sufficient to lead to amikacin release [108].

Another example of microbial responsive nanocarriers is the release of gentamicin out of polymer nanocarriers in contaminated wounds. It has been shown that fluids from *S. aureus*-infected wounds contained thrombin-like molecules, capable of destabilizing thrombin-sensitive linkers that conjugated gentamicin on polymer nanoparticles [154]. A third example is the release of vancomycin out of AuNP functionalized liposomes by α-toxins secreted by *S. aureus*. Alpha-toxins are known for their pore-forming abilities in membranes, and can be used to disrupt liposomal membranes for antibiotic release [155].

3.4 **CONCLUSION**

With multiresistant bacteria emerging worldwide and the shortage of novel antibiotics in the development pipeline, novel approaches for antimicrobial treatment are required. Making use of existing antibiotic molecules, but increasing their delivery toward biofilm cells, is a very promising route to follow. Disrupting biofilms, by enzymatic degradation as well as physical methods, has been shown to greatly enhance antibiotic's efficacy. However, more research should be performed in order to translate these concepts into clinical settings. On the other hand, the potential of nanoparticle-assisted antibiotic delivery has been assessed by many studies, and promising results paved the way for intense worldwide investigation. To achieve maximal output, smart engineering of the NP is essential: NP size, surface charge, chemical composition, attachment of site-directing molecules and responsive linkers are crucial in obtaining elevated antibiotic delivery. Further research should take into account those factors and assess the biofilm penetration ability of the carriers with advanced microscopy techniques such as Single Particle Tracking and Fluorescence Correlation Spectroscopy. Furthermore, more research should focus on the construction of NPs that effectively target biofilm-enclosed cells, while at the same time avoiding interactions with tissues distributed throughout the entire human body. Instead of charge-based interactions, which will not exclusively occur at the biofilm, one should screen for specific targeting agents, directed against precise bacterial determinants, while considering that higher specificity is often accompanied with lower affinity. Also, as the majority of published papers include in vitro data only, assessment of the toxicity and applicability in vivo is urgently needed. In conclusion, rationally designed antibiotic-enclosing NPs have the potential to increase antibiotic delivery, and could be useful in the future treatment of infectious diseases.

REFERENCES

[1] Fukuda DK. WHO | Antimicrobial resistance: global report on surveillance. Geneva: World Health Organization; 2014;257.

[2] Jakab Z, Lönngren T. The bacterial challenge: time to react. A call to narrow the gap between multidrug-resistant bacteria in the EU and the development of new antibacterial agents. ECDC/EMEA Joint Technical report. 2009.

[3]   Lessa FC, Winston LG, McDonald LC. Burden of *Clostridium difficile* infection in the United States. N Engl J Med. 2015;372(24):2369–70.

[4]   Alanis AJ. Resistance to antibiotics: are we in the post-antibiotic era? Arch Med Res 2005;36(6):697–705.

[5]   Chan Margaret. WHO | Global action plan on antimicrobial resistance. Geneva: World Health Organization; 2015.

[6]   Costerton JW, Stewart PS, Greenberg EP. Bacterial biofilms: a common cause of persistent infections. Science. 1999;284(5418):1318–22.

[7]   Bjarnsholt T. The role of bacterial biofilms in chronic infections. APMIS Suppl. 2013;136:1–51.

[8]   Li Y-H, Tian X. Quorum sensing and bacterial social interactions in biofilms. Sensors. 2012;12(3):2519–38.

[9]   Donlan RM, Costerton JW. Biofilms: survival mechanisms of clinically relevant microorganisms. Clin Microbiol Rev 2002;15(2):167–93.

[10]   Davies D. Understanding biofilm resistance to antibacterial agents. Nat Rev Drug Discov 2003;2(2):114–22.

[11]   Van Acker H, Van Dijck P, Coenye T. Molecular mechanisms of antimicrobial tolerance and resistance in bacterial and fungal biofilms. Trends Microbiol 2014;22(6):326–33.

[12]   Stewart PS, Costerton JW. Antibiotic resistance of bacteria in biofilms. Lancet 2001;358(9276):135–8.

[13]   Kim J, Hahn J-S, Franklin MJ, Stewart PS, Yoon J. Tolerance of dormant and active cells in *Pseudomonas aeruginosa* PA01 biofilm to antimicrobial agents. J Antimicrob Chemother 2009;63(1):129–35.

[14]   Picioreanu C, van Loosdrecht MCM, Heijnen JJ. Community structure and co-operation in biofilms Allison DG, Gilbert P, Lappin-Scott HM, Wilson M, editors. Fift-ninth symposium of the society for general microbiology held at the University of Exeter September 2000. Cambridge: Cambridge University Press; 2000.

[15]   Stoodley P, Wilson S, Hall-Stoodley L, Boyle JD, Lappin-Scott HM, Costerton JW. Growth and detachment of cell clusters from mature mixed-species biofilms. Appl Environ Microbiol 2001;67(12):5608–13.

[16]   Flemming H-C, Neu TR, Wozniak DJ. The EPS matrix: the "house of biofilm cells". J Bacteriol 2007;189(22):7945–7.

[17]   Flemming H-C, Wingender J. The biofilm matrix. Nat Rev Microbiol 2010;8(9):623–33.

[18]   Xiong M-H, Bao Y, Yang X-Z, Zhu Y-H, Wang J. Delivery of antibiotics with polymeric particles. Adv Drug Deliv Rev 2014;78:63–76.

[19]   Stoodley P, Boyle JD, DeBeer D, Lappin-Scott HM. Evolving perspectives of biofilm structure. Biofouling. Taylor & Francis Group 1999;14(1):75–90.

[20]   Lewandowski Z, Walser G, Characklis WG. Reaction kinetics in biofilms. Biotechnol Bioeng 1991;38(8):877–82.

[21]   Siegrist H, Gujer W. Mass transfer mechanisms in a heterotrophic biofilm. Water Res 1985;19(11):1369–78.

[22]   Palmer RJ, Sternberg C. Modern microscopy in biofilm research: confocal microscopy and other approaches. Curr Opin Biotechnol 1999;10(3):263–8.

[23]   Cerca N, Gomes F, Pereira S, Teixeira P, Oliveira R. Confocal laser scanning microscopy analysis of *S. epidermidis* biofilms exposed to farnesol, vancomycin and rifampicin. BMC Res Notes 2012;5:244.

[24] Nwaneshiudu A, Kuschal C, Sakamoto FH, Anderson RR, Schwarzenberger K, Young RC. Introduction to confocal microscopy. J Invest Dermatol 2012;132(12):e3.

[25] Lawrence JR, Korber DR, Hoyle BD, Costerton JW, Caldwell DE. Optical sectioning of microbial biofilms. J Bacteriol 1991;173(20):6558–67.

[26] de Beer D, Stoodley P, Roe F, Lewandowski Z. Effects of biofilm structures on oxygen distribution and mass transport. Biotechnol Bioeng 1994;43(11):1131–8.

[27] Stoodley P, Debeer D, Lewandowski Z. Liquid flow in biofilm systems. Appl Environ Microbiol 1994;60(8):2711–6.

[28] de Beer D, Stoodley P, Lewandowski Z. Liquid flow in heterogeneous biofilms. Biotechnol Bioeng 1994;44(5):636–41.

[29] Stewart PS. Diffusion in biofilms. J Bacteriol 2003;185(5):1485–91.

[30] Rani SA, Pitts B, Stewart PS. Rapid diffusion of fluorescent tracers into *Staphylococcus epidermidis* biofilms visualized by time lapse microscopy. Antimicrob Agents Chemother 2005;49(2):728–32.

[31] Stewart PS, Davison WM, Steenbergen JN. Daptomycin rapidly penetrates a *Staphylococcus epidermidis* biofilm. Antimicrob Agents Chemother 2009;53(8):3505–7.

[32] Stone G, Wood P, Dixon L, Keyhan M, Matin A. Tetracycline rapidly reaches all the constituent cells of uropathogenic *Escherichia coli* biofilms. Antimicrob Agents Chemother 2002;46(8):2458–61.

[33] Forier K, Messiaen A-S, Raemdonck K, Nelis H, De Smedt S, Demeester J, et al. Probing the size limit for nanomedicine penetration into *Burkholderia multivorans* and *Pseudomonas aeruginosa* biofilms. J Control Release 2014;195:21–8.

[34] Gordon CA, Hodges NA, Marriott C. Antibiotic interaction and diffusion through alginate and exopolysaccharide of cystic fibrosis-derived *Pseudomonas aeruginosa*. J Antimicrob Chemother 1988;22(5):667–74.

[35] Alkawash MA, Soothill JS, Schiller NL. Alginate lyase enhances antibiotic killing of mucoid *Pseudomonas aeruginosa* in biofilms. APMIS 2006;114(2):131–8.

[36] Nichols WW, Dorrington SM, Slack MP, Walmsley HL. Inhibition of tobramycin diffusion by binding to alginate. Antimicrob Agents Chemother 1988;32(4):518–23.

[37] Kumon H, Tomochika K, Matunaga T, Ogawa M, Ohmori H. A sandwich cup method for the penetration assay of antimicrobial agents through *Pseudomonas* exopolysaccharides. Microbiol Immunol 1994;38(8):615–9.

[38] De Beer D, Srinivasan R, Stewart PS. Direct measurement of chlorine penetration into biofilms during disinfection. Appl Environ Microbiol 1994;60(12):4339–44.

[39] de Beer D, Stoodley P, Lewandowski Z. Measurement of local diffusion coefficients in biofilms by microinjection and confocal microscopy. Biotechnol Bioeng 1997;53(2):151–8.

[40] Szomolay B, Klapper I, Dockery J, Stewart PS. Adaptive responses to antimicrobial agents in biofilms. Environ Microbiol 2005;7(8):1186–91.

[41] Jefferson KK, Goldmann DA, Pier GB. Use of confocal microscopy to analyze the rate of vancomycin penetration through *Staphylococcus aureus* biofilms. Antimicrob Agents Chemother 2005;49(6):2467–73.

[42] Bagge N, Schuster M, Hentzer M, Ciofu O, Givskov M, Greenberg EP, et al. *Pseudomonas aeruginosa* biofilms exposed to imipenem exhibit changes in global gene expression and beta-lactamase and alginate production. Antimicrob Agents Chemother 2004;48(4):1175–87.

[43] Sailer FC, Meberg BM, Young KD. Beta-Lactam induction of colanic acid gene expression in *Escherichia coli*. FEMS Microbiol Lett 2003;226(2):245–9.

[44] Stewart PS. Antimicrobial tolerance in BIOFILMS. Microbiol Spectr 2015;3(3).

[45] Debeer D, Stoodley P. Relation between the structure of an aerobic biofilm and transport phenomena. Water Sci Technol 1995;32(8):11–18.

[46] Lewis K. Riddle of biofilm resistance. Antimicrob Agents Chemother 2001;45(4):999–1007.

[47] Hatch RA, Schiller NL. Alginate lyase promotes diffusion of aminoglycosides through the extracellular polysaccharide of mucoid *Pseudomonas aeruginosa*. Antimicrob Agents Chemother 1998;42(4):974–7.

[48] Lee KY, Mooney DJ. Alginate: properties and biomedical applications. Prog Polym Sci 2012;37(1):106–26.

[49] Tseng BS, Zhang W, Harrison JJ, Quach TP, Song JL, Penterman J, et al. The extracellular matrix protects *Pseudomonas aeruginosa* biofilms by limiting the penetration of tobramycin. Environ Microbiol 2013;15(10).

[50] Billings N, Millan M, Caldara M, Rusconi R, Tarasova Y, Stocker R, et al. The extracellular matrix Component Psl provides fast-acting antibiotic defense in *Pseudomonas aeruginosa* biofilms. PLoS Pathog 2013;9(8):e1003526.

[51] Van Acker H, Van Dijck P, Coenye T. Molecular mechanisms of antimicrobial tolerance and resistance in bacterial and fungal biofilms. Trends Microbiol 2014;22(6):326–33. Elsevier.

[52] Colvin KM, Gordon VD, Murakami K, Borlee BR, Wozniak DJ, Wong GCL, et al. The pel polysaccharide can serve a structural and protective role in the biofilm matrix of *Pseudomonas aeruginosa*. PLoS Pathog 2011;7(1):e1001264.

[53] Stewart PS, Grab L, Diemer JA. Analysis of biocide transport limitation in an artificial biofilm system. J Appl Microbiol 1998;85(3):495–500.

[54] Chen X, Stewart PS. Chlorine penetration into artificial biofilm is limited by a reaction– diffusion interaction. Environ Sci Technol 1996;30(6):2078–83.

[55] Davison WM, Pitts B, Stewart PS. Spatial and temporal patterns of biocide action against *Staphylococcus epidermidis* biofilms. Antimicrob Agents Chemother 2010;54(7):2920–7.

[56] Varela MF, Kumar S. Molecular mechanisms of bacterial resistance to antimicrobial agents Méndez-Vilas A, editor. Microbial pathogens and strategies for combating them: science, technology and education. Spain: Formatex Research Center; 2013. p. 522–34.

[57] Anderl JN, Franklin MJ, Stewart PS. Role of antibiotic penetration limitation in *Klebsiella pneumoniae* biofilm resistance to ampicillin and ciprofloxacin. Antimicrob Agents Chemother 2000;44(7):1818–24.

[58] Zhang T. Modeling of biocide action against biofilm. Bull Math Biol 2012;74(6):1427–47.

[59] Stewart PS, Raquepas JB. Implications of reaction-diffusion theory for the disinfection of microbial biofilms by reactive antimicrobial agents. Chem Eng Sci 1995;50(19):3099–104.

[60] Stewart PS, Roe F, Rayner J, Elkins JG, Lewandowski Z, Ochsner UA, et al. Effect of catalase on hydrogen peroxide penetration into *Pseudomonas aeruginosa* biofilms. Appl Environ Microbiol 2000;66(2):836–8.

[61] Alfonso-Prieto M, Biarnés X, Vidossich P, Rovira C. The molecular mechanism of the catalase reaction. J Am Chem Soc 2009;131(33):11751–61.

[62] Elkins JG, Hassett DJ, Stewart PS, Schweizer HP, McDermott TR. Protective role of catalase in *Pseudomonas aeruginosa* biofilm resistance to hydrogen peroxide. Appl Environ Microbiol 1999;65(10):4594–600.

[63] Kaplan JB. Therapeutic potential of biofilm-dispersing enzymes. Int J Artif Organs 2009;32(9):545–54.

[64] Kaplan JB, Ragunath C, Ramasubbu N, Fine DH. Detachment of Actinobacillus actinomycetemcomitans biofilm cells by an endogenous beta-hexosaminidase activity. J Bacteriol 2003;185(16):4693–8.

[65] Forier K, Raemdonck K, De Smedt SC, Demeester J, Coenye T, Braeckmans K. Lipid and polymer nanoparticles for drug delivery to bacterial biofilms. J Control Release 2014;190:607–23.

[66] Sousa C, Botelho CM, Oliveira R. Nanotechnology applied to medical biofilms control Méndez-Vilas A, editor. Science against microbial pathogens: communicating current research and technological advances. Spain: Formatex; 2011. p. 878–88.

[67] Tetz GV, Artemenko NK, Tetz VV. Effect of DNase and antibiotics on biofilm characteristics. Antimicrob Agents Chemother 2009;53(3):1204–9.

[68] Walencka E, Sadowska B, Rozalska S, Hryniewicz W, Rózalska B. Lysostaphin as a potential therapeutic agent for staphylococcal biofilm eradication. Pol J Microbiol 2005;54(3):191–200.

[69] Thallinger B, Prasetyo EN, Nyanhongo GS, Guebitz GM. Antimicrobial enzymes: an emerging strategy to fight microbes and microbial biofilms. Biotechnol J 2013;8(1):97–109.

[70] Taylor ZD, Navarro A, Kealey CP, Beenhouwer D, Haake DA, Grundfest WS, et al. Bacterial biofilm disruption using laser generated shockwaves Conf proceedingss Annu Int Conf IEEE Eng Med Biol Soc IEEE Eng Med Biol Soc Annu Conf. 201020101028.32

[71] Taraszkiewicz A, Fila G, Grinholc M, Nakonieczna J. Innovative strategies to overcome biofilm resistance. Biomed Res Int 2013;2013:150653.

[72] Xavier JB, Picioreanu C, Rani SA, van Loosdrecht MCM, Stewart PS. Biofilm-control strategies based on enzymic disruption of the extracellular polymeric substance matrix--a modelling study. Microbiology 2005;151(Pt 12):3817–32.

[73] Bjarnsholt T, Ciofu O, Molin S, Givskov M, Høiby N. Applying insights from biofilm biology to drug development - can a new approach be developed? Nat Rev Drug Discov 2013;12(10):791–808.

[74] Carmen JC, Nelson JL, Beckstead BL, Runyan CM, Robison RA, Schaalje GB, et al. Ultrasonic-enhanced gentamicin transport through colony biofilms of *Pseudomonas aeruginosa* and *Escherichia coli*. J Infect Chemother 2004;10(4):193–9.

[75] Messiaen A-S, Nelis H, Coenye T. Investigating the role of matrix components in protection of *Burkholderia cepacia* complex biofilms against tobramycin. J Cyst Fibros 2014;13(1):56–62.

[76] Romeo T, Preston JF. Depolymerization of alginate by an extracellular alginate lyase from a marine bacterium: substrate specificity and accumulation of reaction products. Biochemistry 1986;25(26):8391–6.

[77] Miller MB, Bassler BL. Quorum sensing in bacteria. Annu Rev Microbiol 2001;55:165–99.

[78] Waters CM, Bassler BL. Quorum sensing: cell-to-cell communication in bacteria. Annu Rev Cell Dev Biol 2005;21:319–46.

[79] Parsek MR, Greenberg EP. Sociomicrobiology: the connections between quorum sensing and biofilms. Trends Microbiol 2005;13(1):27–33.

[80] Coenye T. Social interactions in the *Burkholderia cepacia* complex: biofilms and quorum sensing. Future Microbiol 2010;5(7):1087–99.

[81] Brackman G, Coenye T. Quorum sensing inhibitors as anti-biofilm agents. Curr Pharm Des 2015;21(1):5–11.

[82] Brackman G, Coenye T. Inhibition of Quorum Sensing in *Staphylococcus* spp. Curr Pharm Des 2015;21(16):2101–8.

[83] Davies DG, Parsek MR, Pearson JP, Iglewski BH, Costerton JW, Greenberg EP. The involvement of cell-to-cell signals in the development of a bacterial biofilm. Science 1998;280(5361):295–8.

[84] Tomlin KL, Malott RJ, Ramage G, Storey DG, Sokol PA, Ceri H. Quorum-sensing mutations affect attachment and stability of *Burkholderia cenocepacia* biofilms. Appl Environ Microbiol 2005;71(9):5208–18.

[85] Brackman G, Cos P, Maes L, Nelis HJ, Coenye T. Quorum sensing inhibitors increase the susceptibility of bacterial biofilms to antibiotics in vitro and in vivo. Antimicrob Agents Chemother 2011;55(6):2655–61.

[86] Bridier A, Briandet R, Thomas V, Dubois-Brissonnet F. Resistance of bacterial biofilms to disinfectants: a review. Biofouling 2011;27(9):1017–32.

[87] Jabbouri S, Sadovskaya I. Characteristics of the biofilm matrix and its role as a possible target for the detection and eradication of *Staphylococcus epidermidis* associated with medical implant infections. FEMS Immunol Med Microbiol 2010;59(3):280–91.

[88] Zogaj X, Nimtz M, Rohde M, Bokranz W, Römling U. The multicellular morphotypes of Salmonella typhimurium and *Escherichia coli* produce cellulose as the second component of the extracellular matrix. Mol Microbiol 2001;39(6):1452–63.

[89] Stoodley P, Sauer K, Davies DG, Costerton JW. Biofilms as complex differentiated communities. Annu Rev Microbiol 2002;56:187–209.

[90] Qian Z, Stoodley P, Pitt WG. Effect of low-intensity ultrasound upon biofilm structure from confocal scanning laser microscopy observation. Biomaterials 1996;17(20):1975–80.

[91] Dong Y, Chen S, Wang Z, Peng N, Yu J. Synergy of ultrasound microbubbles and vancomycin against *Staphylococcus epidermidis* biofilm. J Antimicrob Chemother 2013;68(4):816–26.

[92] Nishikawa T, Yoshida A, Khanal A, Habu M, Yoshioka I, Toyoshima K, et al. A study of the efficacy of ultrasonic waves in removing biofilms. Gerodontology 2010;27(3):199–206.

[93] Crone S, Garde C, Bjarnsholt T, Alhede M. A novel in vitro wound biofilm model used to evaluate low-frequency ultrasonic-assisted wound debridement. J Wound Care 2015;24(2):64. 66–9, 72.

[94] Yu H, Chen S, Cao P. Synergistic bactericidal effects and mechanisms of low intensity ultrasound and antibiotics against bacteria: a review. Ultrason Sonochem 2012;19(3):377–82.

[95] Williams KA, Clark HA, Allison DG. Use of ultrasound to facilitate antibiotic diffusion through *Pseudomonas aeruginosa* alginate. J Antimicrob Chemother 1995;36(3):463–73.

[96] Johnson LL, Peterson RV, Pitt WG. Treatment of bacterial biofilms on polymeric biomaterials using antibiotics and ultrasound. J Biomater Sci Polym Ed 1998;9(11):1177–85.

[97] Young D, Morton R, Bartley J. Therapeutic ultrasound as treatment for chronic rhinosinusitis: preliminary observations. J Laryngol Otol 2010;124(5):495–9.

[98] Rediske AM, Roeder BL, Brown MK, Nelson JL, Robison RL, Draper DO, et al. Ultrasonic enhancement of antibiotic action on *Escherichia coli* biofilms: an in vivo model. Antimicrob Agents Chemother 1999;43(5):1211–4.

[99] Rediske AM, Roeder BL, Nelson JL, Robison RL, Schaalje GB, Robison RA, et al. Pulsed ultrasound enhances the killing of *Escherichia coli* biofilms by aminoglycoside antibiotics in vivo. Antimicrob Agents Chemother 2000;44(3):771–2.

[100] Carmen JC, Roeder BL, Nelson JL, Ogilvie RLR, Robison RA, Schaalje GB, et al. Treatment of biofilm infections on implants with low-frequency ultrasound and antibiotics. Am J Infect Control 2005;33(2):78–82.

[101] Carmen JC, Roeder BL, Nelson JL, Beckstead BL, Runyan CM, Schaalje GB, et al. Ultrasonically enhanced vancomycin activity against *Staphylococcus epidermidis* biofilms in vivo. J Biomater Appl 2004;18(4):237–45.

[102] Nigri GR, Tsai S, Kossodo S, Waterman P, Fungaloi P, Hooper DC, et al. Laser-induced shock waves enhance sterilization of infected vascular prosthetic grafts. Lasers Surg Med 2001;29(5):448–54.

[103] Alhede M, Kragh KN, Qvortrup K, Allesen-Holm M, van Gennip M, Christensen LD, et al. Phenotypes of non-attached *Pseudomonas aeruginosa* aggregates resemble surface attached biofilm. PLoS One 2011;6(11):e27943.

[104] Pelgrift RY, Friedman AJ. Nanotechnology as a therapeutic tool to combat microbial resistance. Adv Drug Deliv Rev 2013;65(13–14):1803–15.

[105] Jahanshahi M, Babaei Z. Protein nanoparticle: a unique system as drug delivery vehicles. African J Biotechnol 2008;7(25):4926–34.

[106] Brooks BD, Brooks AE. Therapeutic strategies to combat antibiotic resistance. Adv Drug Deliv Rev 2014;78:14–27.

[107] Parveen S, Misra R, Sahoo SK. Nanoparticles: a boon to drug delivery, therapeutics, diagnostics and imaging. Nanomedicine 2012;8(2):147–66.

[108] Meers P, Neville M, Malinin V, Scotto AW, Sardaryan G, Kurumunda R, et al. Biofilm penetration, triggered release and in vivo activity of inhaled liposomal amikacin in chronic *Pseudomonas aeruginosa* lung infections. J Antimicrob Chemother 2008;61(4):859–68.

[109] Chambless JD, Hunt SM, Stewart PS. A three-dimensional computer model of four hypothetical mechanisms protecting biofilms from antimicrobials. Appl Environ Microbiol 2006;72(3):2005–13.

[110] Mugabe C, Azghani AO, Omri A. Liposome-mediated gentamicin delivery: development and activity against resistant strains of *Pseudomonas aeruginosa* isolated from cystic fibrosis patients. J Antimicrob Chemother 2005;55(2):269–71.

[111] Nacucchio MC, Bellora MJ, Sordelli DO, D'Aquino M. Enhanced liposome-mediated activity of piperacillin against staphylococci. Antimicrob Agents Chemother 1985;27(1):137–9.

[112] Turos E, Reddy GSK, Greenhalgh K, Ramaraju P, Abeylath SC, Jang S, et al. Penicillin-bound polyacrylate nanoparticles: restoring the activity of beta-lactam antibiotics against MRSA. Bioorg Med Chem Lett 2007;17(12):3468–72.

[113] Peulen T-O, Wilkinson KJ. Diffusion of nanoparticles in a biofilm. Environ Sci Technol 2011;45(8):3367–73.

[114] Wang L-S, Gupta A, Rotello VM. Nanomaterials for the treatment of bacterial biofilms. ACS Infect Dis 2015; 151021155851001.

[115] Li X, Yeh Y-C, Giri K, Mout R, Landis RF, Prakash YS, et al. Control of nanoparticle penetration into biofilms through surface design. Chem Commun 2015;51(2):282–5.

[116] Duncan B, Li X, Landis RF, Kim ST, Gupta A, Wang L-S, et al. Nanoparticle stabilized capsules for the treatment of bacterial biofilms. ACS Nano 2015;9(8):7775–82.

[117] Jones MN, Kaszuba M, Reboiras MD, Lyle IG, Hill KJ, Song YH, et al. The targeting of phospholipid liposomes to bacteria. Biochim Biophys Acta 1994;1196(1):57–64.

[118] Hill KJ, Kaszuba M, Creeth JE, Jones MN. Reactive liposomes encapsulating a glucose oxidase-peroxidase system with antibacterial activity. Biochim Biophys Acta 1997;1326(1):37–46.

[119] Jones MN, Francis SE, Hutchinson FJ, Handley PS, Lyle IG. Targeting and delivery of bactericide to adsorbed oral bacteria by use of proteoliposomes. Biochim Biophys Acta 1993;1147(2):251–61.

[120] Jones MN, Hill KJ, Kaszuba M, Creeth JE. Antibacterial reactive liposomes encapsulating coupled enzyme systems. Int J Pharm 1998;162(1–2):107–17.

[121] Jones MN, Kaszuba M. Polyhydroxy-mediated interactions between liposomes and bacterial biofilms. Biochim Biophys Acta 1994;1193(1):48–54.

[122] Sanderson NM, Jones MN. Targeting of Cationic Liposomes to Skin-Associated Bacteria. Pestic Sci 1996;46(3):255–61.

[123] Song YH, Jones MN. The interaction of positively charged phospholipid vesicles with bacteria. Biochem Soc Trans 1994;22(3):330S.

[124] Jones MN, Song YH, Kaszuba M, Reboiras MD. The interaction of phospholipid liposomes with bacteria and their use in the delivery of bactericides. J Drug Target 1997;5(1):25–34.

[125] Kaszuba M, Robinson AM, Song Y-H, Creeth JE, Jones MN. The visualisation of the targeting of phospholipid liposomes to bacteria. Colloids Surf B Biointerfaces 1997;8(6):321–32.

[126] Silhavy TJ, Kahne D, Walker S. The bacterial cell envelope. Cold Spring Harb Perspect Biol 2010;2(5):a000414.

[127] Jernberg C, Löfmark S, Edlund C, Jansson JK. Long-term impacts of antibiotic exposure on the human intestinal microbiota. Microbiology 2010;156(Pt 11):3216–23.

[128] Robinson AM, Creeth JE, Jones MN. The use of immunoliposomes for specific delivery of antimicrobial agents to oral bacteria immobilized on polystyrene. J Biomater Sci Polym Ed 2000;11(12):1381–93.

[129] Robinson AM, Creeth JE, Jones MN. The specificity and affinity of immunoliposome targeting to oral bacteria. Biochim Biophys Acta 1998;1369(2):278–86.

[130] Kennedy JF, Palva PMG, Corella MTS, Cavalcanti MSM, Coelho LCBB. Lectins, versatile proteins of recognition: a review. Carbohydr Polym 1995;26(3):219–30.

[131] Strathmann M, Wingender J, Flemming H-C. Application of fluorescently labelled lectins for the visualization and biochemical characterization of polysaccharides in biofilms of *Pseudomonas aeruginosa*. J Microbiol Methods 2002;50(3):237–48.

[132] Kaszuba M, Lyle IG, Jones MN. The targeting of lectin-bearing liposomes to skin-associated bacteria. Colloids Surf B Biointerfaces 1995;4(3):151–8.

[133] Yang K, Gitter B, Rüger R, Albrecht V, Wieland GD, Fahr A. Wheat germ agglutinin modified liposomes for the photodynamic inactivation of bacteria. Photochem Photobiol 2012;88(3):548–56.

[134] Umamaheshwari RB, Jain NK. Receptor mediated targeting of lectin conjugated gliadin nanoparticles in the treatment of *Helicobacter pylori*. J Drug Target 2003;11(7):415–23. discussion 423–4.

[135] Lambert PA. Cellular impermeability and uptake of biocides and antibiotics in Gram-positive bacteria and mycobacteria. J Appl Microbiol 2002;92(Suppl:):46S–54S.

[136] Denyer SP, Maillard J-Y. Cellular impermeability and uptake of biocides and antibiotics in Gram-negative bacteria. J Appl Microbiol 2002;92(s1):35S–45S.

[137] Kohanski MA, Dwyer DJ, Collins JJ. How antibiotics kill bacteria: from targets to networks. Nat Rev Microbiol 2010;8(6):423–35.

[138] Nicolosi D, Scalia M, Nicolosi VM, Pignatello R. Encapsulation in fusogenic liposomes broadens the spectrum of action of vancomycin against Gram-negative bacteria. Int J Antimicrob Agents 2010;35(6):553–8.

[139] Gomes-da-Silva LC, Fonseca NA, Moura V, Pedroso de Lima MC, Simões S, Moreira JN. Lipid-based nanoparticles for siRNA delivery in cancer therapy: paradigms and challenges. Acc Chem Res 2012;45(7):1163–71.

[140] Mugabe C, Halwani M, Azghani AO, Lafrenie RM, Omri A. Mechanism of enhanced activity of liposome-entrapped aminoglycosides against resistant strains of *Pseudomonas aeruginosa*. Antimicrob Agents Chemother 2006;50(6):2016–22.

[141] Halwani M, Mugabe C, Azghani AO, Lafrenie RM, Kumar A, Omri A. Bactericidal efficacy of liposomal aminoglycosides against *Burkholderia cenocepacia*. J Antimicrob Chemother 2007;60(4):760–9.

[142] Beaulac C, Sachetelli S, Lagace J. In-vitro bactericidal efficacy of sub-MIC concentrations of liposome-encapsulated antibiotic against gram-negative and gram-positive bacteria. J Antimicrob Chemother 1998;41(1):35–41.

[143] Sachetelli S, Khalil H, Chen T, Beaulac C, Sénéchal S, Lagacé J. Demonstration of a fusion mechanism between a fluid bactericidal liposomal formulation and bacterial cells. Biochim Biophys Acta 2000;1463(2):254–66.

[144] Beaulac C, Clement-Major S, Hawari J, Lagace J. Eradication of mucoid *Pseudomonas aeruginosa* with fluid liposome- encapsulated tobramycin in an animal model of chronic pulmonary infection. Antimicrob Agents Chemother 1996;40(3):665–9.

[145] Hymes SR, Randis TM, Sun TY, Ratner AJ. DNase inhibits gardnerella vaginalis biofilms in vitro and in vivo. J Infect Dis 2013;207(10):1491–7.

[146] Timko BP, Dvir T, Kohane DS. Remotely triggerable drug delivery systems. Adv Mater 2010;22(44):4925–43.

[147] Horev B, Klein MI, Hwang G, Li Y, Kim D, Koo H, et al. pH-activated nanoparticles for controlled topical delivery of farnesol to disrupt oral biofilm virulence. ACS Nano 2015;9(3):2390–404.

[148] Leung SJ, Romanowski M. Light-activated content release from liposomes. Theranostics 2012;2(10):1020–36.

[149] Huang S-L, MacDonald RC. Acoustically active liposomes for drug encapsulation and ultrasound-triggered release. Biochim Biophys Acta 2004;1665(1–2):134–41.

[150] Pavlukhina S, Lu Y, Patimetha A, Libera M, Sukhishvili S. Polymer multilayers with pH-triggered release of antibacterial agents. Biomacromolecules 2010;11(12):3448–56.

[151] Bibi S, Lattmann E, Mohammed AR, Perrie Y. Trigger release liposome systems: local and remote controlled delivery? J Microencapsul 2012;29(3):262–76.

[152] Fleige E, Quadir MA, Haag R. Stimuli-responsive polymeric nanocarriers for the controlled transport of active compounds: concepts and applications. Adv Drug Deliv Rev 2012;64(9):866–84.

[153] Sánchez M, Aranda FJ, Teruel JA, Espuny MJ, Marqués A, Manresa A, et al. Permeabilization of biological and artificial membranes by a bacterial dir-hamnolipid produced by *Pseudomonas aeruginosa*. J Colloid Interface Sci 2010;341(2):240–7.

[154] Tanihara M, Suzuki Y, Nishimura Y, Suzuki K, Kakimaru Y, Fukunishi Y. A novel microbial infection-responsive drug release system. J Pharm Sci 1999;88(5):510–4.

[155] Pornpattananangkul D, Zhang L, Olson S, Aryal S, Obonyo M, Vecchio K, et al. Bacterial toxin-triggered drug release from gold nanoparticle-stabilized liposomes for the treatment of bacterial infection. J Am Chem Soc 2011;133(11):4132–9.

Chapter 4

Metal Nanoparticles for Microbial Infection

B. Mordorski[1] and A. Friedman[2]

[1]Albert Einstein College of Medicine, Bronx, NY, United States [2]George Washington School of Medicine and Health Sciences, Washington, DC, United States

CHAPTER OUTLINE

4.1 INTRODUCTION

Metal nanoparticles enable the exploitation of unique properties of metals on the nanoscale (1–100 nm) [1]. At such small sizes, many metals have improved physical, chemical, and biological features that make them

Functionalized Nanomaterials for the Management of Microbial Infection.

exceedingly useful for a variety of optical, thermal, chemical sensing, and medicinal applications. Due to their potential to improve diagnostic and therapeutic practices, metal nanoparticles have been explored for use against microbial infection. For example, gold nanoparticles have been used to quantify and identify bacteria in human joint fluid in just over 1 h [2], and wound dressings impregnated with silver nanoparticles have demonstrated potent, broad spectrum activity [3]. These advances are especially important considering the rapid rise of antimicrobial-resistant organisms and stagnating rates of new antibiotic development [4].

In the United States alone, hospital-acquired infections cost thousands of lives and \$125 billion in extra hospital charges annually, and an increasing number of these infections are caused by antimicrobial-resistant organisms [5]. Morbidity and mortality may be worsened by the slow rate of diagnosis associated with conventional microbial cultures, and even newer DNA detection methods may be time-consuming and costly. In addition to delaying necessary antimicrobials and other interventions, patients may be needlessly exposed to antibiotics if empiric therapy is initiated before diagnostic results are available, thereby providing increased opportunity for antimicrobial resistance development. This current paradigm demands a new approach for the development of antimicrobial therapeutics and diagnostic tools for microbial infection.

Metal nanoparticles offer several advantages, many of which arise from their large surface area-to-volume ratio. For example, size reduction for a 10 μm nanoparticle to about 10 nm will increase the contact surface area by 10^9, immensely increasing the interaction between pathogen and nanoparticle surfaces [6]. Additionally, nanoparticles are small relative to most microbes, which often measure on the micron scale or higher end of the nanoscale [7]. In fact, their miniscule structures may only be a few atoms wide with a larger percentage of atoms localized to the surface. Surface atoms have fewer neighboring atoms with which to bond, thereby conferring higher reactivity, and atomic properties at the surface of a material become much more significant [8]. When materials are sized on the nanoscale, new properties emerge that are entirely different from those of the bulk material, as well as the atoms that comprise it. This phenomenon is partly due to the spatial confinement of electrons, which determines the type of motion electrons can execute. When larger materials undergo changes in size and shape, electron confinement is not affected; however, it is affected when size and shape are altered on the nanoscale, thereby creating materials with new properties. This is demonstrated by gold nanospheres whose size and shape can be modulated to reflect all of the colors across the visible spectrum [8–10]. Shape may also be optimized for microbial interaction as demonstrated by silver nanoparticles (Ag-NP) with *Escherichia coli*. Truncated triangular Ag-NP exhibited significantly greater antimicrobial activity compared to spherical or rod-shaped Ag-NP in both liquid media and on agar plates. These findings are attributed to the greater percentage of high atom density (1, 1, 1) facets in truncated triangular Ag-NP compared to spherical or rod-shaped Ag-NP, which confer increased reactivity [6]. Finally, surface characteristics of metal nanoparticles may be modified to optimize pathogen-particle interaction. For instance, capping agents can be applied whose chemical properties confer increased affinity for bacterial, fungal, or viral structures [11,12]. Specific cellular or structural targeting may also be achieved by conjugating metal nanoparticles with a monoclonal antibody, thereby enabling highly specific interactions with pathogenic cells and structures [13,14].

In addition to their advantageous physical properties, metal nanoparticles may be conjugated with established or emerging therapeutic agents that may not otherwise be optimally available in vivo secondary to poor solubility, short half-life, or inability to maintain contact with their biological target. Conjugation may be used to functionalize inert metal nanoparticles, such as gold, although other nanoparticles, such as silver or copper, may be conjugated to enhance their intrinsic antimicrobial activity. By conjugating metal nanoparticles with antimicrobial agents, the density and reactivity of the drug are increased during interaction with microorganisms. Such constructs have demonstrated significant enhancement of antimicrobial activity, including killing of previously resistant organisms, broad spectrum activity against organisms which were previously unaffected by the antimicrobial agent, and inhibition of resistance mechanisms such as efflux pumps [15].

Given these properties, metal nanoparticles may improve antimicrobial therapy in a variety of ways, including enhanced specificity for disease targets and increased bioavailability and bioactivity of conjugated agents. Of special importance in the fight against microbial infection, metal nanoparticles may also circumvent antimicrobial resistance by exerting nonspecific, multi-mechanistic antimicrobial activity and overcoming specific resistance mechanisms such as drug efflux [15] and biofilm formation [16]. Unlike conventional antibiotics, which typically have only one mechanism of action, metal nanoparticles employ multiple mechanisms, as will be further discussed. Therefore, the development of antimicrobial resistance is less likely, as multiple simultaneous mutations would need to occur in the same organism. Resistance may also be thwarted by conjugating multiple antimicrobial agents to a single nanoparticle [17].

While potential for resistance development is a concern for any new or emerging antimicrobial agent, research thus far regarding metal nanoparticles has been reassuring. For example, a minority of bacteria have been identified as less sensitive to Ag-NP and silver ions following long-term exposure, achieved via resistance genes which confer membrane alterations, causing decreased silver uptake and increased efflux [18]. However, expression of these genes does not appear to spread as do resistance genes against conventional antibiotics, and when bacteria are no longer in contact with Ag-NP, expression of resistance genes is lost. This suggests that such genes may decrease evolutionary fitness and are therefore not expressed in the absence of silver [19]. Similar findings have also been demonstrated with copper. For example, bacterial strains isolated from European 50-cent coins were found to exhibit copper resistance; however, when transferred to culture media the bacterial strains were no longer resistant [5]. While microbes have evolved an array of copper protection mechanisms including extracellular sequestration, relative impermeability of cell membranes, scavenging proteins, and copper efflux, these mechanisms only prolong microbial survival; they do not confer resistance [20,21].

The multi-mechanistic antimicrobial activities that protect metal nanoparticles from resistance development are incompletely understood, although several studies have been undertaken to help elucidate the processes involved. For example, studies have indicated that silver nanoparticles primarily exert antimicrobial activity via direct adherence and damage to pathogen membranes and cell walls, as demonstrated by structural alterations, pit formation, and changes in permeability with subsequent cell death. Following membrane disruption, silver ion uptake occurs, leading to perturbation in DNA replication and ATP production [22–26]. It has also been hypothesized that silver ions inhibit membrane-bound enzymes of the respiratory chain, facilitating the generation of reactive oxygen species (ROS) which further damage the cell and increase its porosity [26–28] (Fig. 4.1).

Silver's antimicrobial activities have long been recognized, with centuries of documented use in the treatment of burns and chronic wounds, and silver has been used to make water potable as early as 1000 BC [29]. Likewise, the use of copper to sterilize wounds and drinking water was documented in ancient Egyptian medical texts, as early as 2600–2200 BC [30]. Compared to silver, copper may be a weaker antibacterial agent, although copper seems to have a broader range of antimicrobial activity, particularly against fungi [31]. Currently, copper's antimicrobial properties are utilized in agriculture to control bacterial and fungal diseases, as copper has relatively high toxicity against these pathogens compared to very low toxicity in plants and mammals at the same concentration [32]. Furthermore, the Environmental Protection Agency (EPA) has registered almost 300 different antimicrobial copper and copper alloy surfaces, which

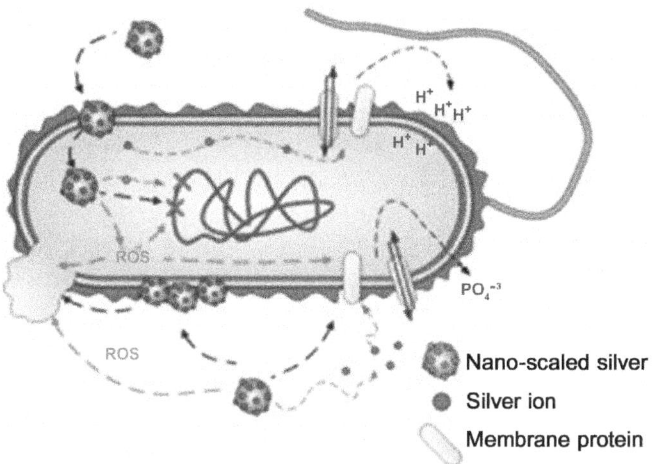

■ **FIGURE 4.1** Silver nanoparticles exhibit multimechanistic antibacterial activity. Silver nanoparticles (1) accumulate at bacterial surfaces, disrupting membrane integrity; (2) inhibit DNA replication; (3) affect membrane proteins and ATP production; and (4) generate reactive oxygen species (ROS). Generated ROS may alter DNA, cell membranes, and membrane proteins. Released silver ions may affect DNA and membrane proteins, as well. Membrane damage leads to cytoplasmic leakage and ultimately bacterial death [22]. Image reproduced from Marambio-Jones C, Hoek EM. A review of the antibacterial effects of silver nanomaterials and potential implications for human health and the environment. J Nanoparticle Res 2010;12(5):1531–51, with permission from Springer.

may be useful in a variety of settings [5]. Like Ag-NP, the antimicrobial activity of copper nanoparticles (Cu-NP) primarily affects microbes via membrane damage, in addition to copper influx, which overwhelms the cells with intracellular copper. Copper ions cause oxidative damage via ROS generated by redox cycling between different copper species. Although DNA fragmentation has also been identified in killed microbial cells, it is likely that this occurs as a secondary event after the cells have already died [5].

Conversely, gold nanoparticles (Au-NP) are known to be inert against microorganisms, as demonstrated by Brown et al. who tested Au-NP against *E. coli*, *Vibrio cholerae*, *Staphylococcus aureus*, *Enterobacter aerogenes*, and *Pseudomonas aeruginosa*, demonstrating no activity via disc diffusion assay or transmission electron microscopy (TEM) [15]. Despite these findings, Au-NP have proven their utility against microbial infection by increasing the activity of antimicrobial substances conjugated to their surfaces, as well as serving as an efficient probe in the detection and quantification of microbial infection. For example, Brown et al. witnessed potent activity of Au-NP against the aforementioned organisms when conjugated with ampicillin, and the activity was significantly greater than that of ampicillin alone [15]. Furthermore, scientists have developed a highly sensitive automated quantification technique for Au-NP at the single particle level [33]. If conjugated with a microbial probe, this assay may be used to improve upon the sensitivity and accuracy of current diagnostic methods by exploiting a phenomenon called surface plasmon resonance, and associated color changes, which will be discussed in further detail [33]. Overall, metal nanoparticles such as silver, copper, and gold have been explored for a variety of applications in the management of microbial infection, which will be reviewed in this chapter.

4.2 SILVER NANOPARTICLES FOR MICROBIAL INFECTION

Silver nanoparticles (Ag-NP) are one of the most promising metal nanoparticles in the fight against microbial infection. They deliver potent, broad-spectrum activity and have the ability to evade resistance development. Ag-NP may be synthesized inorganically, although cost-effective "green synthesis" has also been employed. Recently, eco-friendly processes utilizing several different plants and plant extracts have been developed to produce Ag-NP that are highly effective against microbial pathogens [34–38]. In general, the antimicrobial activity of Ag-NP appears to be size-dependent with smaller-sized Ag-NP less than 10 nm in diameter delivering the most potent effects [18]. Since surface area increases with decreasing size, and Ag-NP's primary killing mechanism requires interaction with microbial surfaces, smaller sizes may facilitate increased contact and greater killing activity [39,40]. Additionally, the increased surface area afforded by smaller particles allows a higher rate of silver ion release, contributing to increased activity, as well [41]. Although the antibacterial activity of Ag-NP has been the most widely investigated, Ag-NP also inhibit fungi [42], viruses [43], and protozoa [44]. Furthermore, Ag-NP have been used to counteract various virulence factors by inhibiting biofilm formation [16,45], neutralizing bacterial toxins [45], and interfering with quorum sensing [46]. Ag-NP have been effectively used in antimicrobial wound dressings [3] and as a coating on medical devices [47]. Moreover, these versatile nanoparticles demonstrate synergy with other antimicrobial agents, killing microbes which were otherwise resistant or unaffected by the drug [48].

4.2.1 Silver Nanoparticles Against Bacterial Infection

Ag-NP have demonstrated potent antimicrobial activity against a variety of microorganisms, including Gram-positive (*S. aureus, Streptococcus pyogenes, Enterococcus faecalis, Bacillus subtilis, Mycobacterium bovis,*

Mycobacterium smegmatis) and Gram-negative bacteria (*E. coli, P. aeruginosa, Acinetobacter baumannii, Salmonella* spp., *V. cholerae*) [26,49–52]. Ag-NP are also effective against drug resistant bacteria, including methicillin-resistant *S. aureus* (MRSA), ampicillin-resistant *E. coli,* and multidrug-resistant *Salmonella enterica* serovar Typhi [49,53].

While Ag-NP are an effective monotherapy, they have also exhibited synergistic effects when conjugated with penicillin G, amoxicillin, erythromycin, clindamycin, and vancomycin for use against *S. aureus* and *E. coli* [48]. Further studies have demonstrated Ag-NP synergy when conjugated with erythromycin, methicillin, chloramphenicol, and ciprofloxacin against *Klebsiella pneumoniae* and *E. aerogenes* [54], as well as with gentamycin, ampicillin, tetracycline and streptomycin against *E. coli, S. aureus,* and *P. aeruginosa* [55]. Another group applied a different approach by mixing gold-silver alloy nanoparticles with kanamycin (no conjugation) to increase its effect against *E. coli.* The alloy nanoparticles were also mixed with ampicillin against sensitive and resistant strains of *S. aureus* and considerably decreased the dosage of antibiotic required for killing in both liquid and solid media. Even though the metal nanoparticles in this case were not physically conjugated to an antimicrobial drug, their combined effect was still greater than the sum of nanoparticle and drug effects alone [56].

In addition to conjugating Ag-NP with conventional antibiotics, they may be combined with new and emerging antimicrobial agents such as chitosan, a polysaccharide derived from chitin, the principal structural element of the crustacean exoskeleton. Chitosan's cationic structure facilitates interaction with negatively charged microbial membranes and cell walls, resulting in osmotic instability, membrane disruption, and leakage of cellular contents. Additionally, chitosan may inhibit mRNA and protein synthesis by binding microbial DNA [57–59]. In one study, hybrid chitosan-Ag-NP demonstrated lower minimum inhibitory and minimum bactericidal concentrations (MIC and MBC) against multiple strains of *S. aureus* compared to silver nanoparticles or chitosan alone [60]. Chitosan has also demonstrated synergy with Ag-NP against MRSA, *P. aeruginosa, Proteus mirabilis,* and *A. baumannii* [61]. Given that chitosan is also active against fungi and viruses, silver-chitosan nanoparticles may be useful against non-bacterial pathogens, as well [62].

Another broad-spectrum antimicrobial agent that has been combined with Ag-NP is titanium dioxide (TiO_2). TiO_2 nanoparticles themselves demonstrate antimicrobial activity, which will be discussed in the chapter "Graphene-Microbial Interactions". A recent study demonstrated the effectiveness of TiO_2-Ag-NP against *E. coli, A. baumannii, P. aeruginosa,*

B. subtilis, M. smegmatis, M. bovis, and *S. aureus,* in addition to fungi such as *Candida albicans* and *Aspergillus* spp. The antimicrobial activity was superior to that of Ag-NP alone, and outperformed the conventional antifungal drug, fluconazole [49]. Hybrid nanoparticles may represent a new paradigm in the fight against resistant bacteria. Not only do they increase the antimicrobial effect, but they can also decrease the potential for resistance development by increasing the number of antimicrobial mechanisms employed [63].

4.2.2 Silver Nanoparticles Against Fungal Infection

Although bacterial "superbugs" receive much of the attention in the discussion of antimicrobial resistance, fungal organisms have exhibited an alarming rise in resistance rates, as well [64]. The antifungal activity of Ag-NP is less studied compared to their antibacterial activity; however, there is significant evidence promoting Ag-NP as a potent antifungal agent. For example, Kim et al. demonstrated Ag-NP activity against 44 strains of six fungal species (*C. albicans, Candida tropicalis, Candida glabrata, Candida parapsilosis, Candida krusei,* and *Trichophyton mentagrophytes*) [42]; Arjun et al. showcased Ag-NP's effectiveness against *Trichophyton rubrum, Aspergillus fumigatus,* and *C. albicans* [65]; and Gajbhiye et al. discovered Ag-NP's usefulness against *Phomaglomerata, Phomaherbarum, Fusarium semitectum, Trichoderma* spp., and *C. albicans,* in addition to demonstrating synergy in combination with fluconazole [66]. Ag-NP also demonstrated superior antifungal activity compared to natamycin against 216 fungal strains from patients with severe keratitis caused by *Fusarium solani, Fusarium verticillioides, Fusarium oxysporum, Aspergillus flavus, A. fumigatus, Aspergillus versicolor, Aspergillus niger,* and *Alternariaalternata* [67]. Although the exact mechanisms of Ag-NP antifungal activity have not been elucidated, studies suggest that they may be similar to mechanisms employed against bacteria, including destruction of membrane integrity and inhibition of normal budding in yeasts [68].

4.2.3 Silver Nanoparticles Against Viral Infection

As with Ag-NP's antifungal activity, research regarding their antiviral effects has remained relatively underdeveloped [64]. However, as antiviral resistance continues to emerge, Ag-NP become increasingly appealing due to their decreased potential for resistance development and broad antiviral activity. Thus far, Ag-NP have demonstrated their effectiveness against hepatitis B (HBV) [69], influenza [70,71], parainfluenza [43], herpes simplex type 1 and 2 (HSV-1 and HSV-2) [43,72], coxsackievirus [73], vaccinia [74], monkeypox [75], and tacaribe virus (TCRV) [76]. Additionally,

Ag-NP exert size-dependent protective activity against human immuno-deficiency virus type 1 (HIV-1) by blocking its binding to CD4 receptors. Only particles 1–10 nm in diameter are able to attach to the virus, and the spatial relationship of bound particles suggests that they may act by binding gp120, a glycoprotein on the viral envelope [77]. In addition to demonstrating optimal viral interaction, smaller-sized Ag-NP may also exert less mammalian cytotoxicity [64]. This was illustrated via incubation of Ag-NP with a hepatoblastoma cell line transfected with HBV, wherein smaller-sized nanoparticles demonstrated increased anti-HBV activity and decreased cytotoxicity against hepatoblastoma cells [69]. In another study, 10 nm Ag-NP had an increased effect against TCRV and decreased cytotoxicity against kidney epithelial cells compared to 25 nm Ag-NP [76].

Ag-NP have also been employed against viruses after being coated with a capping agent, typically a biocompatible substance such as polyvinylpyrrolidone (PVP), citrate, or polyethylene glycol (PEG). Capping agents are known to mute mammalian cytotoxicity of Ag-NP; however, their effect on antimicrobial efficacy remains unclear. For instance, coated Ag-NP were less cytotoxic against renal epithelial cells, while demonstrating significant, albeit decreased activity against TCRV [76]. Conversely, studies involving PVP- or mercaptoethanesulfonate (MESNA)-coated Ag-NP against HIV-1, HSV, and respiratory syncytial virus (RSV) indicate that certain coating agents may actually increase Ag-NP activity via direct interaction with virions [11,12,78]. This has been illustrated by PVP, which enhanced Ag-NP activity against RSV via interaction with glycoproteins on the viral envelope. MESNA-capped Ag-NP are also enhanced by their ability to mimic heparan sulfate, a cellular receptor for HSV, thereby allowing them to compete against binding of HSV virions to host cells. Furthermore, PVP-capped Ag-NP utilized against HIV have been found to bind gp120, preventing CD4-dependent virion binding and subsequent viral entry, and they have effectively inhibited a variety of HIV-1 laboratory, clinical, and resistant strains [11,12,78]. Based on these successful studies, PVP-capped Ag-NP have been used to prevent HIV transmission as a coating on polyurethane condoms [79], as well as a topical vaginal microbicide in a human cervical tissue-based organ culture with simulated in vivo conditions [80]. Both applications demonstrated significant inactivation of HIV [79,80].

4.2.4 **Silver Nanoparticles Against Protozoal Infection**

Lastly, Ag-NP have demonstrated their usefulness against drug-resistant protozoal microorganisms, including those that cause malaria and severe diarrhea. *Plasmodium falciparum*, a malarial species which has achieved nearly 25% resistance against the common antimalarial chloroquine [81], has been successfully inhibited by Ag-NP, including both chloroquine-sensitive and resistant strains [82–84]. Water-borne intestinal protozoa, such as *Giardia lamblia* and *Cryptosporidium parvum*, have also been effectively killed by Ag-NP. This is significant given that *Giardia* and *Cryptosporidium* resist deactivation by conventional preventative measures, such as size-exclusion filters and chlorine treatments [85]. In one study *Cryptosporidium* oocysts were inhibited by Ag-NP at concentrations of 0.005–500 μ gmL^{-1} in a dose-dependent manner. Imaging revealed that Ag-NP caused full-thickness destruction of the oocyst wall, in addition to entering the oocyst and destroying sporozoites [44]. Based on these results, it may be useful to explore Ag-NP impregnation of water filters in future studies. Another group investigated rats infected with *G. lamblia* via nasogastric tube, which were then given Ag-NP, chitosan nanoparticles, or curcumin nanoparticles orally. While all nanoparticles offered therapeutic benefit, Ag-NP was the most effective monotherapy, as measured by fecal parasite shedding and presence of trophozoites in the small intestine, although combination treatment with all three nanoparticles was overall most effective. None of the nanoparticle treatments lead to toxicity as determined by blood chemistries and histopathology, and quantitative analysis of silver in various organs was within acceptable limits for safety [86].

4.2.5 **Medical Applications of Silver Nanoparticles**

Given Ag-NP's broad range of activity against microorganisms, their use has been explored in a variety of different medical applications, including wound dressings, catheters, and orthopedic implants. Wound care is exceedingly important as chronic wounds affect approximately 6.5 million patients in the United States alone, and wound infection remains the most expensive post-surgical complication [87]. In one recent study by Wu et al., Ag-NP were incorporated into nanofibers of bacterial cellulose, an attractive new dressing material which lacks antimicrobial properties [3]. Since Ag-NP were synthesized directly onto cellulose nanofibers, the dressing permitted a slow, sustained release of silver ions without Ag-NP release, decreasing any potential toxicity which may be associated with high concentrations of free Ag-NP. Despite a slower release of silver ions, the dressing effectively killed more than 99% of *E. coli*, *S. aureus*, and *P. aeruginosa* in vitro. Additionally, since bacterial cellulose nanofibers mimic the topographical features of extracellular matrix that is lost from wounds, the dressing facilitated attachment and growth of keratinocytes, which is necessary for reepithelialization, thereby hastening wound closure and reducing opportunity for infection. Furthermore, the Ag-NP nanofiber dressing did not display cytotoxicity against keratinocytes, and it provided an ideal, moist absorbent barrier with high wet strength [3]. Another study investigated polycaprolactone (PCL) nanofiber dressings impregnated with 0.05, 0.25, 0.5 or 1% wt Ag-NP, which effectively killed *S. aureus* and *E. coli* in a dose-dependent manner and displayed excellent mechanical properties with maximal tensile strength achieved at 0.5 wt% Ag-NP [88].

Hydrogel dressings, such as those made from PVP and carrageenan, have been incorporated with Ag-NP, as well. Hydrogels are capable of holding large amounts of fluid without leakage and can donate or absorb water from the wound environment, depending on the hydration status of the tissue [89]. Singh et al. found that Ag-NP incorporation did not alter the favorable fluid properties of hydrogel dressing, and furthermore successfully eradicated *P. aeruginosa, S. aureus, E. coli*, and *C. albicans* isolates from burn patients [89]. Finally, Anisha et al. developed antimicrobial sponges composed of chitosan, hyaluronic acid, and Ag-NP (0.001, 0.005, 0.01, 0.02% wt) for use in diabetic foot ulcers [90]. Sponges containing at least 0.005% Ag-NP demonstrated dose-dependent antibacterial activity against MRSA; however, those with at least 0.01% Ag-NP also showed significant toxicity against human dermal fibroblasts. Thus, the optimal Ag-NP content for these sponges may be between 0.005 and 0.01% wt in order to provide adequate activity against MRSA without significant toxicity. It must also be considered that this group synthesized Ag-NP separately before incorporating into the sponges [90], unlike Wu et al. who synthesized Ag-NP directly onto the dressing, thereby preventing free Ag-NP release into the wound [3]. If Ag-NP had been synthesized directly onto the antimicrobial wound sponges, the optimal range of Ag-NP may have been wider, therefore affording greater potential activity against MRSA.

In addition to wound dressings, Ag-NP have been used to coat plastic catheters, demonstrating sustained release of silver ions over at least 10 days, in addition to successfully killing and inhibiting biofilm formation of *E. coli, Enterococcus* spp., *S. aureus*, coagulase negative *Staphylococcus, P. aeruginosa*, and *C. albicans* [16,46]. Ag-NP conjugated with polymyxin B (polyB-Ag-NP), an antibacterial peptide, have also been used against multi drug-resistant *Vibrio fluvialis* and *P. aeruginosa*. PolyB-Ag-NP exhibited potent antimicrobial activity threefold greater than that of citrate-conjugated Ag-NP, successfully inhibited existing biofilms as well as biofilm formation, and neutralized 97% of endotoxin due to the affinity of polymyxin B. Even after being electroplated onto stainless steel surgical blades, polyB-Ag-NP maintained their antimicrobial properties, thus presenting these nanoparticles as a promising technology against organisms which may colonize medical equipment [45]. The anti-biofilm activity of Ag-NP has been attributed to their small size, thus increasing penetration of virtually impermeable biofilms, and their high surface reactivity confers the necessary potency required to exterminate this highly virulent configuration. Furthermore, Ag-NP have been found to interfere with quorum sensing, an additional virulence factor which also regulates biofilm formation [46].

While Ag-NP have demonstrated effective antimicrobial activity in conjunction with wound dressings, catheters, and surgical blades, they may also be used to prevent microbial infection associated with implanted devices. In orthopedic surgery, implant-associated infection is a major cause of morbidity. Ag-NP have been explored to improve post-surgical outcomes with trauma implants, tumor prostheses, and in combination with hydroxyapatite coatings; however, further in vivo study is necessary to ensure biocompatibility [47]. Thus far, an in vitro study of Ag-NP-embedded external fixator pins demonstrated effective prevention of *Staphylococcus epidermidis* biofilm formation [91], and composite bone grafts laden with Ag-NP have been beneficial against simulated vancomycin-resistant MRSA contamination in rat surgical models [92]. Additionally, a comparison of Ag-NP-loaded bone cement with gentamicin-loaded and plain cement revealed that Ag-NP cement has superior activity against *S. epidermidis*, methicillin-resistant *S. epidermidis* (MRSE), and MRSA [93]. Furthermore, Ag-NP was added to the cement without compromising its mechanical properties or causing cytotoxicity in osteoblast cell cultures [94].

Finally, incorporation of Ag-NP into chitosan-containing hydrocolloids has been studied to prevent *Streptococcus mutans* biofilm formation on dental enamel [95]. Biofilm models were constructed using bovine enamel discs, artificial saliva, and *S. mutans*, which were then immersed in the Ag-NP containing hydrocolloid solution for 1 min on a shaker. Following immersion, the biofilm model demonstrated 0% *S. mutans* cell viability, compared to 36.5% viability seen with the silver fluoride control. Cytotoxic properties of the Ag-NP-hydrocolloid were assessed using chicken eggs, and no cytotoxicity was observed by measuring vascular changes at the chorioallantoic membrane. Furthermore, there was no enamel color change observed following repeated immersion, a finding attributed to lower silver concentrations required to inhibit *S. mutans* when Ag-NP are employed. These findings demonstrate that Ag-NP hydrocolloids may have both antimicrobial and esthetic advantages compared to conventional silver fluoride dental products [95].

4.3 COPPER NANOPARTICLES FOR MICROBIAL INFECTION

Copper is an essential trace element required for the activity of a variety of copper-containing enzymes including lysyl oxidase (collagen cross-linking), tyrosinase (melanin synthesis), and cytochrome *c* oxidase (electron transport chain). While copper metabolism in microorganisms is less understood, some bacteria are known to employ copper uptake ATPase pumps, while Gram-negative bacteria such as *E. coli* may achieve metalation of copper-containing enzymes in the periplasmic space, thereby negating the need to transport copper across the cytoplasmic membrane [5]. Copper nanoparticles (Cu-NP) may be synthesized in a variety of ways, although synthesis from a pure copper metal wire via an inert gas condensation method is common. This involves copper vaporization via thermal evaporation, electron beam evaporation, or sputtering in an inert atmosphere of argon or helium, achieving excellent control over size, distribution, and purity [96]. Like Ag-NP, Cu-NP may also be manufactured via green synthesis, utilizing a variety of plant organisms and extracts [97].

4.3.1 Copper Nanoparticles Against Bacterial Infection

Cu-NP have demonstrated antibacterial activity against a variety of Gram-positive (*S. aureus*, MRSA, *B. subtilis*, *Propionibacterium acnes*, *Micrococcus luteus*) and Gram-negative (*E. coli*, *K. pneumoniae*, *P. aeruginosa*, *Salmonella* spp.) species [97–101]. Cu-NP appear to have a particularly potent effect against *E. coli* [97,98], which has been effectively killed in both liquid and solid growth media [96]. On scanning electron microscopy, *E. coli* treated with Cu-NP has demonstrated irregular forms, as opposed to its usual rod shape, and its cell walls were pitted and destroyed. Furthermore, the cells exhibited intracellular vacuoles with shrinkage of cytoplasmic contents [96] Cu-NP have also inhibited biofilms of *Listeria monocytogenes* and *P. aeruginosa* by reducing their surface area, thickness, and altering their morphology [102,103]. Interestingly, one study aimed to compare the activities of Cu-NP, copper oxide nanoparticles, copper hydroxide nanoparticles, copper microparticles, copper oxide microparticles, and ionic copper against *E. coli* and *Lactobacillus brevis*. The authors found that Cu-NP and copper oxide nanoparticles exhibited stronger antimicrobial effects than their micro-sized counterparts at the same copper concentration. Cu-NP also demonstrated different antimicrobial mechanisms compared to ionic and micro-sized copper, emphasizing that metals and other materials develop novel properties when sized on the nanoscale [104].

In addition to successfully inhibiting bacterial organisms, copper nanoparticles may also be useful against endospores, such as those formed by *Bacillus* and *Clostridium spp.* This is an important feature of Cu-NP, as endospores are especially difficult to eliminate and have been known to resist heat, radiation, desiccation, and denaturing chemicals. In hospitals, *Clostridium difficile* is especially significant, and endospores excreted by infected patients may contaminate surfaces, creating a long-term reservoir for transmission. Despite the challenges presented by endospore formation, *C. difficile* spores are completely inactivated on copper surfaces after 24-48 h [105], and copper nanoparticles kill more than 99% of *Bacillus anthracis* spores within 8 h [106]. These findings may indicate a role for copper nanotechnology against bacteria, as well as robust bacterial endospores.

4.3.2 Copper Nanoparticles Against Fungal Infection

Although the antimicrobial activity of Cu-NP is not as potent as that of Ag-NP, its spectrum of activity may be broader, especially with respect to fungi [63,107]. As previously mentioned, copper is widely used in agriculture to control fungal and bacterial pathogens, and current Cu-NP research has indicated extensive activity against plant pathogenic fungi, including *Fusarium culmorum, F. oxysporum, Fusarium graminearum, F. solani, A. flavus, A. alternata, Penicilliumchrysogenum, Phomadestructiva,* and *Cochlioboluslunatus* [97,108,109]. In fact, Cu-NP were found to be more effective than the commercially available fungicide, bavistin [108]. Due to the ever-increasing number of patients with compromised immunity, fungi have become a major health threat, and fungal infections in nonimmunocompromised patients are also on the rise [31]. Although there is less research regarding Cu-NP against human pathogenic fungi, Cu-NP are noted to be effective against *C. albicans,* and several of the aforementioned species *(F. solani, F. oxysporum, A. flavus, A. alternata, C. lunatus)* may also cause opportunistic infections in human patients [99]. As the antimicrobial activity of Cu-NP is more widely studied, its potential for use in the medical field as a broad spectrum antifungal may become more apparent.

4.3.3 Medical Applications of Copper Nanoparticles

Films or coatings which facilitate controlled release of metal species are exceedingly attractive for a variety of uses, including implants, adhesives, device coatings, and even food and beverage packaging. One study has explored the utility of orthodontic adhesive containing Cu-NP. In vitro testing revealed significant antimicrobial activity of the adhesive against

S. aureus, E. coli, and *S. mutans.* Interestingly, the shear bond strength of the adhesive was actually improved by the addition of Cu-NP, as tested on models using extracted molars. In addition to adhesive strength, esthetics are another important factor in orthodontics, and the addition of Cu-NP to the adhesive did not appear to have adverse effects on color or appearance of the material [110].

Generally speaking, metal/polymer nanocomposites may be useful due the highly dispersed nature and large surface area of metal-releasing particles, as well as the ability to fine-tune their releasing properties for minimization of toxicity and maximization of antimicrobial effect. To meet these needs, one group investigated polymer thin films loaded with Cu-NP [31]. The nanoparticles were electrochemically synthesized in an alkyl ammonium micellar environment which creates Cu-NP of 3.2 ± 1.6 nm in diameter with a stabilizing surfactant shell. Cu-NP were then combined with one of three bio-insoluble polymers, including polyvinylmethyl ketone (PVMK), polyvinyl chloride (PVC), and polyvinylidenefluoride (PVDF), by utilizing an organic solvent capable of dissolving the polymers and dispersing the particles via ultrasonication [31,107]. All three nanocomposites demonstrated significant antimicrobial activity against *Saccharomyces cerevisiae* yeast, molds, and bacteria, including *E. coli, S. aureus* and *L. monocytogenes* [31]. X-ray photoelectron spectroscopy revealed that the Cu-NP surfaces become oxidized, forming a copper oxide shell which further stabilizes the metallic copper core and, upon contact with an aqueous environment, dissolves to release Cu^{2+} ions, but not Cu-NP, at a steady rate over time [107]. The authors found that Cu-NP-PVMK nanocomposites had the strongest biostatic effect, while Cu-NP-PVDF were the least effective. This correlates with the fact that PVDF has a significantly lower Cu-NP loading capacity compared to PVMK and PVC. Subsequent testing with Cu-NP-PVMK nanocomposites demonstrated that increased Cu-NP loading leads to increased release of copper species and increased microbial killing in a predictable manner, which could be modeled with first-order kinetic curves [31]. Therefore, nanocomposite release properties and microbial killing could be closely modulated by controlling Cu-NP loading.

Release properties of nanocomposites were also compared to those of non-stabilized copper sources, including a bulk copper sheet (Cu-bulk) and PVMK containing a water-soluble copper salt ($CuCl_2$-PVMK). $CuCl_2$-PVMK rapidly reached its maximal copper release within the first 2 h. On the other hand, Cu-bulk exhibited first-order kinetics similar to those of Cu-NP loaded nanocomposites, although the plateau was much higher, and neither source of non-stabilized copper demonstrated the potential for control of release kinetics. By contrast, Cu-NP-loaded nanocomposites allowed controlled release at slow, steady rates, and Cu-PVMK and Cu-PVC demonstrated sustained copper release for at least 10 days [31].

4.4 GOLD NANOPARTICLES FOR MICROBIAL INFECTION

Gold nanoparticles (Au-NP) provide a versatile platform for a range of emerging antimicrobial technologies. Although Au-NP alone are inert against microbial organisms, their chemical stability and ability to be conjugated with a wide variety of molecules may be exploited for both diagnostic and therapeutic needs [15]. Advantages may also be derived from their unique optical and photothermal properties, which are attributed to surface plasmon resonance (SPR). This phenomenon occurs in gold particles sized below the electron mean free path, causing free surface electrons to oscillate in unison at a specific frequency, dependent on size and shape. SPR is responsible for Au-NP's powerful absorptive properties, which are strongly coupled to the reflection of light and may also be employed for localized heat generation [9,10].

4.4.1 Gold Nanoparticles for Therapeutic Intervention

To improve upon existing therapeutics, Au-NP may be conjugated with one or more antimicrobial agent, typically resulting in enhanced antimicrobial activity compared to their non-conjugated forms. The improvements associated with Au-NP conjugation may be due to enhanced binding and penetration of the target cell, in addition to increased density and reactivity of the conjugated agent as it interacts with microbial organisms. Au-NP may also be conjugated with antibodies for targeted delivery or with photosensitizing molecules for photodynamic therapy (PDT), and SPR may be utilized to induce hyperthermia leading to microbial lysis. Furthermore, photothermal heat generation may be used to trigger the release of a therapeutic payload, providing yet another mechanism for targeted drug delivery [111].

4.4.2 Gold Nanoparticles for Antimicrobial Drug Delivery

Au-NP have been utilized to improve the efficacy of several antimicrobial drugs, including ampicillin, aminoglycosides, ciprofloxacin, and vancomycin. For example, Brown *et al.* conjugated ampicillin molecules to the surface of Au-NP via their thioether moiety [15]. Subsequent experiments demonstrated significant activity of ampicillin-Au-NP against known multi drug-resistant bacterial strains, including MRSA, *P. aeruginosa*, *E. aerogenes*, and *E. coli*. These results demonstrate the ability of Au-NP to transform a conventional antibiotic into a more powerful, broad spectrum agent, killing both Gram-positive and Gram-negative drug resistant bacteria which otherwise would not be susceptible [15]. Saha et al. also conjugated Au-NP with ampicillin, as well as aminoglycosides streptomycin and kanamycin, and tested the conjugates against both Gram-positive (*S. aureus, M. luteus*) and Gram-negative (*E. coli*) bacteria. All of the conjugates demonstrated enhanced antimicrobial activity compared to the antibiotics alone, especially streptomycin and kanamycin, and all of the antibiotics demonstrated significantly greater heat stability in their Au-NP conjugated forms [112].

Aminoglycoside antibiotics appear to conjugate particularly well with Au-NP, which is attributed to an electrostatic attraction between their amine groups and the nanoparticles. One study demonstrated that Au-NP can even be used as a probe for the quantification of aminoglycoside antibiotics in drug formulations in conjunction with UV-vis spectrophotometry or Fourier transform infrared spectroscopy (FT-IR). Due to Au-NP's strong interaction with aminoglycosides, concentrations were detectable as low as 0.01 µ M [113]. After observing this remarkable affinity, the same group utilized streptomycin, gentamycin and neomycin to create Au-Strep-, Au-Genta- and Au-Neo-NP, respectively. All three conjugates exhibited antibacterial activity against Gram-positive (*S. aureus, M. luteus*) and Gram-negative (*E. coli, P. aeruginosa*) bacteria significantly greater than that of the aminoglycosides alone. The effect was especially pronounced against Gram-negative species, which may be due to differences in architecture of the cell wall. TEM imaging reveals that Au-aminoglycoside nanoparticles act at the cell wall, causing destabilization. While Gram-positive bacteria generally have a thick wall of peptidoglycan, the cell wall of Gram-negative bacteria is much thinner, which may be more easily compromised by gold nanoparticle conjugates. Once compromised, aminoglycosides are free to enter the cells and bind the 30S subunit of bacterial ribosomes, effectively inhibiting bacterial translation [113].

Ciprofloxacin is another antimicrobial drug that has been conjugated with Au-NP. Following physical analysis, Tom et al. determined that Au-NP of 3–4 and 15–20 nm could adsorb an average of 65 and 585 ciprofloxacin molecules, respectively, via interaction with the NH moiety of their piperazine group. Ciprofloxacin-Au-NP demonstrated stability in the dry state, in solution and following re-dispersion, and release rate of the antibiotic was determined by particle size. For example, a higher rate of desorption was seen with smaller Au-NP, and release occurred more readily in basic medium versus pure water, demonstrating the potential for controlled release based on size and chemical environment. Furthermore, the group discovered that ciprofloxacin's fluorescence undergoes a red shift and slight quenching of intensity when conjugated with Au-NP, and fluorescent properties of adsorbed ciprofloxacin did not vary significantly with different-sized nanoparticles [114].

Another group showed that vancomycin-conjugated gold nanoparticles (Au-Vanc-NP) enhanced vancomycin's activity 50-fold against vancomycin-resistant enterococci (VRE). Unexpectedly, Au-Vanc-NP also exhibited significant activity against *E. coli*, a Gram-negative bacterium [115]. Vancomycin inhibits the synthesis of peptidoglycan, a major component of the thick cell walls found in Gram-positive bacteria. Conversely, Gram-negative bacteria have thinner cell walls, and their outer membrane blocks vancomycin entry. Thus, conventional vancomycin is ineffective against Gram-negative bacteria. In this study, TEM of Au-Vanc-NP demonstrated that the nanoparticles bind substrates on *E. coli*'s outer membrane, causing destabilization which may grant vancomycin access to the cell wall [115]. Besides increasing the efficacy of antibiotics, these findings illustrate that Au-NP may be used to widen their spectrum of coverage.

4.4.3 Gold Nanoparticles for Photodynamic Therapy

In addition to antimicrobial drugs, Au-NP have been conjugated with photosensitizing agents for photodynamic therapy (PDT), a treatment modality effective against cancer cells, as well as pathogenic organisms. PDT requires uptake of a photosensitizing molecule which, when struck by a specific wavelength of light, generates singlet oxygen, leading to cell death. However, the effectiveness of many current photosensitizers is suboptimal, and diffusion to non-target body regions may cause phototoxicity. This can be prevented by applying photosensitizers with greater target specificity, as well as administering them at a lower dose [111]. Drawbacks of current photosensitizing agents have lead biomedical researchers to investigate Au-NP as a potential source of improved photosensitizers for PDT.

Thus far, studies have indicated that Au-NP-conjugated photosensitizing molecules yield improved results compared to their nonconjugated form. For example, Gil-Tomas et al. covalently coupled toluidine blue O (TBO) to Au-NP with tiopronin (Au-NP capping agent), and the conjugates demonstrated stability in air, water and PBS. When activated by a beam of 632 nm light, TBO-tiopronin Au-NP experienced a fourfold increase in bactericidal effect against *S. aureus* compared to TBO alone, indicating that lower TBO doses may be used with improved efficacy. Tiopronin-capped Au-NP without TBO did not demonstrate any bactericidal activity [116]. In another study, Au-NP were conjugated with either TBO or methylene blue (MB), and both conjugates demonstrated potent activity against *C. albicans* biofilm and planktonic cells in vitro. Ensuing studies involving cutaneous and oral *C. albicans* infection in BALB/c mice have also demonstrated effective fungal depletion with PDT using either TBO- or MB-Au-NP [117]. These results are promising as Au-NP may offer enhanced PDT against infectious organisms, especially those causing skin and soft tissue infections, which may preclude the need for antibiotic therapy.

In theory, Au-NP-conjugated photosensitizers could be further improved by also conjugating them with specific antibodies targeted to the source of infection. Given the damage that may be done if photosensitizers are not well-localized, pathogen-specific antibody conjugation may provide significant benefits. Targeted PDT has already been explored for cancer therapy by conjugating inert hybrid metal nanoparticles with both photosensitizing molecules and cell-targeting antibodies [118]. Targeted PDT for microbial infection has thus far been explored without the use of nanotechnology, by conjugating photosensitizers directly with pathogen-targeting antibodies for use against MRSA infection [119]. Promising results from both studies indicate that it may only be a matter of time before researchers add infection-targeting antibodies to photosensitizer-conjugated Au-NP for use against bacterial and fungal infections.

4.4.4 Gold Nanoparticles for Photothermal Treatment

Due to Au-NP's photothermal properties, they may be used to generate localized heat, thereby inactivating infectious organisms. Despite the potential usefulness of Au-NP for photothermal therapy, this application has not been extensively studied; however, it has been well studied for cancer therapy [120–122]. The principle involves Au-NP conjugation with a pathogen-specific antibody. Once bound to the target, either pulsed or continuous laser irradiation can be aimed at the area of infection with frequency, fluence, and power that are optimized for the specific size/shape of Au-NP, thereby promoting maximal heat-mediated microbial killing with minimal damage to surrounding tissue [111].

The potential utility of photothermal therapy for microbial infection has been explored by Zharov et al. who used anti-protein A-conjugated Au-NP with an activating laser to demonstrate selective destruction of *S. aureus* in vitro. Bacteria without the presence of nanoparticles were unaffected by the laser [123]. Additional in vitro studies have used anti-PA3-conjugated Au-NP with a near-infrared laser against *P. aeruginosa*, resulting in widespread bacterial death [124].

Further studies regarding photothermal therapy with Au-NP are limited, although some innovative groups have combined Au-NP photothermal therapy with other treatment modalities to yield even more potent effects. For example, Huang et al. loaded polygonal Au-NP with vancomycin and then subjected them to laser irradiation in the presence of Gram-positive (*S. pyogenes, S. aureus*), Gram-negative (*E. coli, E. coli O157:H7, A. baumannii*), and drug-resistant (VRE, MRSA, pan drug-resistant *A. baumannii*) bacteria [125]. For all types of bacteria, greater than 99% death was achieved within 5 min of illumination, which was greater than the microbial killing observed with Au-NP lacking vancomycin conjugation. As before, the apparent effect of vancomycin against Gram-negative bacteria may be due to a compromised external membrane, this time via thermal disruption, thereby allowing access to the cell wall.

In another unique approach, Kuo et al. combined PDT with photothermal therapy by conjugating polyacrylic acid-coated gold nanorods with TBO [126]. Following irradiation with a 633 nm laser for PDT and an 808 nm laser for photothermal therapy, the nanorods effectively exterminated MRSA cells, to an even greater extent than when PDT was used alone. Finally, Radt et al. have demonstrated that photothermal properties may be utilized for targeted drug release at the site of infection. Their group designed nanostructured capsules of polyelectrolyte multilayers with incorporated Au-NP, which were loaded with lysozymes. Upon irradiation, the capsules burst, releasing their payload and successfully killing *Micrococcus lysodeikticus* in suspension [127]. Although these novel applications have been featured in a limited number of studies, the results thus far are encouraging and demonstrate the potential for significant future benefits (Fig. 4.2).

4.4.5 **Gold Nanoparticles for Diagnostic Applications**

Reflective properties of Au-NP may be used for several diagnostic applications in the field of medicine, including pathogen identification. Although conventional culturing methods are very accurate, clinicians must routinely wait 24 h or more to obtain results, which may lead to treatment delays or cause patients to receive unnecessary antibiotics, thereby increasing risks of side effects and providing opportunities for antimicrobial resistance development [111]. To expedite the identification of bacterial pathogens, some medical facilities utilize polymerase chain reaction (PCR); however, these methods are expensive, requiring sophisticated equipment and species-specific DNA primers which hinder their use [128].

Conversely, Au-NP are inexpensive, they provide rapid, highly accurate results, and are easy to use. Various microbial species may be detected

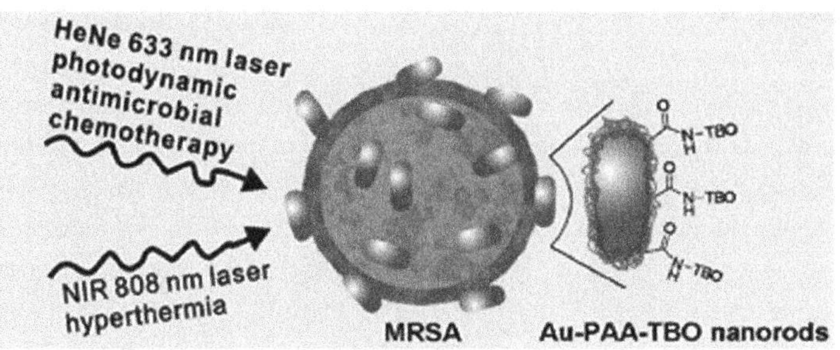

■ **FIGURE 4.2**   Gold nanorods for combination photodynamic-photothermal therapy. Gold nanorods were coated with polyacrylic acid (PAA) and conjugated with a photosensitizing molecule, toluidine blue O (TBO) [126]. Nanorods were then irradiated with a 633 nm helium-neon (HeNe) laser, as well as an 808 nm near-infrared (NIR) laser. This technique applied a combination of photodynamic and photothermal therapy, respectively, eliminating MRSA to an even greater extent than photodynamic therapy alone. *Image reproduced from Kuo WS, Chang CN, Chang YT, Yeh CS. Antimicrobial gold nanorods with dual-modality photodynamic inactivation and hyperthermia. Chem Commun 2009;32:4853–5, with permission from Royal Society of Chemistry.*

using Au-NP by exploiting their tendency to aggregate and cause an associated color change. For example, spherical Au-NP, which reflect red light in suspension, can aggregate to reflect blue light due to a shift in peak SPR. Scientists can create two unique types of Au-NP (e.g., type A and type B) via modification with two different oligonucleotides, thereby functionalizing them to bind two different regions of target DNA from a given microorganism. In suspension, both types of Au-NP reflect red. In the presence of the microorganism, the nanoparticles bind DNA, and Au-NP types A and B subsequently aggregate with each other, reflecting a blue color [129].

Using this method, Elghanian et al. utilized Au-NP to detect as few as 10 femtomoles of an analyte oligonucleotide [129]. Ensuing work by Storhoff et al. enabled the detection of 50 femtomolar MRSA DNA following a brief 1-hour hybridization [130]. Aggregation properties have also been utilized by Baptista et al. who conjugated Au-NP with an oligonucleotide sequence complementary to DNA of *Mycobacterium tuberculosis* [131]. In this case, *M. tuberculosis* was detected via preservation of Au-NP's red color following the addition of sodium chloride (NaCl). Since Au-NP aggregate in the presence of NaCl, turning the sample blue, a persistent red color indicates binding of *M. tuberculosis* DNA, which prevents NaCl-induced aggregation [131].

In addition to functionalizing Au-NP with oligonucleotides, antibodies may be used as well, as demonstrated by the sensitive and specific detection of *S. aureus* using antibodies against Staphylococcal protein A (SpA), a cell wall protein. Huang et al. utilized anti-SpA antibodies with two distinct specificities, immobilizing one type of antibody on a membrane and conjugating the other to Au-NP. As sample flowed along the membrane, SpA was bound, which then bound Au-NP detector reagent, producing a bright red color. Results were rapid and observed by the naked eye, yielding 100% sensitivity for 306 *S. aureus* strains and 96–100% specificity for 44 non-*S. aureus* strains [132]. In addition to obvious potential advantages at the point of care, this assay may prove useful for infection control in the food industry. The assay was assessed by inoculating twelve processed food samples with 0.9, 1.2, 2.4, and 6 CFU g^{-1} of *S. aureus*, and all contaminated samples yielded positive assay results [132].

In medical practice, Au-NP may be useful for rapid detection and typing of bacteria in human joint fluid. This application could be used to diagnose and guide treatment for periprosthetic joint infection, one of the most dreaded complications associated with arthroplasty. In this setting, rapid detection is ideal to ensure prompt intervention and minimal damage, along with typing for proper choice of antibiotic regimen. Unfortunately, conventional cultures are time-consuming and associated with a high false-negative rate, thereby delaying intervention. Conversely, alternate methods such as PCR are tedious and associated with a high false-positive rate. To meet this diagnostic need, Chang et al. designed an automated microfluidic platform using Au-NP for rapid detection of live bacteria within 30 min, with an additional 40 min required for typing of captured bacteria via accelerated PCR on the microfluidic chip [2]. The platform demonstrated rapid results with a low limit of detection and high accuracy, thereby presenting another potential application for Au-NP.

In addition to bacterial detection, Au-NP have been employed for viral diagnosis. For example, Jeong et al. developed a microwell plate-based multiplex immunoassay for detection of antibodies against HIV, HBV, and hepatitis C (HCV). Detection of all three viruses was achieved simultaneously without sacrificing sensitivity, accuracy or reproducibility, in contrast to conventional ELISA assays which do not allow multiplexing and therefore require separate tests [133]. The Au-NP immunoassay involves capture of target antibodies by specific antigens on DNA-encoded Au-NP. Then, target antibodies may be quantified using RNA probes that produce RNase H-mediated fluorescent signals. This immunoassay may enable efficient, simultaneous detection for three different viruses, which cause serious infections and frequently occur concomitantly (Fig. 4.3).

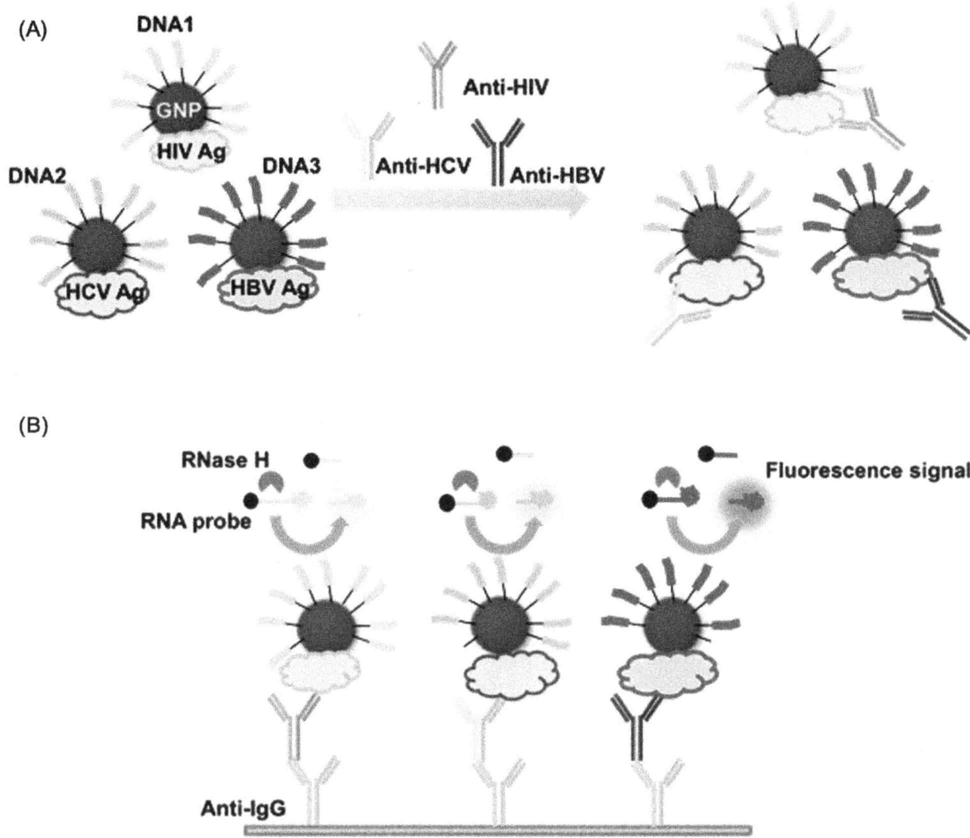

■ **FIGURE 4.3** (A) Gold nanoparticles for microwell plate-based multiplex immunoassay. Gold nanoparticles enabled efficient, simultaneous detection of HIV, HBV, and HCV immunoglobulins [133]. Target antibodies were captured by binding specific antigens on DNA-encoded gold nanoparticles. (B) The antibodies were then quantified using RNA probes that produced RNase H-mediated fluorescent signals. *Image reproduced from Jeong MS, Ahn DR. A microwell plate-based multiplex immunoassay for simultaneous quantitation of antibodies to infectious viruses. Analyst 2015;140(6):1995–2000, with permission from Royal Society of Chemistry.*

4.5 **SAFETY CONSIDERATIONS**

Metal nanoparticles are an emerging technology that holds great promise for a variety of medical and nonmedical fields. In particular, their unique properties may be harnessed to fight microbial infection, as outlined in this chapter. From our discussion thus far, it is clear that many unknowns remain with respect to metal nanoparticles, regarding their potential uses, as well as possible modes of toxicity. Obviously, any metal—silver, copper, gold—can cause toxicity at a high enough concentration, yet there are also ranges in which these metals are acceptable [134]. When it comes to metal nanoparticles, determining this range may be more difficult since we must consider the released metal species, as well as the nanoparticles themselves, unless they are stabilized, for example in an external composite [31]. Furthermore, metal nanoparticles come in a variety of sizes, shapes, and surface functionalizations, making generalizations difficult. For example, if 5 nm silver nanospheres are proven safe at a particular concentration, this may not necessarily translate to 60 nm silver nanorods, or 25 nm nanospheres conjugated with a conventional antibiotic [76]. Additionally, data are not yet available which demonstrate metal nanoparticles' ultimate fate in vivo, including whether they may be chemically modified, excreted, or accumulate in specific tissues.

Perhaps the greatest setback is that there are currently no standardized methods for determining the toxicity of metal nanoparticles. The studies reported in the literature are widely varied and have utilized a range of models and techniques. For example, the toxicity of different types of Ag-NP have been individually studied in vitro using hepatoblastoma cells [69], renal epithelial cells [76], and keratinocytes [3], all under different laboratory conditions. Yet another study assessed Ag-NP using the "hen's egg test" which measures the acute effects of a substance on the small vessels and proteins of the chorioallantoic membrane of chicken eggs [95]. This method is quick and simple, and has been officially accepted in France and Germany for identifying irritating substances, although it is unclear whether such a method may be useful for assessing metal nanoparticle toxicity. Another, more thorough investigation of Ag-NP assessed various concentrations of 7–20 nm spherical particles against fibroblasts and hepatocytes, which were also examined on TEM. The images revealed dark, electron-dense aggregates in the mitochondria and cytoplasm; however, morphology remained unaltered in these cells following 24-h exposure. The cells were found to have significantly higher GSH levels compared to unchallenged cells, as well as less lipid peroxidation, indicating that, despite Ag-NP entry, these cells may possess antioxidant compensatory mechanisms which enable tolerance and prevent changes in morphology at the tested concentrations [135].

Given the current available data, it is too soon to draw conclusions regarding metal nanoparticle toxicity. However, moving forward, researchers should seek standardized methodologies for toxicological assessment, taking into account different sizes, shapes, concentrations, and methods of synthesis. Future practices should also consider different routes of administration (e.g., topical, oral, parenteral), acute versus chronic exposures, and in vivo models should be further developed. Although metal nanoparticles are not yet ready to be used in humans, development of more sophisticated toxicological assessment tools may help identify which applications are associated with acceptable levels of toxicity. In the future, this may allow us to safely capitalize on the medical benefits of metal nanoparticles as outlined in the preceding chapter.

4.6 CONCLUSION

The growing presence of antimicrobial resistance among bacterial, fungal, viral, and protozoal pathogens calls for new methods of diagnosis and treatment for microbial infection. Antimicrobial effects of metal nanoparticles are enhanced by their large surface area-to-volume ratio, small size, and increased surface reactivity. They may also be functionalized to improve upon existing antimicrobial agents, and surface plasmon resonance enables their use for light-based therapy. Importantly, metal nanoparticles may circumvent antimicrobial resistance development by exerting nonspecific, multimechanistic antimicrobial activity and thwarting specific resistance mechanisms such as drug efflux, quorum sensing, and biofilm formation. Furthermore, the physical properties of gold nanoparticles allow faster, more sensitive detection of antigens and antibodies associated with microbial infection, which in turn may lead to earlier administration of appropriate interventions. Overall, metal nanoparticles provide a vast array of applications for the identification and eradication of microbial pathogens, which could potentially be of enormous benefit to public health. However, metal nanoparticles are still an emerging technology. Further studies are necessary to better understand the range of possible applications, and standardized methods for determining metal nanoparticle toxicity are urgently needed to determine which practices may be safe for use in patients.

REFERENCES

[1] Kim BY, Rutka JT, Chan WC. Nanomedicine. New Engl J Med 2010; 363(25):2434–43.

[2] Chang W-H, Wang C-H, Lin C-L, Wu J-J, Lee MS, Lee G-B. Rapid detection and typing of live bacteria from human joint fluid samples by utilizing an integrated microfluidic system. Biosensors Bioelectronics 2015;66:148–54.

[3] Wu J, Zheng Y, Song W, Luan J, Wen X, Wu Z, et al. In situ synthesis of silver-nanoparticles/bacterial cellulose composites for slow-released antimicrobial wound dressing. Carbohydr Polym 2014;102:762–71.

[4] Huh AJ, Kwon YJ. "Nanoantibiotics": a new paradigm for treating infectious diseases using nanomaterials in the antibiotics resistant era. J Controlled Release 2011;156(2):128–45.

[5] Grass G, Rensing C, Solioz M. Metallic copper as an antimicrobial surface. Appl Environ Microbiol 2011;77(5):1541–7.

[6] Pal S, Tak YK, Song JM. Does the antibacterial activity of silver nanoparticles depend on the shape of the nanoparticle? A study of the gram-negative bacterium *Escherichia coli*. Appl Environ Microbiol 2007;73(6):1712–20.

[7] Landriscina A, Rosen J, Friedman AJ. Biodegradable chitosan nanoparticles in drug delivery for infectious disease. Nanomedicine 2015;10(10):1609–19.

[8] Eustis S, El-Sayed MA. Why gold nanoparticles are more precious than pretty gold: noble metal surface plasmon resonance and its enhancement of the radiative and nonradiative properties of nanocrystals of different shapes. Chem Soc Rev 2006;35(3):209–17.

[9] El-Sayed MA. Some interesting properties of metals confined in time and nanometer space of different shapes. Acc Chem Res 2001;34(4):257–64.

[10] Alfano RR, Ni X, Zevallos M.Changing skin-color perception using quantum and optical principles in cosmetic preparations. Google Patents;2013.

[11] Sun L, Singh AK, Vig K, Pillai SR, Singh SR. Silver nanoparticles inhibit replication of respiratory syncytial virus. J Biomed Nanotechnol 2008;4(2):149–58.

[12] Baram-Pinto D, Shukla S, Perkas N, Gedanken A, Sarid R. Inhibition of herpes simplex virus type 1 infection by silver nanoparticles capped with mercaptoethane sulfonate. Bioconjug Chem 2009;20(8):1497–502.

[13] Janát-Amsbury M, Ray A, Peterson C, Ghandehari H. Geometry and surface characteristics of gold nanoparticles influence their biodistribution and uptake by macrophages. Eur J Pharmaceutics Biopharm 2011;77(3):417–23.

[14] Zhou J, Shum K-T, Burnett JC, Rossi JJ. Nanoparticle-based delivery of RNAi therapeutics: progress and challenges. Pharmaceuticals 2013;6(1):85–107.

[15] Brown AN, Smith K, Samuels TA, Lu J, Obare SO, Scott ME. Nanoparticles functionalized with ampicillin destroy multiple-antibiotic-resistant isolates of *Pseudomonas aeruginosa* and *Enterobacter aerogenes* and methicillin-resistant *Staphylococcus aureus*. Appl Environ Microbiol 2012;78(8):2768–74.

[16] Roe D, Karandikar B, Bonn-Savage N, Gibbins B, Roullet J-B. Antimicrobial surface functionalization of plastic catheters by silver nanoparticles. J Antimicrob Chemother 2008;61(4):869–76.

[17] Pelgrift RY, Friedman AJ. Nanotechnology as a therapeutic tool to combat microbial resistance. Adv Drug Deliv Rev 2013;65(13):1803–15.

[18] Knetsch ML, Koole LH. New strategies in the development of antimicrobial coatings: the example of increasing usage of silver and silver nanoparticles. Polymers 2011;3(1):340–66.

[19] Lara HH, Ayala-Núñez NV, Turrent LdCI, Padilla CR. Bactericidal effect of silver nanoparticles against multidrug-resistant bacteria. World J Microbiol Biotechnol 2010;26(4):615–21.

[20] Elguindi J, Wagner J, Rensing C. Genes involved in copper resistance influence survival of *Pseudomonas aeruginosa* on copper surfaces. J Appl Microbiol 2009;106(5):1448–55.

[21] Santo CE, Taudte N, Nies DH, Grass G. Contribution of copper ion resistance to survival of *Escherichia coli* on metallic copper surfaces. Appl Environ Microbiol 2008;74(4):977–86.

[22] Marambio-Jones C, Hoek EM. A review of the antibacterial effects of silver nanomaterials and potential implications for human health and the environment. J Nanopart Res 2010;12(5):1531–51.

[23] Feng Q, Wu J, Chen G, Cui F, Kim T, Kim J. A mechanistic study of the antibacterial effect of silver ions on *Escherichia coli* and *Staphylococcus aureus*. J Biomed Mater Res 2000;52(4):662–8.

[24] Yamanaka M, Hara K, Kudo J. Bactericidal actions of a silver ion solution on *Escherichia coli*, studied by energy-filtering transmission electron microscopy and proteomic analysis. Appl Environ Microbiol 2005;71(11):7589–93.

[25] Sondi I, Salopek-Sondi B. Silver nanoparticles as antimicrobial agent: a case study on *E. coli* as a model for Gram-negative bacteria. J Colloid Interface Sci 2004;275(1):177–82.

[26] Huang K-S, Chang S-C, Yang C-H, Wang C-Y. Advances in bio-hybrid nanostructures with anti-pathogenic activity. Curr Med Chem 2014;21(29):3323–32.

[27] Bragg P, Rainnie D. The effect of silver ions on the respiratory chain of *Escherichia coli*. Can J Microbiol 1974;20(6):883–9.

[28] McDonnell G, Russell AD. Antiseptics and disinfectants: activity, action, and resistance. Clin Microbiol Rev 1999;12(1):147–79.

[29] Richard J, Spencer B, McCoy L, Carina E, Washington J, Edgar P. Acticoat versus Silverlon: the truth. J Burns Surg Wound Care 2002;1(1):11–19.

[30] Dollwet H, Sorenson J. Historic uses of copper compounds in medicine. Trace Elem Med 1985;2(2):80–7.

[31] Cioffi N, Torsi L, Ditaranto N, Tantillo G, Ghibelli L, Sabbatini L, et al. Copper nanoparticle/polymer composites with antifungal and bacteriostatic properties. Chem Mater 2005;17(21):5255–62.

[32] Cha J-S, Cooksey DA. Copper resistance in *Pseudomonas syringae* mediated by periplasmic and outer membrane proteins. Proc Natl Acad Sci USA 1991;88(20):8915–9.

[33] Xu X, Li T, Xu Z, Wei H, Lin R, Xia B, et al. Automatic enumeration of gold nanomaterials at the single-particle level. Anal Chem 2015;87(5):2576–81.

[34] Gavade N, Kadam A, Suwarnkar M, Ghodake V, Garadkar K. Biogenic synthesis of multi-applicative silver nanoparticles by using Ziziphus Jujuba leaf extract. Spectrochim Acta Part A 2015;136:953–60.

[35] Rajkuberan C, Sudha K, Sathishkumar G, Sivaramakrishnan S. Antibacterial and cytotoxic potential of silver nanoparticles synthesized using latex of *Calotropis gigantea* L. Spectrochim Acta Part A 2015;136:924–30.

[36] Emmanuel R, Palanisamy S, Chen S-M, Chelladurai K, Padmavathy S, Saravanan M, et al. Antimicrobial efficacy of green synthesized drug blended silver nanoparticles against dental caries and periodontal disease causing microorganisms. Mater Sci Eng: C 2015;56:374–9.

[37] Gogoi N, Babu PJ, Mahanta C, Bora U. Green synthesis and characterization of silver nanoparticles using alcoholic flower extract of Nyctanthes arbortristis and in vitro investigation of their antibacterial and cytotoxic activities. Mater Sci Eng: C 2015;46:463–9.

[38] Mapara N, Sharma M, Shriram V, Bharadwaj R, Mohite K, Kumar V. Antimicrobial potentials of *Helicteres isora* silver nanoparticles against extensively drug-resistant (XDR) clinical isolates of *Pseudomonas aeruginosa*. Appl Microbiol Biotechnol 2015:1–13.

[39] Liu H-L, Dai SA, Fu K-Y, Hsu S-H. Antibacterial properties of silver nanoparticles in three different sizes and their nanocomposites with a new waterborne polyurethane. Int J Nanomed 2010;5:1017.

[40] Shameli K, Ahmad MB, Jazayeri SD, Shabanzadeh P, Sangpour P, Jahangirian H, et al. Investigation of antibacterial properties silver nanoparticles prepared via green method. Chem Cent J 2012;6(1):73.

[41] Liu J, Sonshine DA, Shervani S, Hurt RH. Controlled release of biologically active silver from nanosilver surfaces. ACS Nano 2010;4(11):6903–13.

[42] Kim K-J, Sung WS, Moon S-K, Choi J-S, Kim JG, Lee DG. Antifungal effect of silver nanoparticles on dermatophytes. J Microbiol Biotechnol 2008;18(8):1482–4.

[43] Gaikwad S, Ingle A, Gade A, Rai M, Falanga A, Incoronato N, et al. Antiviral activity of mycosynthesized silver nanoparticles against herpes simplex virus and human parainfluenza virus type 3. Int J Nanomed 2013;8:4303.

[44] Cameron P, Gaiser BK, Bhandari B, Bartley PM, Katzer F, Bridle H. Silver nanoparticles decrease the viability of *Cryptosporidium parvum* oocysts. Appl Environ Microbiol 2015 AEM. 02806-15.

[45] Lambadi PR, Sharma TK, Kumar P, Vasnani P, Thalluri SM, Bisht N, et al. Facile biofunctionalization of silver nanoparticles for enhanced antibacterial properties, endotoxin removal, and biofilm control. Int J Nanomed 2015;10:2155.

[46] Singh BR, Singh BN, Singh A, Khan W, Naqvi AH, Singh HB. Mycofabricated biosilver nanoparticles interrupt *Pseudomonas aeruginosa* quorum sensing systems. Sci Rep 2015:5.

[47] Brennan S, Fhoghlú CN, Devitt B, O'Mahony F, Brabazon D, Walsh A. Silver nanoparticles and their orthopaedic applications. Bone Joint J 2015;97(5): 582–9.

[48] Shahverdi AR, Fakhimi A, Shahverdi HR, Minaian S. Synthesis and effect of silver nanoparticles on the antibacterial activity of different antibiotics against *Staphylococcus aureus* and *Escherichia coli*. Nanomed: Nanotechnol Biol Med 2007;3(2):168–71.

[49] Martinez-Gutierrez F, Olive PL, Banuelos A, Orrantia E, Nino N, Sanchez EM, et al. Synthesis, characterization, and evaluation of antimicrobial and cytotoxic effect of silver and titanium nanoparticles. Nanomed: Nanotechnol Biol Med 2010;6(5):681–8.

[50] Ghosh S, Kaushik R, Nagalakshmi K, Hoti S, Menezes G, Harish B, et al. Antimicrobial activity of highly stable silver nanoparticles embedded in agar–agar matrix as a thin film. Carbohydr Res 2010;345(15):2220–7.

[51] Krishnaraj C, Jagan E, Rajasekar S, Selvakumar P, Kalaichelvan P, Mohan N. Synthesis of silver nanoparticles using *Acalypha indica* leaf extracts and its antibacterial activity against water borne pathogens. Colloids Surf B: Biointerfaces 2010;76(1):50–6.

[52] Devi LS, Joshi S. Antimicrobial and synergistic effects of silver nanoparticles synthesized using soil fungi of high altitudes of Eastern Himalaya. Mycobiology 2012;40(1):27–34.

[53] Shrivastava S, Bera T, Roy A, Singh G, Ramachandrarao P, Dash D. Characterization of enhanced antibacterial effects of novel silver nanoparticles. Nanotechnology 2007;18(22):225103.

[54] Bawaskar M, Gaikwad S, Ingle A, Rathod D, Gade A, Duran N, et al. A new report on mycosynthesis of silver nanoparticles by *Fusarium culmorum*. Curr Nanosci 2010;6(4):376–80.

[55] Bonde S, Rathod D, Ingle A, Ade R, Gade A, Rai M. Murraya koenigii-mediated synthesis of silver nanoparticles and its activity against three human pathogenic bacteria. Nanosci Methods 2012;1(1):25–36.

[56] Dos Santos MM, Queiroz MJ, Baptista PV. Enhancement of antibiotic effect via gold: silver-alloy nanoparticles. J Nanopart Res 2012;14(5):1–8.

[57] Ma Y, Zhou T, Zhao C. Preparation of chitosan–nylon-6 blended membranes containing silver ions as antibacterial materials. Carbohydr Res 2008;343(2):230–7.

[58] Sanpui P, Murugadoss A, Prasad PD, Ghosh SS, Chattopadhyay A. The antibacterial properties of a novel chitosan–Ag-nanoparticle composite. Int J Food Microbiol 2008;124(2):142–6.

[59] Banerjee M, Mallick S, Paul A, Chattopadhyay A, Ghosh SS. Heightened reactive oxygen species generation in the antimicrobial activity of a three component iodinated chitosan– silver nanoparticle composite. Langmuir 2010;26(8):5901–8.

[60] Potara M, Jakab E, Damert A, Popescu O, Canpean V, Astilean S. Synergistic antibacterial activity of chitosan–silver nanocomposites on *Staphylococcus aureus*. Nanotechnology 2011;22(13):135101.

[61] Huang L, Dai T, Xuan Y, Tegos GP, Hamblin MR. Synergistic combination of chitosan acetate with nanoparticle silver as a topical antimicrobial: efficacy against bacterial burn infections. Antimicrob Agents Chemother 2011;55(7):3432–8.

[62] Kong M, Chen XG, Xing K, Park HJ. Antimicrobial properties of chitosan and mode of action: a state of the art review. Int J Food Microbiol 2010;144(1):51–63.

[63] Blecher K, Nasir A, Friedman A. The growing role of nanotechnology in combating infectious disease. Virulence 2011;2(5):395–401.

[64] Rai M, Kon K, Ingle A, Duran N, Galdiero S, Galdiero M. Broad-spectrum bioactivities of silver nanoparticles: the emerging trends and future prospects. Appl Microbiol Biotechnol 2014;98(5):1951–61.

[65] Arjun TV, Bholay A. Biosynthesis of silver nanoparticles and its antifungal activities. J Environ Res Develop 2012;7(1A).

[66] Gajbhiye M, Kesharwani J, Ingle A, Gade A, Rai M. Fungus-mediated synthesis of silver nanoparticles and their activity against pathogenic fungi in combination with fluconazole. Nanomed: Nanotechnol Biol Med 2009;5(4):382–6.

[67] Xu Y, Gao C, Li X, He Y, Zhou L, Pang G, et al. In vitro antifungal activity of silver nanoparticles against ocular pathogenic filamentous fungi. J Ocular Pharmacol Therapeut 2013;29(2):270–4.

[68] Kim K-J, Sung WS, Suh BK, Moon S-K, Choi J-S, Kim JG, et al. Antifungal activity and mode of action of silver nano-particles on *Candida albicans*. Biometals 2009;22(2):235–42.

[69] Lu L, Sun R, Chen R, Hui C-K, Ho C-M, Luk JM, et al. Silver nanoparticles inhibit hepatitis B virus replication. Antivir Ther 2008;13(2):253.

[70] Mehrbod P, Motamed N, Tabatabaian M, Estyar RS, Amini E, Shahidi M, et al. In vitro antiviral effect of "Nanosilver" on influenza virus. DARU J Pharmaceut Sci 2009;17(2):88–93.

[71] Xiang D-X, Chen Q, Pang L, Zheng C-l. Inhibitory effects of silver nanoparticles on H1N1 influenza A virus in vitro. J Virol Methods 2011;178(1):137–42.

[72] Sun RW-Y, Chen R, Chung NP-Y, Ho C-M, Lin C-LS, Che C-M. Silver nanoparticles fabricated in Hepes buffer exhibit cytoprotective activities toward HIV-1 infected cells. Chem Commun 2005;40:5059–61.

[73] Ben Salem AN, Zyed R, Lassoued MA, Nidhal S, Sfar S, Mahjoub A. Plant-derived nanoparticles enhance antiviral activity against coxsakievirus B3 by acting on virus particles and vero cells. Digest J Nanomater Biostruct 2012;7(2):737–44.

[74] Trefry JC, Wooley DP. Silver nanoparticles inhibit vaccinia virus infection by preventing viral entry through a macropinocytosis-dependent mechanism. J Biomed Nanotechnol 2013;9(9):1624–35.

[75] Rogers JV, Parkinson CV, Choi YW, Speshock JL, Hussain SM. A preliminary assessment of silver nanoparticle inhibition of monkeypox virus plaque formation. Nanoscale Res Lett 2008;3(4):129–33.

[76] Speshock JL, Murdock RC, Braydich-Stolle LK, Schrand AM, Hussain SM. Interaction of silver nanoparticles with Tacaribe virus. J Nanobiotechnol 2010;8(1):19.

[77] Elechiguerra JL, Burt JL, Morones JR, Camacho-Bragado A, Gao X, Lara HH, et al. Interaction of silver nanoparticles with HIV-1. J Nanobiotechnol 2005; 3(6):1–10.

[78] Lara HH, Ayala-Nuñez NV, Ixtepan-Turrent L, Rodriguez-Padilla C. Mode of antiviral action of silver nanoparticles against HIV-1. J Nanobiotechnol 2010;8(1):1–8.

[79] Fayaz AM, Ao Z, Girilal M, Chen L, Xiao X, Kalaichelvan P, et al. Inactivation of microbial infectiousness by silver nanoparticles-coated condom: a new approach to inhibit HIV-and HSV-transmitted infection. Int J Nanomed 2012;7:5007.

[80] Lara HH, Ixtepan-Turrent L, Garza-Treviño EN, Rodriguez-Padilla C. Research PVP-coated silver nanoparticles block the transmission of cell-free and cell-associated HIV-1 in human cervical culture. J Nanobiotechnol 2010;13:8.

[81] Fall B, Pascual A, Sarr FD, Wurtz N, Richard V, Baret E, et al. *Plasmodium falciparum* susceptibility to anti-malarial drugs in Dakar, Senegal, in 2010: an ex vivo and drug resistance molecular markers study. Malar J 2013;12:107.

[82] Panneerselvam C, Ponarulselvam S, Murugan K. Potential anti-plasmodial activity of synthesized silver nanoparticle using *Andrographis paniculata* Nees (Acanthaceae). Arch Appl Sci Res 2011;3(6):208–17.

[83] Ponarulselvam S, Panneerselvam C, Murugan K, Aarthi N, Kalimuthu K, Thangamani S. Synthesis of silver nanoparticles using leaves of *Catharanthus roseus* Linn. G. Don and their antiplasmodial activities. Asian Pacific J Trop Biomed 2012;2(7):574–80.

[84] Murugan K, Shri K, Barnard D.Green synthesis of silver nanoparticles from botanical sources and their use for control of medical insects and malaria parasites.2013.

[85] Korich D, Mead J, Madore M, Sinclair N, Sterling CR. Effects of ozone, chlorine dioxide, chlorine, and monochloramine on *Cryptosporidium parvum* oocyst viability. Appl Environ Microbiol 1990;56(5):1423–8.

[86] Said D, Elsamad L, Gohar Y. Validity of silver, chitosan, and curcumin nanoparticles as anti-Giardia agents. Parasitol Res 2012;111(2):545–54.

[87] Sen CK, Gordillo GM, Roy S, Kirsner R, Lambert L, Hunt TK, et al. Human skin wounds: a major and snowballing threat to public health and the economy. Wound Repair Regener 2009;17(6):763–71.

[88] Augustine R, Kalarikkal N, Thomas S. Electrospun PCL membranes incorporated with biosynthesized silver nanoparticles as antibacterial wound dressings. Appl Nanosci 2015:1–8.

[89] Singh D, Singh A, Singh R. Polyvinyl pyrrolidone/carrageenan blend hydrogels with nanosilver prepared by gamma radiation for use as an antimicrobial wound dressing. J Biomater Sci Polym Ed 2015;26(17):1269–85.

[90] Anisha B, Biswas R, Chennazhi K, Jayakumar R. Chitosan–hyaluronic acid/nano silver composite sponges for drug resistant bacteria infected diabetic wounds. Int J Biol Macromol 2013;62:310–20.

[91] Furkert FH, Sörensen JH, Arnoldi J, Robioneck B, Steckel H. Antimicrobial efficacy of surface-coated external fixation pins. Curr Microbiol 2011;62(6):1743–51.

[92] Zheng Z, Yin W, Zara JN, Li W, Kwak J, Mamidi R, et al. The use of BMP-2 coupled–Nanosilver-PLGA composite grafts to induce bone repair in grossly infected segmental defects. Biomaterials 2010;31(35):9293–300.

[93] Alt V, Bechert T, Steinrücke P, Wagener M, Seidel P, Dingeldein E, et al. An in vitro assessment of the antibacterial properties and cytotoxicity of nanoparticulate silver bone cement. Biomaterials 2004;25(18):4383–91.

[94] Prokopovich P, Leech R, Carmalt CJ, Parkin IP, Perni S. A novel bone cement impregnated with silver–tiopronin nanoparticles: its antimicrobial, cytotoxic, and mechanical properties. Int J Nanomed 2013;8:2227.

[95] Freire PL, Stamford TC, Albuquerque AJ, Sampaio FC, Cavalcante HM, Macedo RO, et al. Action of silver nanoparticles towards biological systems: cytotoxicity evaluation using hen's egg test and inhibition of *Streptococcus mutans* biofilm formation. Int J Antimicrob Agents 2015;45(2):183–7.

[96] Raffi M, Mehrwan S, Bhatti TM, Akhter JI, Hameed A, Yawar W, et al. Investigations into the antibacterial behavior of copper nanoparticles against *Escherichia coli*. Ann Microbiol 2010;60(1):75–80.

[97] Shende S, Ingle AP, Gade A, Rai M. Green synthesis of copper nanoparticles by *Citrus medica* Linn.(Idilimbu) juice and its antimicrobial activity. World J Microbiol Biotechnol 2015;31(6):865–73.

[98] Ramyadevi J, Jeyasubramanian K, Marikani A, Rajakumar G, Rahuman AA. Synthesis and antimicrobial activity of copper nanoparticles. Mater Lett 2012;71:114–6.

[99] Usman MS, El Zowalaty ME, Shameli K, Zainuddin N, Salama M, Ibrahim NA. Synthesis, characterization, and antimicrobial properties of copper nanoparticles. Int J Nanomed 2013;8:4467.

[100] Betancourt-Galindo R, Reyes-Rodriguez P, Puente-Urbina B, Avila-Orta C, Rodríguez-Fernández O, Cadenas-Pliego G, et al. Synthesis of copper nanoparticles by thermal decomposition and their antimicrobial properties. J Nanomater 2014;2014:10.

[101] Gopinath M, Subbaiya S, Selvam MM, Suresh D. Synthesis of copper nanoparticles from Nerium oleander leaf aqueous extract and its antibacterial activity. Int J Curr Microbiol Appl Sci 2014;3(9):814–8.

[102] LewisOscar F, MubarakAli D, Nithya C, Priyanka R, Gopinath V, Alharbi NS, et al. One pot synthesis and anti-biofilm potential of copper nanoparticles (CuNPs) against clinical strains of *Pseudomonas aeruginosa*. Biofouling 2015;31(4):379–91.

[103] Ghasemian E, Naghoni A, Rahvar H, Kialha M, Tabaraie B. Evaluating the effect of copper nanoparticles in inhibiting *Pseudomonas aeruginosa* and *Listeria monocytogenes* biofilm formation. Jundishapur J Microbiol 2015;8(5).

[104] Kaweeteerawat C, Chang CH, Roy KR, Liu R, Li R, Toso D, et al. Cu nanoparticles have different impacts in *Escherichia coli* and *Lactobacillus brevis* than their microsized and ionic analogues. ACS Nano 2015;9(7):7215–25.

[105] Weaver L, Michels H, Keevil C. Survival of *Clostridium difficile* on copper and steel: futuristic options for hospital hygiene. J Hospital Infect 2008;68(2):145–51.

[106] Pandey P, Packiyaraj MS, Nigam H, Agarwal GS, Singh B, Patra MK. Antimicrobial properties of CuO nanorods and multi-armed nanoparticles against B. anthracis vegetative cells and endospores. Beilstein J Nanotechnol 2014;5(1):789–800.

[107] Cioffi N, Torsi L, Ditaranto N, Sabbatini L, Zambonin PG, Tantillo G, et al. Antifungal activity of polymer-based copper nanocomposite coatings. Appl Phys Lett 2004;85(12):2417–9.

[108] Kanhed P, Birla S, Gaikwad S, Gade A, Seabra AB, Rubilar O, et al. In vitro antifungal efficacy of copper nanoparticles against selected crop pathogenic fungi. Mater Lett 2014;115:13–17.

[109] Ghasemian E, Naghoni A, Tabaraie B, Tabaraie T. In vitro susceptibility of filamentous fungi to copper nanoparticles assessed by rapid XTT colorimetry and agar dilution method. J Med Mycol 2012;22(4):322–8.

[110] Argueta-Figueroa L, Scougall-Vilchis RJ, Morales-Luckie RA, Olea-Mejía OF. An evaluation of the antibacterial properties and shear bond strength of copper nanoparticles as a nanofiller in orthodontic adhesive. Aust Orthod J 2015;31(1):42–8.

[111] Pissuwan D, Cortie CH, Valenzuela SM, Cortie MB. Functionalised gold nanoparticles for controlling pathogenic bacteria. Trends Biotechnol 2010;28(4):207–13.

[112] Saha B, Bhattacharya J, Mukherjee A, Ghosh AK, Santra CR, Dasgupta AK, et al. In vitro structural and functional evaluation of gold nanoparticles conjugated antibiotics. Nanoscale Res Lett 2007;2(12):614–22.

[113] Grace AN, Pandian K. Antibacterial efficacy of aminoglycosidic antibiotics protected gold nanoparticles—a brief study. Colloids Surf A: Physicochem Eng Aspects 2007;297(1):63–70.

[114] Tom RT, Suryanarayanan V, Reddy PG, Baskaran S, Pradeep T. Ciprofloxacin-protected gold nanoparticles. Langmuir 2004;20(5):1909–14.

[115] Gu H, Ho P, Tong E, Wang L, Xu B. Presenting vancomycin on nanoparticles to enhance antimicrobial activities. Nano Lett 2003;3(9):1261–3.

[116] Gil-Tomás J, Tubby S, Parkin IP, Narband N, Dekker L, Nair SP, et al. Lethal photosensitisation of *Staphylococcus aureus* using a toluidine blue O–tiopronin–gold nanoparticle conjugate. J Mater Chem 2007;17(35):3739–46.

[117] Sherwani MA, Tufail S, Khan AA, Owais M. Gold nanoparticle-photosensitizer conjugate based photodynamic inactivation of biofilm producing cells: potential for treatment of *C. albicans* infection in BALB/c mice. PLoS One 2015;10(7).

[118] Narsireddy A, Vijayashree K, Irudayaraj J, Manorama SV, Rao NM. Targeted in vivo photodynamic therapy with epidermal growth factor receptor-specific peptide linked nanoparticles. Int J Pharm 2014;471(1):421–9.

[119] Embleton ML, Nair SP, Cookson BD, Wilson M. Antibody-directed photodynamic therapy of methicillin resistant *Staphylococcus aureus*. Microb Drug Resist 2004;10(2):92–7.

[120] Hirsch LR, Stafford R, Bankson J, Sershen S, Rivera B, Price R, et al. Nanoshell-mediated near-infrared thermal therapy of tumors under magnetic resonance guidance. Proc Natl Acad Sci USA 2003;100(23):13549–54.

[121] O'Neal DP, Hirsch LR, Halas NJ, Payne JD, West JL. Photo-thermal tumor ablation in mice using near infrared-absorbing nanoparticles. Cancer Lett 2004;209(2):171–6.

[122] Huang X, El-Sayed IH, Qian W, El-Sayed MA. Cancer cell imaging and photothermal therapy in the near-infrared region by using gold nanorods. J Am Chem Soc 2006;128(6):2115–20.

[123] Zharov VP, Mercer KE, Galitovskaya EN, Smeltzer MS. Photothermal nanotherapeutics and nanodiagnostics for selective killing of bacteria targeted with gold nanoparticles. Biophys J 2006;90(2):619–27.

[124] Norman RS, Stone JW, Gole A, Murphy CJ, Sabo-Attwood TL. Targeted photothermal lysis of the pathogenic bacteria, *Pseudomonas aeruginosa*, with gold nanorods. Nano Lett 2008;8(1):302–6.

[125] Huang W-C, Tsai P-J, Chen Y-C. Functional gold nanoparticles as photothermal agents for selective-killing of pathogenic bacteria. Nanomedicine 2007;2(6):777–87.

[126] Kuo W-S, Chang C-N, Chang Y-T, Yeh C-S. Antimicrobial gold nanorods with dual-modality photodynamic inactivation and hyperthermia. Chem Commun 2009;32:4853–5.

[127] Radt B, Smith TA, Caruso F. Optically addressable nanostructured capsules. Adv Mater 2004;16(23-24):2184–9.

[128] Babalola OO. Molecular techniques: an overview of methods for the detection of bacteria. Afr J Biotechnol 2004;2(12):710–3.

[129] Elghanian R, Storhoff JJ, Mucic RC, Letsinger RL, Mirkin CA. Selective colorimetric detection of polynucleotides based on the distance-dependent optical properties of gold nanoparticles. Science 1997;277(5329):1078–81.

[130] Storhoff JJ, Marla SS, Bao P, Hagenow S, Mehta H, Lucas A, et al. Gold nanoparticle-based detection of genomic DNA targets on microarrays using a novel optical detection system. Biosensors Bioelectronics 2004;19(8):875–83.

[131] Baptista PV, Koziol-Montewka M, Paluch-Oles J, Doria G, Franco R. Gold-nanoparticle-probe–based assay for rapid and direct detection of *Mycobacterium tuberculosis* DNA in clinical samples. Clin Chem 2006;52(7):1433–4.

[132] Huang S-H. Gold nanoparticle-based immunochromatographic assay for the detection of *Staphylococcus aureus*. Sensors Actuators B: Chem 2007;127(2):335–40.

[133] Jeong M-S, Ahn D-R. A microwell plate-based multiplex immunoassay for simultaneous quantitation of antibodies to infectious viruses. Analyst 2015;140(6):1995–2000.

[134] James G. Water Treatment, 4th ed. Cleveland, OH: CRC Press; 1971.

[135] Arora S, Jain J, Rajwade J, Paknikar K. Interactions of silver nanoparticles with primary mouse fibroblasts and liver cells. Toxicol Appl Pharmacol 2009;236(3):310–8.

Chapter 5

Lipid-Based Nanopharmaceuticals in Antimicrobial Therapy

N. Škalko-Basnet[1] and Ž. Vanić[2]

[1]*Drug Transport and Delivery Research Group, Department of Pharmacy, Faculty of Health Sciences, University of Tromsø The Arctic University of Norway, Tromsø, Norway* [2]*Department of Pharmaceutical Technology, Faculty of Pharmacy and Biochemistry, University of Zagreb, A. Kovačića 1, Zagreb, Croatia*

Functionalized Nanomaterials for the Management of Microbial Infection.

5.1 **NANOPHARMACEUTICALS**

The rapid development in the field of nanotechnology in the past forty years resulted in even faster advancements in the application of nanotechnology in medical and biomedical fields. Nanosized delivery systems, carrying active molecules and often referred to as nanopharmaceuticals, offer the potential to:

1. provide protection to associated drug molecule;
2. enhance drug absorption via enhanced diffusion through epithelial barriers;
3. modify the drug release and distribution profile, especially assuring a sustained release;
4. improve intracellular delivery; and
5. achieve drug targeting [1].

What is particularly interesting is the possibility to tailor the nanocarrier in such a manner as to modulate the immune system and assure improved therapeutic outcome [2]. Various nanosized delivery systems can fulfill these desired features to different extents. Most importantly, a safe and pharmacologically efficient carrier needs to be biocompatible and biodegradable [3]. Ideally, the carrier will provide a high drug payload in a sustained delivery manner; prevent the "burst release" of the encapsulated drug before reaching the target site, therefore reducing the damage to healthy tissue; be inert and protective toward the encapsulated/incorporated drug and able to reach the target site [3].

Although we have still not fully reached the Paul Erlich concept of "magic bullet" and drug targeting, some superior nanosized formulations in cancer therapy, such as Doxil® for example, are indicators of research focus being on the right track. Regarding the anti-infective therapy, the development of Ambisome® justified the use of lipid-based nanocarriers to deliver a highly toxic drug with limited side effects [4].

A detailed review of nanocarrier-based delivery of various antibiotics is provided by Abed and Couvreur [3]. The authors suggested that liposomes in particular are able to increase the therapeutic index of antibiotics via systemic administration, target the infected intracellular compartments and allow combination therapy [3]. A structured overview of the features and advantages of liposomes as nanocarriers for antibiotics delivery is provided by Drulis-Kawa and Dorotkiewicz-Jach [5].

5.2 **LIPID-BASED NANOPHARMACEUTICALS**

Lipid-based nanocarriers comprise mostly phospholipids, cholesterol (CHOL) and triglycerides. The lipid sources can be from natural origin and increasingly synthesized by tailored design. Many are approved as pharmaceutical products by the governing agencies, which remains the major advantage over the polymeric particles of similar size [6]. The basic characteristics of most commonly studied nanosystems built up of lipids are provided in this chapter. More detailed characteristics of the systems are described in the following sections, followed by an overview of lipid-based delivery systems for various antimicrobials classified according to route of administration.

5.2.1 **Liposomes**

Liposomes are phospholipid vesicles consisting of bilayers built up by amphiphilic lipids enclosing aqueous space (Fig. 5.1). They have been extensively studied as nanosized delivery systems for various routes of drug administration [7] with several products successfully reaching the market.

Liposomes are biodegradable nanocarriers in which a hydrophobic molecule can be successfully incorporated within their liposomal bilayers and hydrophilic molecules within their interior aqueous space (Fig. 5.1A). Recently, Kubitschke et al. [9] added another dimension to liposomes as drug carriers, by suggesting liposomes derived from vase-shaped cavitands, which are receptor-like molecules that wrap around "guest" compounds. The cavitands can encapsulate different guest molecules and present them at high densities at the liposome surface, an approach which should be utilized in drug delivery, including antimicrobial delivery. The first obstacle for the wider use of liposomes in drug delivery was their limited half-life *in vivo*, since liposomes are quickly removed from the bloodstream by immune cells unless their surface is modified to avoid opsonization [7]. By chemically attaching polymer molecules to the lipids' polar heads (Fig. 5.1B), the "stealth" liposomes were designed and are nowadays widely utilized in marketed liposome-based products. PEG prevents the attachment of plasma proteins that label foreign matter for removal by immune cells. The possibility of modifying the liposomal surface leads to further tailoring of vesicles through chemical modification of their lipids' head groups. Targeting specific receptors could be achieved by binding of peptides/ligands attached at the distal end of the polymers, able to interact with the receptors on specific cells and tissues (Fig. 5.1B). Further development of liposomes as nanocarriers was the design of liposome–DNA complexes such as for example the multilamellar structure in which single layers of DNA molecules are inserted between lipid bilayers (Fig. 5.1C). Recently, a novel modification of liposomes in a form of covalent incorporation of cavitands into phospholipids was proposed (Fig. 5.1D). When self-assembled into liposomes, the amphiphilic cavitands can encapsulate guests on three distinct length scales: ångström-sized molecules can be trapped in the vase-like hydrophobic cavity of each cavitand; molecules a few nanometers can be caught in the hydrophobic bilayer of the liposome; and molecules up to 100nm in size, and perhaps even up to several hundred nanometers, can be encapsulated in the liposome's interior. Intriguingly, this means that the different-sized guests can be entrapped in locations that have distinct

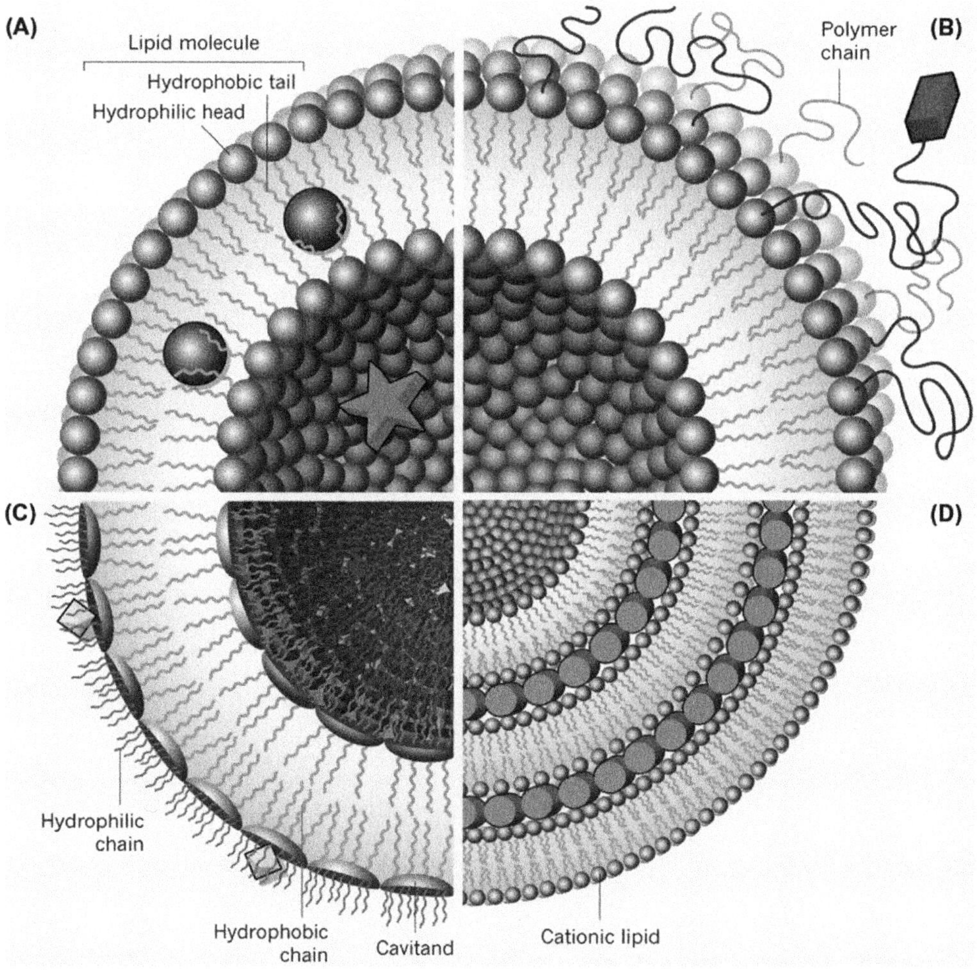

■ **FIGURE 5.1** The evolution of liposomes. (A) Simple liposomes are vesicles that have a shell consisting of a lipid bilayer. (B) In 'stealth' liposomes developed for drug-delivery applications, the lipid bilayer contains a small percentage of polymer lipids. (C) Most cationic liposome–DNA complexes have an onion-like structure, with sandwiched between cationic membranes. (D) Liposomes in which the bilayer assembles from cavitands—vase-shaped molecules—with attached hydrophobic and hydrophilic chains. *From [8] Safinya CR, Ewert KK. Materials chemistry: liposomes derived from molecular vases. Nature 2012;489:372–4, with permission from Nature Publishing Group.*

dimensionalities within the cavitand, in the bilayer and three-dimensional in the liposome's center [8]. More details on liposomal delivery of antimicrobials are provided in following sections.

Liposomes are often classified into conventional liposomes, deformable (elastic) liposomes, ethosomes, penetration enhancer-embodying liposomes (PEVs), and invasomes [10]. In brief, deformable liposomes are highly fluid vesicles composed of lipids and an edge activator and introduced by [11] as a superior system for skin penetration. Ethosomes are soft phospholipid vesicles developed by Touitou et al. [12] to improve treatment of skin diseases. In addition to lipid and water, ethosomes contain high ethanol content (up to 45% v/v). They are in general smaller in size and able to penetrate into deeper skin layers [10]. PEVs contain penetration enhancer such as propylene glycol or diethylenegycolmonoethyl ether; if the penetration enhancer is terpen, this type of liposomes is referred to as invasomes. More details are available in our recent review [10].

5.2.2 Solid Lipid Nanoparticles (SLNs) and Nanostructured Lipid Carriers (NLCs)

SLNs are made up of a solid lipid core stabilized by surfactants and are rather amorphous structures in which bilayers are not distinguished (Fig. 5.2). SLNs are biodegradable and biocompatible, relatively easy to scale up and offer more means for sterilization as compared to liposomes. They can be seen as "hybrid" nanodevices between polymeric nanoparticles and liposomes [3]. SLNs were designed as a superior lipid-based

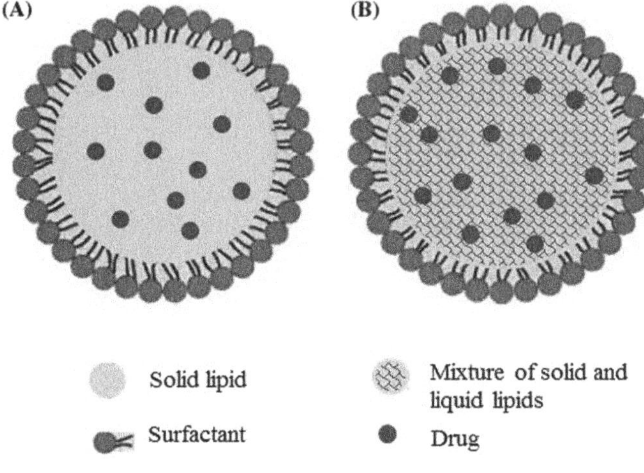

■ **FIGURE 5.2** Schematic drawing of SLN (A) and NLC (B).

system able to improve long-term stability, better incorporation efficacy for lipophilic drugs and scaling-up features as compared to liposomes [13], and are superior in cytotoxicity in comparison to polymeric nanoparticles [14]. To overcome the limitations related to relatively low loading capacity for nonlipophilic drugs, stability issues related to highly ordered crystalline lipid structures and drug expulsion, a new generation of lipid-based particles was developed, the nanostructured lipid carriers (NLCs; Fig. 5.2).

5.2.3 **Nanostructured Lipid Carriers (NLCs)**

Unlike SLNs, NLCs are made of solid and liquid lipids. The surfactants used to stabilize them are the same as in SLNs [15]. Liquid lipids are included in order to prevent the formation of the arranged lipid matrix that caused stability issues, as mentioned earlier. These lipid arrangements within NLCs allow a biphasic pattern of drug release with an initial burst effect followed by a sustained drug release [16]. The examples of NLCs with various antimicrobials are provided in the following sections.

There are several other types of lipid-based nanopharmaceuticals, which have been studied to a lesser extent [10] and are not discussed in detail in this chapter. Lipid micelles as a nanosized drug delivery system have been relatively less studied than the polymeric micelles. They are monolayered nanostructures (up to 35 nm) composed of polyethylene glycol (PEG) conjugated phospholipids or lysolipids, which are formed at lipid concentrations above critical micellar concentration of the lipid in aqueous medium [17]. Those lipid-core micelles have been loaded with mostly with hydrophobic anticancer drugs [18].

5.3 **ANTIMICROBIAL THERAPY OF INTEREST**

Current antimicrobial therapy is suffering from increased drug resistance among pathogens, which limits the therapy's success. To address this alarming antibacterial resistance, research efforts have been focused on two pipelines; search for new antimicrobials and development of novel drug carriers able to assure improved performance of currently available antimicrobials [5]. The second line is especially interesting considering that the patents for many of the valuable antimicrobials have expired and the development of nanomedicine based on those antimicrobials is attractive for pharmaceutical companies. Moreover, after the "golden era" in discoveries of novel antibiotic classes, a gap of nearly 35 years in antibiotic discovery suggests that searching for means to improve antibiotic delivery, rather than "wonder antibiotics", might be more feasible [3]. In addition, the number of immunocompromised patients is rapidly increasing as a consequence of aged society, cancer treatment or human immunodeficiency virus (HIV) infections [1].

Nanotechnology in medicine, or nanomedicine as it is often referred to, offers almost unlimited potential to exploit delivery of antimicrobials to infection sites, intracellularly [1,3]. Intracellular infections, such as tuberculosis, are of specific interest and challenge [19]. They are characterized by pathogens that shelter mostly in the mononuclear phagocyte system (MPS), more often called the reticuloendothelial system (RES). This concept of the "hijacked" reservoir allows pathogens to form secondary infectious foci and is responsible for recurrent systemic infections [3]. Fig. 5.3 summarizes the concept of the treatment of intracellular bacteria with nanoantibiotics [3].

Another increasingly serious problem in current antimicrobial therapy is the treatment of biofilms. Treating biofilm infections is one of the major challenges still lacking confirmed efficient therapeutic treatment [20]. Biofilms are matrix-enclosed, complex and differentiated bacterial communities adherent to inert or biological surfaces in which the bacterial cells

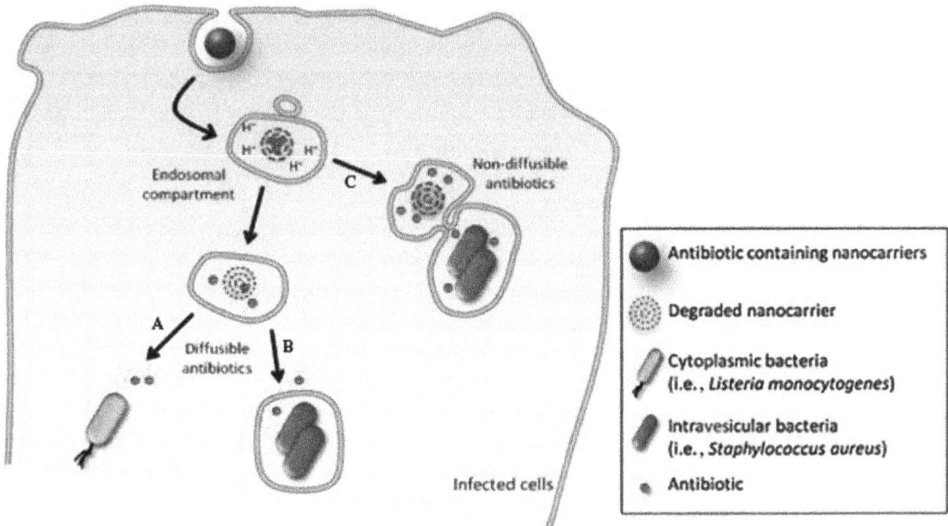

■ **FIGURE 5.3**   Intracellular delivery of antibiotics to treat intracellular infections. Following cell internalization through one or several endocytic pathways, nanocarriers should be destabilized and/or degraded by hydrolytic enzymes and/or acidic pH to allow the antibiotic to be released. A diffusible antibiotic will easily escape from endosomes and reach cytoplasmic bacteria (A) or vesicles containing bacteria (B) to exert antimicrobial activity. In some cases, fusion between endosomes (with captured nanocarriers + antibiotic) and vesicles containing bacteria (C) are needed in order to allow non-diffusible antibiotics to exert their antimicrobial activity. *From [3] Abed N, Couvreur P. Nanocarriers for antibiotics: a promising solution to treat intracellular bacterial infections. Int J Antimicrob Agents 2014;43:485–96, with permission from Elsevier.*

start producing an extracellular matrix and organizing themselves together in densely packed clusters. Upon maturation, individual cells or biofilm fragments can start colonizing new surfaces assuring the spread of biofilm. Moreover, the biofilm bacteria can communicate and alter each other's phenotypes enabling adaptation growth of bacteria and causing a serious therapeutic challenge. One of the criteria used to confirm that the infection involves biofilm is the fact that the infection is not eradicated by classical antimicrobial therapy [21]. It is estimated that over 60% of bacterial infections in humans involve biofilms [22]. Nanomedicine, especially targeted to treat chronic wound infections and slow-healing wounds, tries to address and overcome these related obstacles by designing and developing advanced wound dressings [23].

Facing the urgent need to overcome the resistance and tolerance against antimicrobial agents used in the treatment of biofilms leads to search for nanopharmaceuticals-based antimicrobial therapy, especially lipid- and polymer-based nanosystems. The biocompatibility of lipids and selected polymers, the versatility of nanomaterials and possibility of their surface modification/manipulation, together with potential for drug targeting and stimuli-responsive drug release, open the possibilities for targeted design of functionalized nanocarriers [24]. Biofilm-caused recalcitrant infections cannot be cured by classical antibiotic therapy based on conventional drug dosage forms. Lipid nanocarriers able to fuse with the bacterial outer membrane, deliver the antimicrobial agent directly to the bacterial cells, and protect the sensitive agents, could offer significant advantages. In addition, the targeting approach can assure the maximized therapeutic effects and enable minimizing the unwanted side effects [22].

Drug delivery by lipid-based nanopharmaceuticals such as fusogenic liposomes is considered to be one of the most promising approaches to overcoming biofilm resistance [22].

An alternative novel approach in antimicrobial therapy can be offered by antimicrobial photodynamic inactivation (PDI); the therapy is based on the use of nontoxic dyes and visible light to generate reactive oxygen species that selectively kill microbial cells, and often relies on nanocarriers to enhance the therapeutic outcome [25]. PDI efficacy can be optimized by loading the photosensitizer (PS, dye) into lipid nanocarriers, mostly liposomes [26]. It is worth noting that a serious drawback of this approach is the limited tissue penetration of external light delivered from conventional light sources resulting in limited therapy success in deep infections. Recently, chemiluminescent photodynamic antimicrobial therapy (CPAT)

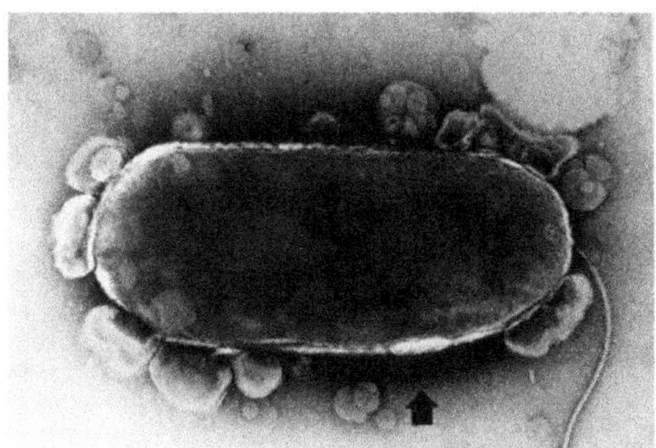

■ **FIGURE 5.4** Interaction of fluidosomes with *Pseudomonas aeruginosa 429* cells as observed by negative staining with PTA. Magnification: × 50561. *From [28] Sachetelli S, Khalil H, Chen T, Beaulac C, Sénéchal S, Lagacé J. Demonstration of a fusion mechanism between a fluid bactericidal liposomal formulation and bacterial cells. Biochim Biophys Acta 2000;1443:254–66, with permission from Elsevier.*

using light emitted as a result of the internal chemiluminescent reaction of liminal to excite PS has been proposed as an improved approach [27]. CPAT allows PDI in the absence of a conventional light source to activate the PS, enabling the treatment of deeper infections.

5.4 POTENTIAL OF NANOPHARMACEUTICALS IN TREATMENT OF MICROBIAL INFECTIONS

Nanomedicine is expected to enhance the antimicrobial effect of drugs by means through which nanosystems fuse with bacterial cell walls (Fig. 5.4).

Nanosystems may cause the disruption of cell wall (membrane) via direct adsorption or oxidative stress production, leading to the protein denaturation and ultimately cell death. As the bacterial walls possess the net negative charge, a positively charged nanosystem may interact with bacteria through the electrostatic attractions. The inclusion of CHOL in lipid-based nanosystems may contribute to the accumulation of antimicrobials on the biofilm interface via nonspecific adhesion [29]. Nanosystems able to adhere onto biofilms can diffuse across the water-filled pores in biofilm.

Nano-antibiotics are considered a promising approach to overcome the (mucus) drug barrier and prolong the antibiotic retention at the infected site. Nanocarriers employed to deliver antibiotics provide an opportunity to assure controlled and sustained release at a closer proximity to the infection target.

Liposomes are one of the most studied nanosystems for various diseases, including antimicrobial therapy. Numerous reports have stated that the encapsulation of antibiotics in liposomes lowers the minimal inhibitory concentrations (MIC) for clinically relevant biofilm forming organisms and/or lower minimal biofilm inhibitory concentrations (MBIC) compared to non nanocarrier-associated antibiotics [22].

The pharmacological profile of liposomally associated drugs will be influenced by the interplay between the characteristics of the drug molecule and tailored characteristics of liposomes as drug carrier. Readers are referred to an extensive review on pharmacokinetic profiles of drug release from liposomes provided by Drummond and coworkers [30]. How fast the liposomally associated drug will be cleared from the blood will be affected by:

1. the clearance of the carrier itself
2. dissociation of the complexed, entrapped, or lipid-membrane solubilized drug from liposome
3. rate of clearance and metabolism of free drug once released from liposomes [30].

Non-PEGylated liposomes, upon intravenous administration, follow saturable nonlinear pharmacokinetics with relatively fast clearance at lower doses [31]. PEGylated liposomes, on the other hand, display log-linear pharmacokinetics over a wider dose range; however, repeated administration and very low doses complicate the predictability of pharmacokinetics [32].

PEGylated liposomes have been extensively utilized in anticancer therapy and for more details on their characteristics, the readers are referred for example to reviews by Allen et al. [31] and Gabizon et al. [33]. PEGylation does not only improve the *in vivo* stability of liposomes, but also offer additional advantages, such as the possibility of using a wide variety of lipid compositions, use of unsaturated phospholipids, charged lipids, or the preparation of liposomes in the absence of CHOL. Most importantly, PEGylated liposomal drugs exhibited more predictable pharmacokinetics compared to non-PEGylated liposomal drugs [30].

Liposomal physicochemical properties can be modified/optimized by selecting the right type of lipids, lipid ratios, manipulating the size and size distributions, surface charge, pH, and/or temperature sensitivity and rigidity of liposomal bilayers [5]. We summarized the approaches in optimization of liposomal characteristics with respect to the features of liposomes as delivery system in Table 5.1. However, the choice of liposomal characteristics will also be influenced by the drug to be entrapped/incorporated into liposomes.

Table 5.1 Liposomal Characteristics Affecting the Performance of Liposomally Associated Antimicrobials

Liposomal Characteristics	Expected Effect	Remarks
Size	Smaller liposomes lead to better biodistribution and prolonged half-life	Clearance of PEGylated liposomes is less affected by size Lower drug load in smaller liposomes
The net surface charge 1. Neutral 2. Positive 3. Negative	In general, charged lipids lead to faster clearance	Cationic lipids contribute to system's toxicity
Surface-modification 1. PEG 2. Antibody 3. Polymer 4. Receptor-targeted ligand	1. Prolonged half-life 2. Targeting via antibodies and ligands can be achieved 3. Polymer-coating can assure mucoadhesiveness and improve retention time	Reports on immunogenicity of PEGylated liposomes upon repeated administration
Lipid composition 1. Membrane fluidity 2. Inclusion of CHOL 3. pH-sensitive 4. Fusogenic	1. Highly saturated lipids lead to faster release (absence of CHOL) 2. Stabilizes membrane, decreases opsonization 3. Intracellular targeting 4. Improved antibacterial activity	Effect not applicable in PEGylated liposomes

CHOL, cholesterol; PEG, polyethylene glycol.

Liposomal formulations are expected to assure prolonged half-life of antimicrobials [30]. Liposomal formulations, in comparison to free drugs, were superior in treatment of various infections; liposomes enabled the increase of the safe dose concentration due to reduced toxicity (improved therapeutic index), all of which is of great importance in fungal and parasite infections [1].

However, although liposomes offer many advantages, all liposomal formulations must satisfy the quality control and stability requirements; often the chemical stability of liposomal formulation can cause concern in respect to the stability of both the drug and the lipid. Physical stability, especially potential vesicle fusion and aggregation, and liposomal membrane permeability, which may lead to the unpredictable drug release, need to be taken into consideration and fulfilled [34].

5.4.1 Fusogenic Liposomes

In respect to antibacterial therapy, fusogenic liposomes are particularly attractive. Liposomal content can be directly delivered to cytosol after the fusion between the target cell and liposome; targeting non-internalizing receptors. Fusogenicliposomes are built of lipids that make bilayers more fluid and prone to fusion. Most commonly studied lipids for fusogenic activity are phosphatidylethanolamine (PE) derivatives such as 1,2-dioleoyl-sn-glycero-3-phosphoethanolamine (DOPE) [35]. Liposomes coated with, for example, Sendai virus coat-proteins rely on the fusion of viral components and the cells for their fusogenic activity. Although viral targeting is efficient, it might be nonspecific and raises safety concerns [26].

Most studies on liposomal antibiotics were focused on aminoglycosides, quinolones, polypeptides, and beta-lactams. The studies tried to prove the superiority of liposomes in enhancing antibacterial activity and improving pharmacokinetics and reducing toxicity. The most promising *in vivo* results confirming the super pharmacokinetics of liposomal antibiotics were reported for aminoglycosides, quinolones and polymyxin B [5,36–40]. The plasma circulation time of liposomal antibiotics can be further improved by using PEGylated liposomes [38].

Very promising results were reported on the enhanced activity of liposomal antibiotics against intracellular pathogens. Drulis-Kawa and Dorotkiewics-Jach [5] prepared an overview of liposomal antibiotics designed for tuberculosis treatment.

Similarly, extracellular pathogens were the target of many studies involving the liposomally associated antimicrobials with relative success in therapy [5].

Forier and colleagues [22] prepared an extensive overview of all liposomal formulations developed and tested in the context of drug delivery to biofilms.

5.4.2 Solid Lipid Nanoparticles (SLNs) and Nanostructured Lipid Carriers (NLCs)

Several researchers confirmed the feasibility of dry powder inhalable SLN and NLC incorporated into microparticles, reconstructed into nanodispersions after delivery to the airway fluid [34]. However, the potential and advantages of SLNs over liposomes need to be confirmed in *in vivo* studies. A triple combination of antibiotics rifampicin, isoniazid, and pyrazinamide loaded into SLN and delivered via inhalation as a single dose in tuberculosis-infected guinea pigs was proven to be successful [41].

Currently there are no human data available on use of SLNs and NLCs, as most of the research on these vesicles is in preclinical stage [34].

5.5 POTENTIAL OF LIPID-BASED NANOPHARMACEUTICALS FOR ANTIMICROBIAL THERAPY RELATED TO ROUTE OF ADMINISTRATION

In order to address the specificity of the lipid-based delivery via different routes of drug administration, we divided the overview of the efficacy of lipid-based nanopharmaceuticals by the route of nanocarrier's entrance onto/into the body. Due to the large number of available research articles and limited space, we tried to present the representative examples of the systems and drugs. The selection is based on the authors' opinion; however, relevant reviews in the literature are cited and readers are referred to these for deeper insight.

5.5.1 Systemic Infections (Parenteral Route)

The challenges associated with systemic drug delivery are primarily related to possible inactivation of the antimicrobial on the route to the microbial target, and side effects of the antimicrobials. In addition, it is known that some bacteria can survive inside the phagocytic cells, form biofilms and ultimately develop resistance [42]. Encapsulation of aminoglycosides in liposomes confirmed the superior effect in treatment of bacterial infections [42]. Intravenous administration of liposomal amikacin in *Mycobacterium avium* mice infection model resulted in increased drug concentration in the liver and spleen, with limited drug concentration in the lungs [43]. Long-circulating sterically stabilized liposomes with amikacin in the same model were shown being able to eliminate bacteria from infected organs, decrease the treatment period and prevent relapse of bacterial infection [44]. Intravenously administered conventional liposomes loaded with gentamicin exhibited prolonged plasma half-life as compared with the free drug [45]. The potential of liposomal antibiotics to treat challenging intracellular mycobacterial infections was confirmed by liposomally associated rifabutin against *M. avium* [46].

Intraperitoneal injection of liposomal vancomycin reduced the infection in murine systemic infection caused by methicillin-resistant *Staphylococcus aureus* (MRSA) in kidney and spleen by two to three log of magnitude [47].

Intravenous administration of cationic liposomes with vancomycin or ciprofloxacin was effective in reducing infection in rabbit bone osteomyelitis caused by *S. aureus* [48]. The efficiency of the therapy was attributed to prolonged circulation time assured by liposomes.

A more advanced approach using receptor targeting was applied in mannose-grafted liposomes containing amphotericin B. The approach was based on the fact that infected macrophages overexpress mannose receptors that bind and internalize mannose-terminated glycoproteins. Mannose-grafted liposomes were superior to conventional liposomes and free drug against *Leishmania donovani* [49].

5.5.2 **Systemic Infections (Oral Route)**

The oral route remains the easiest mode of drug administration in the treatment of systemic infections. It is also highly acceptable by patients and most of the available formulations are relatively affordable. However, lipid-based nanopharmaceuticals destined for oral administration face a specific challenge in the instability of lipids in the harsh environment of gastrointestinal tract. Many attempts have been performed to prepare lipid nanopharmaceuticals able to survive enzymatic attacks. Recently, Pantze et al. [50] developed a lipid-based liposomal carrier system for the oral delivery of peptides. The lipid bilayer of standard egg-phosphatidylcholine (PC)/CHOL liposomes were stabilized by gelatin. The gelatin also works as a thickening agent by forming a matrix in which liposomes are embedded. Matrix liposomes containing 20% gelatin showed an approximately linear release of liposomes which might be further exploited in the delivery of antimicrobials. Interestingly, pharmacokinetics of liposomal delivery of amphotericin B administered orally confirmed improved therapeutic outcome and prophylaxis of systemic fungal infections [51]. SLNs exhibited significant antibacterial activity after oral administration in antituberculosis therapy [41].

5.5.3 **Skin Infections**

The skin microbiome is comprised of diverse populations of bacteria, viruses, archaea, fungi, mutes, and anthropodes. It is affected by endogenous hosts, exogenous environments, and topographical location [29]. Details on the skin microbiome, including bacteria, fungi, viruses, archaea, and small arthropods colonizing the skin surface, are provided by Kong and Segre [52].

Antimicrobials have been applied epicutaneously in the treatment of various skin infections such as acne (e.g., *Propionibacterium acnes*) [53], persistent bacterial infections, MRSA infections, superficial fungal infections of skin and nails, and infections of prosthetic devices implanted into the skin (*Staphylococcus epidermidis*) [58]. To successfully treat skin infections, it is of great importance to understand the skin property as a complex barrier to penetration of particles into/through skin [55]. In dermatology, the property of the nanocarrier which is expected to have the biggest impact on its penetration potential seems to be the "softness" and rigidity of carrier's membrane rather than the actual size [56]. Although no consensus on the penetration of nanoparticles into/through the skin has been reached, it is safe to postulate that only very small particles (size below 10nm) can readily penetrate into the skin whereas the penetration of other nanocarriers will be affected by their characteristics, as summarized by DeLouise [57]. A recent review on nanopharmaceuticals for the treatment of skin infections is provided by Aljuffali and coauthors [58].

Since the pioneering paper by Mezei and Gulasekharam [59], liposomes have been widely studied as nanocarriers for skin administration [60]. The first nanodermatological product to reach the market was Pevaryl Lipogel (containing econazole) in 1988 [60]. Liposomes were intensively studied in the treatment of various skin infections, such as acne, skin and wound infections [54]. Ethosomes have been shown able to enhance the antimicrobial activity of many drugs [10]. SLNs and NLCs were able to assure controlled release of topically administered clotrimazole [61]. An overview of phospholipid-based nanocarriers tested clinically and in animal studies is provided in the Table 5.2.

5.5.4 **Vaginal Infections**

The majority of nanopharmaceuticals evaluated for vaginal drug delivery were intended for the prevention of sexually transmitted diseases (microbicides) and the local treatment of vaginal infections [75,76]. Design of an effective vaginal formulation requires proper balancing of numerous parameters affecting the rate and extent of vaginal absorption such as anatomical and physiological conditions, physicochemical properties of drug, and the properties of formulation alone [77,78].

Among lipid-based nanosystems, liposomes have been predominantly explored, although the use of SLNs and NLCs has been reported as well. The potential of liposomes as an advanced vaginal drug delivery system was recognized relatively late (1997) in comparison to their use via other

Table 5.2 Lipid-Based Nanopharmaceuticals for the Treatment of Skin Infections: Selected Clinical and *In Vivo* Studies

Drug	Type of Lipid Nanosystem	Study	Findings	References
Clindamycin-HCl	*Conventional liposomes*	Clinical	Significantly decreased number of acne lesions compared to standard formulation	[62]
			Almost fourfold greater effectiveness in reducing open comedones in comparison to the drug solution	[63]
Erythromycin	*Ethosomes*	*In vivo* (mice)	Improved healing of *Staphylococcus aureus*-induced deep dermal infections	[64]
Neomycin sulfate	*Deformable liposomes*	*In vivo* (rats)	Complete eradication of *Staphylococcus* infections after 7 days	[65]
Daptomycin	*Deformable liposomes*	*In vivo* (mice)	Maintained effective therapeutic concentrations for several hours, notably inhibited bacterial growth and injury-induced biofilms	[66]
Econazole	*Conventional liposomes*	Clinical	Increased concentration of antifungal in *stratum corneum*, facilitated fast onset of action; shorten duration of treatment than with conventional formulation	[67]
Econazole nitrate	*SLNs*	*In vivo* (humans)	Faster penetration of the drug through *stratum corneum* after 1 h and enhanced diffusion in deeper skin layers after 3 h in comparison to control formulation	[68]
Amphotericin B	*Ethosomes*	*In vivo* (rats)	Improved antifungal activity against *C. albicans*; no irritation observed	[69]
Terbinafine-HCl	*SLNs*	*In vivo* (rats)	Reduced fungal burden of *C. albicans* in shorter time than with commercial product	[70]
Povidone-iodine	*Conventional liposomes*	Clinical	Increased epithelialization and better wounds healing than with chlorhexidine gauze	[71]
		Clinical	Faster and complete healing of the burn wounds in comparison to silver-sulfadiazine cream; good cosmetic results and local tolerability	[72]
		Clinical (Phase III)	Considerably improved epithelialization, reduced transplant losses; significantly improved healing, especially in smokers	[73]
Acyclovir	*Ethosomes*	Clinical	Efficient treatment of HSV infections	[74]

HSV, herpes simplex virus; NLCs, nanostructured lipid carriers; SLNs, solid lipid nanoparticles.

routes of administration [79]. Different types of liposomes, varying in size and surface properties, have been explored including conventional, elastic and mucoadhesive liposomes. We have classified the systems according to the targeted infections.

5.5.4.1 Bacterial and Fungal Infections

Liposomes

The earliest research in this area was focused on the *in vitro* and *in situ* stability of conventional liposomes containing antimicrobial drugs (clotrimazole, metronidazole, or chloramphenicol) in the media mimicking vaginal conditions in pre- and post-menopausal women. The lower pH and the presence of cow vaginal mucosa affected the stability of liposomes resulting in a higher drug release rate in the first 6h [80]. Since the liquid nature of liposomes limits their direct vaginal administration, increasing viscosity by embedding liposomes into compatible hydrogels has been recommended as a means to overcome this deficiency [81–83]. By incorporating liposomes in bioadhesive Carbopol gels, a viscosity suitable for vaginal administration was achieved simultaneously enabling prolonged drug release and improved stability of all the examined liposomes. Even after 24h of incubation in vaginal fluid simulant (pH 4.5) more than 30% of the originally entrapped clotrimazole, 50% of metronidazole, or 40% of chloramphenicol were retained in the gels [82,84]. The extended release of clotrimazole from proliposomes in simulated vaginal conditions has been confirmed by Ning et al. [85,86]. Increased antifungal activity (*in vivo*, rats) achieved by clotrimazole-loaded proliposomes was justified by better drug retention in vaginal mucosa in comparison to standard clotrimazole formulation. In addition, no morphological changes of vaginal tissue have been associated with proliposomes [85]. Similar findings have also been reported later with ciclopiroxolamine-loaded liposomes. Gel formulation of liposomes composed of hydrogenated lipids significantly extended release of the drug, while gel-base assured the good mucoadhesivity on sheep vaginal tissue [87]. Kang et al. suggested thermosensitive gel with amphotericin B-loaded cationic liposomes for the topical treatment of fungal infection. Cytotoxicity study on HEK293 cell line demonstrated lower toxicity of the gel formulation compared to the free drug; however, further studies are required to prove the antifungal effect, release profile and retention of formulation at the site of application [88].

To enhance delivery of antimicrobials deeper into the vaginal mucosa and infected epithelial cells our groups suggested the application of elastic liposomes. Deformable liposomes composed of egg lecithin and sodium deoxycholate (SDCh) as an edge activator (85:15, mass ratio) were able to deliver more metronidazole across *in vitro* model barriers than conventional liposomes, which retained the drug on the epithelial surface. However, the low entrapment of the drug into deformable liposomes is considered as limitation in spite of the increased penetration potential [89]. Therefore, to increase metronidazole loading into the elastic liposomes, so-called deformable propylene glycol liposomes were developed. Presence of propylene glycol improved solubility of metronidazole enabling higher entrapment efficiency and additionally increased bilayer elasticity of deformable liposomes. Compared to conventional liposomes-in-gel, release of the drugs (clotrimazole and metronidazole) from elastic liposomes was significantly enhanced [90].

Advanced vaginal delivery of metronidazole can be also achieved by the use of mucoadhesive pectin- and chitosan-containing liposomes, i.e., pectosomes and chitosomes [91,92]. The presence of mucoadhesive polymer on the surface of the vesicles is expected to prolong the retention time inside the vagina [91,92].

A challenging approach for vaginal delivery of liposomes with antimicrobial drugs includes preparation of tablets from preliposome powder. Selection of an appropriate excipient, added to preliposome powders before tableting, permits adequate drug release profile and retention at the site of administration. In this context, chitosan-based preliposome tablets allowed the highest encapsulation of metronidazole in *in situ* formed liposomes and assured the mucoadhesiveness of the system [93].

The chitosan-coated liposomes with clotrimazole allowed better retention of clotrimazole in vaginal tissue (*ex vivo* from pregnant sheep) and facilitated sustained drug release in comparison to control formulation [94]. Moreover, due to the cationic nature, chitosan exhibits antimicrobial activity and the ability to disrupt bacterial biofilms to some extent. Considering its biodegradability, chitosan-containing formulations have great potential in the treatment of bacterial vaginosis and candidiasis as well [95].

Curcumin or resveratrol, are natural-origin compounds (polyphenols) with many therapeutic properties including anti-inflammatory, antioxidant, immunomodulatory, antioxidant, antineoplastic, and antimicrobial [96,97]. Entrapment of curcumin and curcuminoides into conventional soy phosphatidylcholine liposomes resulted in improved anti-inflammatory activities in comparison to controls [98]. A promising pharmacokinetic outcome was achieved when curcumin was applied in liposomes to vaginal mucosa (*ex vivo*). Depending on the liposomal size, the concentration of curcumin in different layers of vaginal tissue was significantly higher as compared to curcumin solution [99]. Further increase of curcumin permeability through the artificial and isolated bovine mucus was obtained by the coating of curcumin liposomes with bioadhesive polymers, chitosan and Carbopol [100]. A very recent study by Jøraholmen et al. [101] reports on the potentials of chitosan-coated resveratrol liposomes as a safe topical treatment for vaginal inflammations and infections. Resveratrol has been confirmed to possess fungicidal, antibacterial, and antiviral effects [97,102,103], but its low solubility and poor bioavailability limit its clinical uses. Although the actual antimicrobial effect has not been determined, this approach opens the possibility for administration of new antimicrobials able to combat pathogen resistance and escape serious side effects related to current antimicrobial drugs [101].

Solid Lipid Nanoparticles (SLNs) and Nanostructured Lipid Carriers (NLCs)

To the best of our knowledge, up to now, there are only two studies indicating the potential of SLNs and NLCs for the topical therapy of infectious vaginitis. A polyoxyethylene-40 (PEG-40) stearate acrylate SLNs have been explored for vaginal delivery of ketoconazole and clotrimazole. Research based on physicochemical characterization and antifungal activity indicated the potential applicability for the treatment of candidiasis [104]. Clotrimazole has also been incorporated into NLCs embedded in Poloxamer 407 gel. *Ex vivo* permeation studies on pig vaginal mucosa showed localization of the drug in mucosa with antifungal activity fourfold higher than the commercial product. *In vitro* studies on HeLa cells confirmed the low toxicity profile resulting in a cell viability of 77.2% indicating suitability of the formulation for vaginal administration [105].

5.5.4.2 Treatment of Viral Infections

Liposomes

One of the first reports on the vaginal application of liposomes is associated to antiviral therapy. In the preliminary clinical testing of liposomes containing interferon alpha, Foldvari and Moreland [79] proved the potential of topical application of liposomes for the treatment of human papillomavirus (HPV) infections. Cervical lesions were completely resolved at the end of therapy in the female patient, while in the male patient application of liposomal interferon decreased the number of originally present lesions. Our group suggested the vaginal delivery of mucoadhesive liposomal gels with acyclovir for the treatment of *herpes simplex virus* (HSV) infections. Surface properties of liposomes have been shown to influence the stability of liposomes in simulated vaginal conditions (Fig. 5.5). The investigations were based on *in vitro* studies, in simulated vaginal conditions. Incorporation of liposomes into 1% Carbopol 974P gel in the concentration

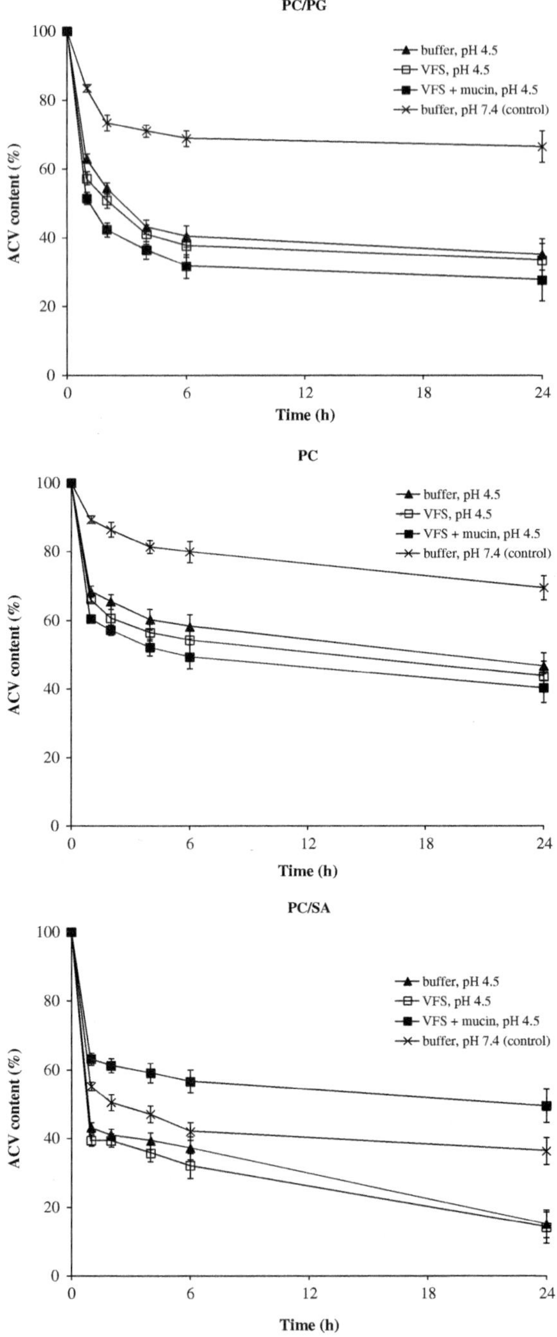

■ **FIGURE 5.5** *In vitro* stability of negatively charged (PC/PG), neutral, (PC) and positively charged (PC/SA) acyclovir (ACV) liposomes in different media simulating vaginal conditions. Liposomes, free from the unencapsulated drug, were incubated in different testing media at 37°C. Samples were taken at certain time intervals and the concentration of ACV still present in liposomes determined spectrophotometrically. The values denote the average of three preparations ± S.D. *From [106] Pavelić Ž, Škalko-Basnet N, Filipović-Grčić J, Martinac A, Jalšenjak I. Development and in vitro evaluation of a liposomal vaginal delivery system for acyclovir. J Controlled Release 2005;106:34–43, with permission from Elsevier.*

of 10% (w/w) enabled suitable retention at the administration site and contributed to prolonged and controlled release profile in simulated vaginal conditions [106].

A novel challenging approach suggested for the treatment of viral infections requires delivery of small interfering RNA (siRNA) through the mucus barrier. To allow effective siRNA delivery, Wu et al. [107] developed a biodegradable alginate scaffold system containing muco-inert PEGylated lipoplexes. The formulation was capable of assuring the extended vaginal presence of lipoplexes *in vivo*, and facilitated the delivery of siRNA into the vaginal epithelium.

5.5.4.3 Prophylaxis of Viral Infections

Liposomes

The abilities of liposomes to solubilize poorly soluble drugs have been utilized in topical microbicide development for the prevention of sexually transmitted viral infections. Several *in vivo* animal and clinical studies demonstrated the advantages of liposome formulations; however, up to now none of the liposome formulations reached the market.

Liposome-based microbicides offer less vaginal irritation in comparison to conventional formulations. The liposomal gel formulation caused less irritation of rabbit vaginal tissue than nonoxynol-9 [108]. When efficacy of MC1220 liposomes incorporated in hydrogels has been evaluated in rhesus macaque model, incomplete antiviral defense has been demonstrated. Surprisingly, the highest tested dose proved only partial protection, probably as a result of low aqueous solubility of MC1220 affected by pH [109]. Compared to conventional gel formulation, liposomal octylglycerol demonstrated better efficacy against HIV, HSV and *Neisseria gonnorrhoeae*. The safety of the liposomal formulation has been confirmed *in vitro* on Lactobacilli, *ex vivo* on human ectocervical tissue and *in vivo* on macaque model [110].

The microbicide effect can be also achieved by specially designed "empty" liposomes (without incorporated active substance). By optimizing the ratio of natural and synthetic lipids in liposomes and varying their physicochemical properties, Malavia et al. [111] developed liposomes for prevention of HIV infection.

A promising mode for preventing HIV infection presents vaginal administration of vaccines. Gupta et al. [112] suggested the liposome-base rods (lyophilized liposomal gel) with HIV-1 envelope glycoprotein CN54gp140 for vaginal immunization. The rods were designed to revert to gel form upon intravaginal application. However, according to the latest literature research, no animal studies on the efficacy of developed formulation are reported.

Solid Lipid Nanoparticles (SLNs)

Alukda et al. [113] suggested the polylysine-heparin functionalized SLNs for microbicide delivery. Even though relatively low encapsulation efficiency of tenofovir was achieved, SLNs were able to enhance cellular uptake of hydrophobic microbicide, and were well tolerated during testing in the *in vitro* model of vaginal epithelium for 48 h [113]. Table 5.3 provides an overview over lipid-based nanopharmaceuticals tested for treatment of vaginal infections.

5.5.5 **Lung/Pulmonary Infections**

Respiratory infections are commonly treated by parenterally administered high doses of antimicrobials, resulting in rather severe side effects. In an ideal case, targeted drug delivery to the airways with minimal side effects of the drugs should be achieved via inhaled antimicrobials. Liposomal formulations can be delivered to the lung in aerosol form, enabling the delivery of drug by inhalation of either nebulized droplets or dry powder particles [42]. Currently, only tobramycin and aztreonam solutions are approved in nebulized form in the United States. The dry powder formulations of tobramycin (Tobi® podhaler) and colistinmethanesulfonate (Colobreathe®) are approved in both the United States and Europe [114]. Nebulized tobramycin (Tobi®) and aztreonam (Cayston®) are rapidly cleared from the lungs and are administered in multiple doses to assure required concentrations. Liposomal formulations are expected to assure sustained release and prolong the clearance time. Liposomal ciprofloxacin formulations provided zero order release kinetics and assured prolonged antibiotic residence time in the lungs, as conformed on isolated perfused lung [115]. Moreover, modifications of liposomal surface are expected to lead to targeted drug therapy. Liposomes reaching the respiratory tract will be engulfed and degraded by the macrophages, a feature of great importance in intracellular drug therapy [116]. The macrophages expressing mannose receptors can be additionally targeted by modification of liposomes with mannose units [117]. An extensive review on the lipid-based formulations in preclinical and clinical studies was provided by di Cipolla and colleagues [34].

Technological advancements in mesh nebulizers are expected to assist this goal [114]. Encapsulation/incorporation of antimicrobials in nanocarriers is one of the emerging strategies for treating chronic lung infections, particularly those caused by *Pseudomonas aeruginosa* [118].

Table 5.3 Summary of Selected Lipid-Based Nanosystems for Topical Treatment of Vaginal Infections

Nanosystem		Drug/Active Compound	Formulation Type	Type of Study	Comments	References
Type	Ingredients					
Liposomes	HSPC:CHOL:stearic acid (2:1:1.4, molar ratio)	*Interferon α*	Suspension (non-mucoadhesive)	Clinical (preliminary)	Decreased number of HPV lesions (humans)	[79]
	Soy lecithin (powder)	*Clotrimazole, metronidazole, chloramphenicol*	Suspension (non-mucoadhesive)	*In vitro, ex vivo*	Stability of liposomes affected by the vaginal pH (pre- and post-menopause)	[80]
	EPC:EPG (9:1, molar ratio)	*Clotrimazole, metronidazole, chloramphenicol*	Mucoadhesive gel	*In vitro*	Increased stability of liposomes in mimicked vaginal conditions; sustained release of all the drugs; compatibility of liposomes with Carbopol gel	[82,84]
	EPC; EPC:EPG (9:1); EPC:SA (9:3); (molar ratios)	*Acyclovir*	Mucoadhesive gel	*In vitro*	Lipid composition influenced the stability of liposomes in simulated vaginal conditions; sustained and controlled release of the drug from liposomal gel	[106]
	DOPE:DOTAP: CHOL (4:5:1, molar ratio)	*Amphotericin B*	Thermosensitive gel	*In vitro*	Increased stability and reduced toxicity in comparison to the free drug.	[88]
	SPC, 0.1–0.5% chitosan	*Clotrimazole*	Suspension (mucoadhesive)	*In vitro, ex vivo*	Increased vaginal tissue retention and reduced penetration (pregnant sheep)	[94]
	SPC, 0.50% (w/w) pectin; SPC, 0.05–0.1% (w/w) chitosans	*Metronidazole*	Suspension (mucoadhesive)	*In vitro*	Significantly enhanced entrapment of the drug in both pectosomes and chitosomes	[92]
	EPC:SDCh (17:3, w/w) EPC:EPG (18:2, w/w)	*Metronidazole*	Suspension (non-mucoadhesive)	*In vitro*	Increased permeation of the drug on Caco-2 cell monolayer from deformable (EPC/SDCh) liposomes	[89]
	SPC(75): SDCh:propylene glycol (17:3:100, w/w/w)	*Metronidazole-clotrimazole*	Mucoadhesive gel	*In vitro*	Improved encapsulation efficiency in elastic deformable propylene glycol liposomes; sustained release profile; compatibility of the Carbopol gel with elastic lipsomes	[90]

(Continued)

Table 5.3 Summary of Selected Lipid-Based Nanosystems for Topical Treatment of Vaginal Infections (Continued)

Nanosystem		Drug/Active Compound	Formulation Type	Type of Study	Comments	References
Type	Ingredients					
	H-PC:CHOL (2:1, molar ratio)	*MC-1220*	Mucoadhesive gel	*In vitro*, in vivo (rabbits)	High encapsulation efficiency; improved stability during storage. Good tolerability	[108]
				In vivo (macaque monkeys)	Partial antiviral protection	[109]
	EPC	*Octylglycerol*	Mucoadhesive/ thermosensitive gels	*In vitro*, *in vivo* (macaque monkeys)	Improved efficacy against HSV, HIV and *N. gonorrhoeae*; confirmed safety in in vivo animal model; improved storage stability	[110]
	EPC:CHOL:SA (2:1:1) EPC:CHOL:DMPG (2:1:1) EPC:CHOL (2:1) (molar ratios)	*CN54gp140*	Mucoadhesive rods	*In vitro*	Good encapsulation efficiency; *in situ* formation of gel with suitable mucoadhesiveness.	[112]
Pre-liposomes	SPC(75)/mannitol (1:2, molar ratio)	*Metronidazole*	Tablets (non- and mucoadhesive excipients)	*In vitro*	*In situ* formation of liposomes in vaginal environment; sustained release profile	[93]
Pro-liposomes	EPC/CHOL/sorbitol (6:4:100, w/w/w)	*Clotrimazole*	Powder	*In vitro*, *in vivo*	Formation of liposomes upon addition of water; longer term reduction of yeasts; well tolerability	[85]
SLNs	PEG-40 stearate PEG-40 stearate acrylate	*Ketoconazole*, *Clotrimazole*	Suspension (non-mucoadhesive)	*In vitro*	More extended release with PEG-40 stearate acrylate; efficacy in promoting transcellular transport	[104]
NLCs	Tristearin/tricaprin 2:1 (w/w), Pluronic F68	*Clotrimazole*	Thermosensitive gel	*In vitro*, *ex vivo*	Increased antifungal activity; low toxicity profile	[105]

CHOL, cholesterol; CN54gp140, recombinant HIV-1 clade-C trimeric gp140 envelope protein; DOPE, dioleoylphosphatidylethanolamine, DMPG, dimiristoylphosphatidylglycerol; DOTAP, dioleoyltrimethylammoniumpropan; EPC, egg phosphatidylcholine; EPG, egg phosphatidylglycerol; H-PC, hydrogenated egg phosphatidylcholine; HSPC, hydrogenated soy phosphatidylcholine; NLCs, nanostructured lipid carriers; PC, phosphatidylcholine; SA, stearylamine; SDCh, sodium deoxycholate; SLNs, solid lipid nanoparticles; SPC, soy phosphatidylcholine; SPC(75), soy lecithin with 75% phosphatidylcholine.

Pseudomonas aeruginosa is known to form biofilms, an additional challenge in the therapy [119,120]. Various liposomal formulations for antibacterials were tested, such as for example liposomal azithromycin [121], sodium colistimethate loaded solid lipid nanoparticles and nanostructured lipid carriers [122]. For example, tobramycin, a cationic aminoglycoside, is often prescribed to treat *P. aeruginosa* biofilms despite its known adherence to the mucus and extracellular polymeric substances (EPS) matrix and short retention time in infected lungs (less than 2h) [123]. Therefore, conventional tobramycin therapy relies on at least 25-fold concentration above MIC twice daily, increasing the risk of systemic toxicity [124]. To address this low tobramycin exposure in the vicinity of the biofilms, various nano-antibiotic strategies have been suggested [118]. The pioneering work has been performed by Lagacé et al. [125] and Omri et al. [126] who evaluated the effects of liposomal tobramycin both *in vitro* and *in vivo*. Later work focused on liposomes the with ability to permeate the *P. aeruginosa* biofilms and codeliver antibiotics with another antimicrobial agents, for example metals [118]. Two proprietary liposomal antibiotics, namely ciprofloxacin (Lipoquin™ [127,128]) and amikacin (Arikace™ [129]) have reached Phase II clinical trials for cystic fibrosis and noncystic fibrosis bronchiectasis. Both formulations are designed for one daily administration. An overview of Arikace (inhaled liposomal amikacin) and inhaled liposomal ciprofloxacin clinical studies is provided by Cipolla and colleagues [34].

Hadinoto and Cheow [118] summarized the research findings related to liposomal antibiotics for pulmonary delivery tested in *in vitro* and *in vivo* conditions and warned that many studies were performed in the absence of mucus and very few studies focused on the mucus-liposome interactions. Only recently, the superiority of liposomal antibiotics was tested against the dormant biofilm form of *P. aeruginosa* [130]. It seems that the affinity of liposomes to adsorb onto cell membranes is the predominant feature behind the higher antipseudomonal efficacy of liposomal antibiotics in *in vitro* conditions [118]. However, the possible interactions between liposomal lipids and some antibiotics should not be neglected as they may inhibit antimicrobial activity of liposomally associated antibiotics [131]. It was suggested that the fluidity of the liposomal membrane plays a dominant role in antipseudomonal activity *in vivo* [132]. Following this line of research is the use of fusogenic liposomes and the addition of calcium ions [133]. The dipalmitoylphosphatidylcholine (DPPC) and 1,2-dimyristoyl-sn-glycero-3-phosphorylglycerol sodium salt (DMPG) liposomes with tobramycin composed of fluid lipid membranes were named Fluidosomes™ [132]. The liposomal surface charge is equally contributory to *in vivo* activity of liposomal antibiotic. The anionic liposomes exhibited longer lung retention time and faster tobramycin release than the neutrally charged liposomes, probably due to lesser aggregation [132]. The Fluidosomes™ were confirmed to be non-immunogenic [134]. However, the nebulization as a means of Fluidosomes™ delivery resulted in a large drug loss [135], which was overcome by the use of dry powder Fluidosomes™ prepared by freeze-drying [136]. The inclusion of CHOL in liposome compositions improves the stability of liposomes, such as in nebulized liposomal amikacin (Arikace®) and liposomal ciprofloxacin (Lipoquiin®), currently in clinical trials [128,129]. Stealth liposomes were also tested for the pulmonary delivery [36].

However, the presence of mucus should not be neglected when optimizing the liposomal composition for superior pulmonary delivery. Most of liposomes do not penetrate through mucus pores and might not be able to reach the biofilm colonies. Meers et al. [137] demonstrated an effective mucus penetration by DPPC:CHOL liposomes carrying amikacin and confirmed by florescence imaging the penetration of liposomes into the biofilm extracellular polymeric substances (EPS) matrix. Neutral liposomes are suggested to be superior to charged liposomes if their size is optimized [118]. However, the possible leakage of liposomally associated antibiotic occurring while diffusing through the mucus can also contribute to the prolonged retention of liposomal antibiotic in the lungs and the sustained release of the antibiotic still retained within liposomes [138]. The co-encapsulation of metals such as bismuth and gallium to enhance the antipseudomonal efficacy of liposomal antibiotics has been exploited as an alternative path to improved antipseudomal therapy, resulting in increased efficacy and reduced cytotoxicity. The coencapsulation of antibiotics and antibacterial metals resulted in synergistic effects [118,139].

Spray-drying technique has been employed to manufacture spray-dried formulations of proliposomes (dry powders which can form liposomes upon contact with water) [140]. The choice of proliposome constituents will affect both the release profile and the aerosolization performance, which needs to be further exploited in optimization of inhaled liposomal formulations [114].

In addition to inhaled liposomal antibacterials, research has also been extended to the inhaled antivirals and antifungals [114]. The majority of the work has been focused on nebulizing the commercially available formulations of amphotericin B such as Fungizone®, Ambisome®, Abelect®, and Amphotec®, as summarized by Kuiper and Ruijgrok [141]. A promising delivery device seems to be Atomiser NL9M nebulizer [142].

SLNs prepared as inhalable sustained drug delivery systems were found to have potential in tuberculosis treatment [143].

5.5.6 **Local Therapy of Eye Infections**

During the last decades a considerable number of researchers have evaluated the applicability of lipid-based nanopharmaceuticals as a superior delivery system in respect to prolonged preocular residence of various antimicrobial drugs and improved drug bioavailability. Lipid nanosystems can be instilled in liquid form as eye drops and thereof reduce the discomfort associated with the administration of semisolid ointments. The nanosystems are considered to be patient friendly due to their less frequent administration, not only extending the duration of drug retention in the extraocular portion but preventing the blurred vision often associated with the ointment form [144,145]. Up to date, liposomes are the most widely investigated lipid-based nanosystem for the treatment of eye infections. Their effectiveness depends on various parameters such as surface charge, rigidity, lipid composition, encapsulation efficiency and release rates, similarly as for liposomes administered via other routes. Numerous studies have reported that the positively charged liposomes have a higher binding affinity for the corneal surface than the neutral or negatively charged vesicles, which is explained by the interaction between positively charged liposomes and the polyanionic corneal and conjunctival mucoglycoproteins [146–148]. Furthermore, the release rate of the drug from liposomes was found to be affected by the pH and vesicle surface charge. For example, by increasing the ratio of positively charged stearylamine in liposome membranes, the release rate of liposomal acyclovir was faster. On the contrary, for negatively charged liposomes, the release rate was greater at lower dicetylphosphate content [149]. Other parameters such as the lipid composition and phase transition temperature of lipids can also influence the drug release profile [150].

The advantages of liposomes in improving the ocular bioavailability of antivirals have been proven by several research groups. Fresta et al. [148] have shown better encapsulation of acyclovir in both positively and negatively charged liposomes than in neutral DPPC liposomes. *In vivo* (rabbit model) evaluation has proven that the most suitable liposomes have been composed of DPPC, CHOL, and dimethyl dioctadecyl glycerol bromide (7:4:1, molar ratio); this formulation delivered a greater amount of the drug into the aqueous humor than acyclovir solution. Similar findings have been reported by Law's group. Positively charged liposomes provided a higher extent of acyclovir absorption in vivo. The drug deposition in the cornea was greater with positively charged than with negatively charged liposomes and free acyclovir as a consequence of binding cationic liposomes to the corneal surface [151]. In another *in vivo* study (rabbits), the efficacy of positively charged liposomes has been compared to acyclovir ointment. In spite of the much higher dose of acyclovir in ointment (1.5 *vs* 0.18 mg in liposomes), the area under the curve (AUC) produced by the ointment was only 1.6 times greater than the AUC of liposome dispersion [152].

SLNs and NLCs have also been explored as a means to improve ocular delivery of acyclovir. Physicochemical evaluation demonstrated NLCs as the formulation of choice due to the higher encapsulation efficiency than in SLNs and faster permeation of the drug through the excised bovine cornea [153]. However, the research was based only on *in vitro* evaluation and up to now, there are no *in vivo* data to confirm the efficacy of suggested nanosystem.

Chronic ocular infections such as conjunctivitis and bacterial keratitis necessitate high drug concentration at the site of infection. Their treatment requires frequent eye drop administration that may cause drug resistance and also decrease patient compliance. To minimize the precorneal drainage and increase the residence time, liposomal hydrogels have been proposed as advanced delivery systems [145]. Bioadhesive poly(vinyl alcohol)- and polymethacrylic acid-based gels have been evaluated as vehicles for liposomes containing ciprofloxacin to minimize tear-driven dilution in the conjunctival sac [150]. Both tested hydrogels ensured a steady and prolonged drug release. Lipid composition and phase transition of lipids building liposomes influenced ciprofloxacin release. Prolonged release of the drug was demonstrated with DPPC liposomes as compared to lecithin liposomes, probably as a result of more rigid bilayers produced by synthetic lipids containing only saturated fatty acid with a transition temperature of 41°C [150]. Superiority of liposomal ciprofloxacin has been confirmed in *in vivo* studies on rabbits. Amongst 18 liposomal formulations, differing in the lipid composition and liposome surface charge, the highest relative bioavailability was achieved for liposomes composed of DMPC [146].

Utilizing chitosan in liposome formulations was reported to be beneficial in improving the precorneal residence time, as an alternative approach. Mucoadhesive chitosan-coated liposomes enabled 1.74-fold higher permeability of ciprofloxacin across the isolated rabbit cornea and increased antibacterial activity against *P. aeruginosa* and *S. aureus* in comparison to the free drug [154].

Although fungal infections of the eye are less common than infections by bacteria or viruses, these infections are usually severe and may lead to loss of vision. Effectiveness of topical therapy depends on the concentration of antifungal achieved in the target ocular tissue, which is determined by the drug molecular mass, concentration of the applied drug and its penetration potential [155,156]. To enhance ocular delivery of fluconazole, liposomal formulation (2 mg/mL) was developed and evaluated on *Candida keratitis* model in rabbits. A significantly better response in terms of ulcer and hypopyonsize has been obtained using fluconazole-loaded liposomes than a solution of the free drug. Complete healing was achieved at the end of the third week in 19 rabbits (86.4%) treated with liposomes and nine rabbits (50%) treated with the drug solution [157].

In another study, SLNs comprising Compritol 888 ATO and PEG 600 matrix were evaluated for ocular delivery of ketoconazole. Presence of PEG allowed better drug incorporation (70%) and improved permeation into the eye. Pharmacokinetic studies with ketoconazole NLCs demonstrated 2.5- and 1.6-fold higher bioavailability (AUC) in aqueous and humor, respectively, in comparison to the drug suspension [158]. Gatifloxacin was loaded in SLNs to improve antibacterial activity [159].

5.5.7 Periodontal Infections

Periodontal biofilms are known to cause acute and chronic gingivitis and periodontitis. The pathogenic microorganisms in the biofilm will also be supported by various genetic and environmental factors aggravating the disease conditions. The control and treatment of bacterial biofilms is of great importance [160,161]. More than 500 distinct microbial species can be found in dental plaque. Although most of the research was focused on bacterial microflora, researchers have implicated *Herpes* viruses and *Candida albicans* species as contributors to bacterial biofilms [162]. Current options used in delivery of local antibiotics fail to completely clear the bacteria filling dentinal tubules and exhibit limited therapeutic efficacy [163]. The local treatment of periodontal diseases with sophisticated and multifunctional nanomedicine is expected to improve long-term healing through: (1) removal of the biofilms; (2) successful delivery of compounds for host inflammatory response modulation; and (3) regeneration of periodontal tissue, which is often not achieved by the conventional periodontal therapy [161]. The liposome-based delivery systems able to penetrate deeply into the tubules, even reaching the terminal ends, were confirmed to have potential in *ex vivo* studies on extracted human teeth [161,163].

Negatively charged liposomes were found to be most resistant to the components of saliva, therefore superior to other types of liposomes [164].

However, Longo et al. [165] suggested that aluminum chloride phthalocyanine (AICIPc) entrapment in cationic liposomes results in efficient reduction of the microbial load from bacterial cultures in the oral cavity. The clinical study confirmed 82% reduction of total bacteria after photodynamic therapy based on liposomal AICIPc and safety of developed formulation.

To assure prolonged residence time in oral cavity, liposomes can be incorporated into mucoadhesive ointments based on polymethyl methacrylate [166].

5.6 **TOXICITY**

Lipid-based delivery systems are expected to be safe due to their biocompatibility and biodegradability. Their use should assure reduction in the therapeutic toxicity associated with antimicrobials. However, the drug's toxicity will be in the best case reduced, but should be taken into the consideration when deciding on the risk to benefit therapeutic outcome. Lipid toxicity has been neglected in many early research efforts. It is known that many lipids exhibit dose-dependent toxicity in exceedingly high doses; however most of the approved lipid-based formulations contain low lipid content. Cationic lipids have been shown to promote non-specific binding to erythrocytes, lymphocytes and endothelial cells, as well as activation of the human complement system. On the other hand, unprotected surface negative charge on lipid molecules act as a binding site for opsonization [6]. Although hydrophilic PEG can shield the surface charge and decrease the system's immunogenicity, recent indications suggest that PEG may generate complement activation products in serum and antibody production [167,168]. If these indications are confirmed, the safety of PEGylated products needs to be re-evaluated, particularly in chronic administration [6]. Although several *in vivo* studies confirmed that positively charged (cationic) liposomes improved ocular bioavailability of the poorly soluble antimicrobial drugs in comparison to neutral and negatively charged liposomes, the toxicity studies have not been performed to confirm the safety of developed nanosystems. We assume that optimal formulations could be achieved by fine-tuning the ratio of cationic and neutral lipids in these cases.

Long-term toxicity/safety of lipid-based nanopharmaceuticals, especially those for the treatment of chronic infections, has to be determined. The potential long-term toxicity of repeatedly administered nanosized lipid particles remains an issue necessary to be addressed and was the basis for the establishment of the Nanodermatology Society in 2010 aiming to promote a deeper understanding of the scientific and medical aspects of nanotechnology related to skin therapy.

5.7 CURRENT LIMITATIONS

In spite of extensive efforts to develop a superior therapy of infections based on the progress in nanotechnology and related areas, up to now, nanomedicine-based antibiotic therapy failed to completely eradicate intracellular bacterial infections [3]. Whether some of the promising lines of research will indeed succeed in the next several years remains to be seen. There are several limitations which need to be addressed by a multidisciplinary approach; the scientific challenges of better understanding of the fate of nanopharmaceuticals in the body can be solved by joint efforts based on the combined expertise in various disciplines and support by pharmaceutical and biotechnological industries. The costs of the newly developed nanoformulations versus the benefits obtained by the new formulations are often the topic of public concerns and remain to be a societal challenge. In close correlation is the public skepticism expressed in many Western societies and is related to chemical properties that may become toxic at "nano" level [1]. Those challenges require close dissemination of the findings and building up the trust between academia, industry and the public. The process of approval of new nanoformulations by various governing agencies in different countries should be synchronized and addressed internationally.

The fact that many *in vivo* studies failed to confirm the prolonged half-life of liposomes and lipid particles needs to be re-evaluated. Often, the low drug load provided by small particles might be the reason behind lack of the success of many formulations. The pressure to prove the superiority of a novel system might lead to rejection and closure of some research pipelines even before all scientific questions are addressed. For example, for pulmonary delivery, liposomal antibiotics face the bottleneck of loss during their preparation, storage, inhalation, mucus interactions and limited physicochemical stability [118], challenges the pharmaceutical industry needs to face. Similarly, the sterilization methods suitable for liposomal antibiotics must avoid heat, irradiation, and chemical agents due to the sensitivity of lipids and are mostly limited to mechanical filtration; from an industrial point of view, this represents an unfavorable choice [3]. Often the experimental manufacturing methods face a challenge in scaling up, which should bring new technological advancements in the area. Finally, the specific protocols for future clinical trials involving liposomal formulations should be discussed and optimized [42].

5.8 PERSPECTIVES

Many unanswered questions and challenges open exciting opportunities for further investigations. The marketing failure of Exubera (inhaled insulin) led to stagnation in research on inhaled formulations destined for systemic action [34]. However, the possibility of delivering lipid-based formulations to treat systemic disorders after inhalation therapy should be, in our opinion, revised. Polymyxin B, tobramaycin, and amikacin, as well as amphotericin B, as the examples of the success of liposomal antimicrobial nanosystems could serve as a base for further optimization [42]. Cationic biodegradable lipids of natural origin, such as chitosan, have great potential in improved antimicrobial topical therapy of lipid-based nanosystems due to (1) cationic nature contributing to increased antimicrobial effect and (2) mucoadhesiveness, allowing prolonged retention of the formulation/drug at the site of application. Although we have not discussed in detail the chitosan-containing lipid-based formulations, their potential remains to be further evaluated, especially in the treatment of biofilms. An interesting approach in improved amphotericin delivery via the pulmonary route are NanoDisk complexes of apolipoprotein A-1 and phospholipids developed by Lypro Biosciences [34]. Recently, antibiotic delivery by liposomes made from prokaryotic microorganisms was proposed. Membranes of prokaryotes obtained from bacteria with limited growth requirements (low-costs systems) were used to prepare liposomes with penicillin G which exhibited superior antibiotic activity [169]. Looking into nature to learn and mimic is one of the most promising lines in biomimetics, which needs to be exploited in depth in the future development of superior delivery systems for various antimicrobials.

5.9 CONCLUSIONS

Lipid-based nanocarriers for antimicrobial therapy are here to stay. They offer an opportunity to design and develop tailored nanocarriers with superior features, which, in comparison to polymeric nanoparticles, are considered safer. However, there is a need for further technological advancements in respect to the manufacturing methods. In addition, deeper insight into the barriers faced by the drug administered via different routes of administration needs to be addressed and dealt with in order to optimize the nanopharmaceuticals.

REFERENCES

[1] Couvreur P, Vauthier C. Nnaotechnology: inteligent design to treat complex disease. Pharm Res 2006;23:1417–50.

[2] Riehemann K, Schneider SW, Luger TA, Godin B, Ferrari M, Fuchs H. Nanomedicine – challenge and perspectives. Angewandte Chem Int Ed 2009;48:872–97.

[3] Abed N, Couvreur P. Nanocarriers for antibiotics: a promising solution to treat intracellular bacterial infections. Int J Antimicrob Agents 2014;43:485–96.

[4] Sykes R. Towards the magic bullet. Int J Antimicrob Agents 2000;14:1–12.

[5] Drulis-Kawa Z, Dorotkiewics-Jach A. Liposomes as delivery systems for antibiotics. Int J Pharm 2010;387:187–98.

[6] Lim SB, Banarjee A, Önyüksel H. Improvement of drug safety by the use of lipid-based nanocarriers. J Controlled Release 2012;163:34–45.

[7] Torchilin VP. Recent advances with liposomes as pharmaceutical carriers. Nat Rev Drug Discovery 2005;4:145–60.

[8] Safinya CR, Ewert KK. Materials chemistry: liposomes derived from molecular vases. Nature 2012;489:372–4.

[9] Kubitschke J, Javor S, Rebek Jr. J. Deep cavitand vesicles – multicompartment hosts. Chem Commun 2012;48:9251–3.

[10] Vanić Ž, Holsæter A-M, Škalko-Basnet N. (Phospho)lipid-based nanosystems for skin administration. Curr Pharm Des 2015;21:4174–92.

[11] Cevc G, Blume G. Lipid vesicles penetrate into intact skin owing to the transdermal osmotic gradients and hydration force. Biochim Biophys Acta 1992;1104:226–32.

[12] Touitou E, Dayan N, Bergelson L, Godin B, Eliaz M. Ethosomes – novel vesicular carriers for enhanced delivery: characterization and skin penetration studies. J Controlled Release 2000;65:403–18.

[13] Mehnert W, Mäder K. Solid lipid nanoparticles production characterization and applications. Adv Drug Delivery Rev 2001;47:165–96.

[14] Müller RH, Maaßen S, Weyhers H, Specht F, Lucks JS. Cytotoxicity of magnetite-loaded polylactide, polylactide/glycolide particles and solid lipid nanoparticles. Int J Pharm 1996;138:85–94.

[15] Müller RH, Petersen RD, Hommoss A, Pardeike J. Nanostructured lipid carriers (NLC) in cosmetic dermal products. Adv Drug Delivery Rev 2007;59:522–30.

[16] Zhang X, Liu J, Qiao H, Liu H, Ni J, Zhang W, et al. Formulation optimization of dihydroartemisinin nanostructured lipid carrier using response surface methodology. Powder Technol 2010;197:120–8.

[17] Constantidinis PP, Chaubal MV, Shorr R. Advances in lipid nanodispersions for parenteral drug delivery and targeting. Adv Drug Delivery Rev 2008;60:757–67.

[18] Torchilin VP. Micellar nanocarriers: pharmaceutical perspectives. Pharm Res 2007;24:1–16.

[19] Armstead AL, Li B. Nanomedicine as an emerging approach against intracellular pathogens. Int J Nanomed 2011;6:3281–93.

[20] Hall-Stoodley L, Stoodley P. Evolving concepts in biofilm infections. Cell Microbiol 2009;11:1034–43.

[21] Parsek MR, Singh PK. Bacterial biofilms: an emerging link to disease pathogenesis. Annu Rev Microbiol 2003;57:677–701.

[22] Forier K, Raemdonck K, De Smedt SC, Demeester J, Coenye T, Braeckmans K. Lipid and polymeric nanoparticles for drug delivery to bacterial biofilms. J Controlled Release 2014;190:607–23.

[23] Hurler J, Škalko-Basnet N. Advancements in burn therapy: promise of nanomedicine McLaughlin ES, Paterson AO, editors. Burns: prevention, causes and treatment. New York, NY: Nova Science Publishers; 2012. p. 39–63.

[24] Parveen S, Misra R, Sahoo SK. Nanoparticles: a boon to drug delivery, therapeutics, diagnostics and imaging. Nanomed: Nanotechnol, Biol Med 2012;8:147–66.

[25] Yin R, Agrawal T, Khan U, Gupta GK, Rai V, Huang Y-Y, et al. Antimicrobial photodynamic inactivation in nanomedicine: small light strides against bad bugs. Nanomedicine (London) 2015;10:2379–404.

[26] Derycke ASL, de Witte PAM. Liposomes for photodynamic therapy. Adv Drug Delivery Rev 2004;56:17–30.

[27] Nakonechny F, Firer MA, Nitzan Y, Nisnevitch M. Intracellular antimicrobial therapy: a novel technique for efficient eradication of pathogenic bacteria. Photochem Photobiol 2010;86:1350–5.

[28] Sachetelli S, Khalil H, Chen T, Beaulac C, Sénéchal S, Lagacé J. Demonstration of a fusion mechanism between a fluid bactericidal liposomal formulation and bacterial cells. Biochim Biophys Acta 2000;1443:254–66.

[29] Aljuffali IA, Huang C-H, Fang J-Y. Nanomedical strategies for targeting skin microbiomes. Curr Drug Metabol 2015;16:255–71.

[30] Drummond DC, Noble CO, Hayes ME, Park JW, Kirpotin DB. Pharmacokinetics and in vivo drug release rates in liposomal nanocarrier development. J Pharm Sci 2008;97:4696–740.

[31] Allen TA, Hansen CB, Lopes de Menezes DE. Pharmacokinetics of long-circulating liposomes. Adv Drug Delivery Rev 1995;16:267–84.

[32] Ishida T, Harada M, Wang XY, Ichihara M, Irimura K, Kiwada H. Accelerated blood clearance of PEGylated liposomes following preceding liposome injection: effects of lipid dose and PEG surface-density and chain length of the first-dose liposomes. J Controlled Release 2005;105:305–17.

[33] Gabizon A, Shmeeda H, Barenholz Y. Pharmacokinetics of PEGylated liposomal doxorubicin. Clin Pharmacokinet 2003;42:419–36.

[34] Cipolla D, Shekunov B, Blanchard J, Hickey A. Lipid-based carriers fro pulmonary products: preclinical development and case studies in humans. Adv Drug Delivery Rev 2014;75:53–80.

[35] Nicolosi D, Scalia M, Nicolosia VM, Pignatello R. Encapsulation in fusogenic liposomes broadens the spectrum of action of vancomycin against Gram-negative bacteria. Int J Antimicrob Agents 2010;35:553–8.

[36] Bakker-Woudenberg IAJM, Ten Kate MT, Guo L, Working P, Mouton JW. Ciprofloxacin in polyethylene glycol-coated liposomes: efficacy in rat models of acute and chronic *Pseudomonas aeruginosa* infection. Antimicrob Agents Chemother 2002;46:2575–81.

[37] Omri A, Suntres ZE, Shek PN. Enhanced activity of liposomal polymyxin B against *Pseudomonas aeruginosa* in a rat model of lung infection. Biochem Pharmacol 2002;64:1407–13.

[38] Schiffelers R, Storm G, Bakker-Woudenberg I. Liposome-encapsulated aminoglycosides in pre-clinical and clinical studies. J Antimicrob Chemother 2001;48:333–44.

[39] Marier J-F, Brazier JL, Lavigne J, Ducharme MP. Liposomal tobramycin against pulmonary infections of *Pseudomonas aeruginosa*: a pharmacokinetic and efficacy study following single and multiple intratracheal administrations in rats. J Antimicrob Chemother 2003;52:247–52.

[40] Marier J-F, Lavigne J, Ducharme MP. Pharmacokinetics and efficiencies of liposomal and conventional formulations of tobramuycin after intratracheal

administration in rats with pulmonary *Burkholderia cepacia* infection. Antimicrob Agents Chemother 2002;46:3776–81.

[41] Pandey R, Sharma S, Khuller GK. Oral solid lipid nanoparticle-based antitubercular chemotherapy. Tuberculosis 2005;85:415–20.

[42] Alhariri M, Azghani A, Omri A. Liposomal antibiotics for the treatment of infectious diseases. Exp Opin Drug Delivery 2013;10:1515–32.

[43] Leitzke S, Bucke W, Borner K, Müller R, Hahn H, Ehlers S. Rationale for and efficacy of prolonged-interval treatment using liposome-encapsulated amikacin in experimental *Mycobaterium avium* infection. Antimicrob Agents Chemother 1998;42:459–61.

[44] de Steenwinkel JEM, van Vianen W, ten Kate MT, Verbrugh HA, van Agtmael MA, Schiffelers RM, et al. Targeted drug delivery to enhance efficacy and shorten treatment duration in disseminated *Mycobacterium avium* infection in mice. J Antimicrob Chemother 2007;60:1064–73.

[45] Swenson CE, Stewart KA, Hammett JH, Fitzsimmons WE, Ginsberg RS. Pharmacokinetics and *in vivo* activity of liposome-encapsulated gentamicin. Antimicrob Agents Chemother 1990;34:235–40.

[46] Gaspar MM, Cruz A, Penha AF, Reymao J, Sousa AC, Eleuterio CV, et al. Rifabutin encapsulated in liposomes exhibits increased therapeutic activity in a model of disseminated tuberculosis. Int J Antimicrob Agents 2008;31:37–45.

[47] Sande L, Sanchez M, Montes J, Wolf AJ, Morgan MA, Omri A, et al. Liposomal encapsulation of vancomycin improves killing of methicillin-resistant *Staphylococcus aureus* in a murine infection model. J Antimicrob Chemother 2012;67:2191–4.

[48] Kadry AA, Al-Suwayeh SA, Abd-Allah ARA, Bayomi MA. Treatment of experimental osteomyelitis by liposomal antibiotics. J Antimicrob Chemother. 2004;54:1103–8.

[49] Rathore A, Jain A, Gulbake A, Shilpi S, Khare P, Jain A, et al. Mannosylated liposomes bearing Amphotericin B for effective management of visceral Leishmaniasis. J Liposome Res 2011;21:333340.

[50] Pantze SF, Parmentier J, Hofhaus G, Fricker G. Matrix liposomes: a solid liposomal formulation for oral administration. Eur J Lipid Sci Technol 2014;116:1145–54.

[51] Gershkovich P, Wasan EK, Lin M, Sivak O, Leon CG, Clement JG, et al. Pharmacokinetics and biodistribution of amphotericin B in rats following oral administration in a novel lipid-based formulation. J Antimicrob Chemother 2009;64:101–8.

[52] Kong HH, Segre J,A. Skin microbiome: looking back to move forward. J Invest Dermatol 2012;132:933–9.

[53] Castro GA, Ferreira LAM. Novel vesicular and particulate drug delivery systems for topical treatment of acne. Exp Opin Drug Delivery 2008;5:665–79.

[54] Basnet P, Škalko-Basnet N. Nanodelivery systems for improved topical antimicrobial therapy. Curr Pharm Des 2013;19:7237–43.

[55] Prow TW, Grice JE, Lin LL, Faye R, Butler M, Becker W, et al. Nanoparticles and microparticles for skin drug delivery. Adv Drug Delivery Rev 2011;63:470–91.

[56] Kristl J, Teskač K, Ahlin P. Current vie won nanosized solid lipid carriers for drug delivery to the skin. J Biomed Nanotechnol 2010;6:529–42.

[57] DeLouise LA. Applications of nanotechnology in dermatology. J Invest Dermatol 2012;132:964–75.

[58] Aljuffali IA, Huang C-H, Fang J-Y. Nanomedical strategies for targeting skin microbiomes. Curr Drug Metabol 2015;16:255–71.

[59] Mezei M, Gulasekharam V. Liposomes- a selective drug delivery system for the topical route of administration. lotion dosage form. Life Sci 1980;26:1473–7.

[60] Korting HC, Schäfer-Korting M. Carriers in the topical treatment of skin disease. in drug deliverySchäfer-Korting M, editor. Handbook of experimental pharmacology, 197. Berlin: Springer-Verlag; 2010. p. 435–7.

[61] Souto EB, Wissing SA, Barbosa CM, Müller RH. Development of a controlled release formulation based on SLN and NLC for topical clotrimazole therapy. Int J Pharm 2004;278:71–7.

[62] Škalko N, Čajkovac M, Jalšenjak I. Liposomes with clindamycin hydrochloride in the therapy of Acne vulgaris. Int J Pharm 1992;85:97–101.

[63] Honzak L, Šentjurc M. Development of liposome encapsulated clindamycin for treatment of acne vulgaris. Pflügers Arch Eur J Physiol 2000;440:R44–5.

[64] Godin B, Touitou E, Rubinstein E, Athamna A, Athamna M. A new approach for treatment of deep skin infections by an ethosomal antibiotic preparation: an in vivo study. J Antimicrob Chemother 2005;55:989–94.

[65] Darwhekar G, Jain DK, Choudhary A. Elastic liposomes for delivery of neomycin sulphate in deep skin infection. Asian J Pharm Sci 2012;7:230–40.

[66] Li C, Zhang X, Huang X, Wang X, Liao G, Chen Z. Preparation and characterization of flexible nanoliposomes loaded with daptomycin, a novel antibiotic, for topical skin therapy. Int J Nanomed 2013;8:1285–92.

[67] Naeff R. Feasibility of topical liposome drugs produced on an industrial scale. Adv Drug Delivery Rev 1996;18:343–7.

[68] Sanna V, Gavini E, Cossu M, Rassu G, Giunchedi P. Solid lipid nanoparticles (SLN) as carriers for topical delivery of econazole nitrate: in-vitro characterization, ex-vivo and in-vivo studies. J Pharm Pharmacol 2007;59:1057–64.

[69] Kaur L, Jain SK, Manhas RK, Sharma D. Nanoethosomal formulation for skin targeting of amphotericin B: an in vitro and in vivo assessment. J Liposome Res 2014;25:294–307.

[70] Vaghasiya H, Kumar A, Sawant K. Development of solid lipid nanoparticles based controlled release system for topical delivery of terbinafine hydrochloride. Eur J Pharm Sci 2013;49:311–22.

[71] Vogt PM, Hauser J, Rossbach O, Bosse B, Fleischer W, Steinau HU, et al. Polyvinyl pyrrolidone-iodine liposome hydrogel improves epithelialization by combining moisture and antisepis. a new concept in wound therapy. Wound Repair Regener 2001;9:116–22.

[72] Homann HH, Rosbach O, Moll W, Vogt Peter M, Germann G, et al. A liposome hydrogel with polyvinyl-pyrrolidone iodine in the local treatment of partial thickness burn wounds. Ann Plast Surg 2007;59:423–7.

[73] Vogt PM, Reimer K, Hauser J, Rossbach O, Steinau HU, Bosse B, et al. PVP-iodine in hydrosomes and hydrogel – a novel concept in wound therapy leads to enhanced epithelialisation and reduced loss of skin grafts. Burns 2006;32:698–705.

[74] Touitou E, Godin B. Vesicular carriers for enhanced delivery through the skin. In: Touitou E, Godin B, Barry BW, editors. Enhancement in drug delivery. Boca Raton, FL: CRC Press; 2006. p. 255–301.

[75] Vanić Ž, Škalko-Basnet N. Nanopharmaceuticals for improved topical vaginal therapy: can they deliver? Eur J Pharm Sci 2013;50:29–41.

[76] Vanić Ž, Škalko-Basnet N. Mucosal nanosystems for improved topical drug delivery: vaginal route of administration. J Drug Delivery Sci Technol 2014;24:435–44.

[77] Okada K, Hillery AM. Vaginal drug delivery Hillery AM, Lloyd AW, Swarbrick J, editors. Drug delivery and targeting for pharmacists and pharmaceutical sciences. London: Taylor & Francis; 2001. p. 301–28.

[78] Machado RM, Palmeira-de-Oliveira A, Gaspar C, Martinez-de-Oliveira J, Palmeira-de-Oliveira R. Studies and methodologies on vaginal drug permeation. Adv Drug Delivery Rev 2015;92:14–26.

[79] Foldvari M, Moreland A. Clinical observations with topical liposome-encapsulated interferon alpha for the treatment of genital papillomavirus infections. J Liposome Res 1997;7:115–26.

[80] Pavelić Ž, Škalko-Basnet N, Jalšenjak I. Liposomes for treatment of vaginal infections. Eur J Pharm Sci 1999;8:345–51.

[81] Pavelić Ž, Škalko-Basnet N, Schubert R. Liposomal gels for vaginal drug delivery. Int J Pharm 2001;219:139–49.

[82] Pavelić Ž, Škalko-Basnet N, Jalšenjak I. Liposomal gel with chloramphenicol: characterization and in vitro release. Acta Pharm 2004;54:319–30.

[83] Pavelić Ž, Škalko-Basnet N, Schubert R, Jalšenjak I. Liposomal gels for vaginal drug deliveryDüzgünes N, editor. Liposomes (Part D) in methods in enzymology, Vol. 287. San Diego, CA: Elsevier Academic Press; 2004. p. 287–99.

[84] Pavelić Ž, Škalko-Basnet N, Jalšenjak I. Characterization and in vitro evaluation of bioadhesive liposome gels for the local treatment of vaginitis. Int J Pharm 2005;301:140–8.

[85] Ning M-Y, Guo Y-Z, Pan H-Z, Yu H-M, Gu Z-W. Preparation and evaluation of proliposomes containing clotrimazole. Chem Pharm Bull 2005;53:620–4.

[86] Ning M, Guo Y, Pan H, Chen X, Gu Z. Preparation, in vitro and in vivo evaluation of liposomal/niosomal gel delivery systems for clotrimazole. Drug Dev Ind Pharm 2005;31:375–83.

[87] Karimunnisa S, Atmaram P. Mucoadhesive nanoliposomal formulation for vaginal delivery of an antifungal. Drug Dev Ind Pharm 2013;39:1328–37.

[88] Kang J-W, Davaa E, Kim Y-T, Park J-S. A new vaginal delivery system of amphotericin B: a dispersion of cationic liposomes in a thermosensitive gel. J Drug Targeting 2010;18:637–44.

[89] Vanić Ž, Hafner A, Bego M, Škalko-Basnet N. The characterization of various deformable liposomes with metronidazole. Drug Dev Ind Pharm 2013;39:481–8.

[90] Vanić Ž, Hurler J, Ferderber K, Golja Gašparović P, Škalko-Basnet N, Filipović-Grčić J. Novel vaginal drug delivery system: deformable propylene glycol liposomes-in-hydrogel. J Liposome Res 2014;24:27–36.

[91] Andersen T, Bleher S, Flaten GE, Tho I, Mattsson S, Škalko-Basnet N. Liposomal delivery system enhances anti-inflammatory properties of curcumin. Chitosan in mucoadhesive drug delivery: focus on local vaginal therapy. Mar Drugs 2015;13:222–36.

[92] Andersen T, Vanić Ž, Flaten GE, Mattsson S, Tho I, Škalko-Basnet N. Pectosomes and chitosomes as delivery systems for metronidazole: the one-pot preparation method. Pharmaceutics 2013;5:445–56.

[93] Vanić Ž, Planinšek O, Škalko-Basnet N, Tho I. Tablets of pre-liposomes govern in situ formation of liposomes: concept and potential of the novel drug delivery system. Eur J Pharm Biopharm 2014;88:443–54.

[94] Jøraholmen MW, Vanić Ž, Tho I, Škalko-Basnet N. Chitosan-coated liposomes for topical vaginal therapy: assuring localized drug effect. Int J Pharm 2014;472:94–101.

[95] Kandimalla KK, Borden E, Omtri RW, Boyapati SP, Smith M, Lebby K, et al. Ability of chitosan gels to disrupt bacterial biofilms and their application in the treatment of bacterial vaginosis. J Pharm Sci 2013;102:2096–101.

[96] Strimpakos AS, Sharma RA. Curcumin: preventive and therapeutic properties in laboratory studies and clinical trials. Antioxid Redox Signaling 2008;10:511–45.

[97] Chan MM. Antimicrobial effect of resveratrol on dermatophytes and bacterial pathogens of the skin. Biochem Pharmacol 2002;63:99–104.

[98] Basnet P, Hussain H, Tho I, Škalko-Basnet N. Liposomal delivery system enhances anti-inflammatory properties of curcumin. J Pharm Sci 2012;101:598–609.

[99] Berginc K, Škalko-Basnet N, Basnet P, Kristl A. Development and evaluation of an in vitro vaginal model for assessment of drug's biopharmaceutical properties: curcumin. AAPS PharmSciTech 2012;13:1045–53.

[100] Berginc K, Suljaković S, Škalko-Basnet N, Kristl A. Mucoadhesive liposomes as new formulation for vaginal delivery of curcumin. Eur J Pharm Biopharm 2014;87:40–6.

[101] Jøraholmen MW, Škalko-Basnet N, Acharya G, Basnet P. Resveratrol-loaded liposomes for topical treatment of the vaginal inflammation and infections. Eur J Pharm Sci 2015;79:112–21.

[102] Docherty JJ, Fu MM, Hah JM, Sweet TJ, Faith SA, Booth T. Effect of resveratrol on herpes simplex virus vaginal infection in the mouse. Antiviral Res 2005;67:155–62.

[103] Houille B, Papon N, Boudesocque L, Bourdeaud E, Besseau S, Courdavault V, et al. Antifungal activity of resveratrol derivatives against *Candida* species. J Nat Prod 2014;77:1658–62.

[104] Cassano R, Ferrarelli T, Vittoria Mauro M, Cavalcanti P, Picci N, Trombino S. Preparation, characterization and in vitro activities evaluation of solid lipid nanoparticles based on PEG-40 stearate for antifungal drugs vaginal delivery. Drug Deliv 2016;23:1047–56.

[105] Ravani L, Esposito E, Bories C, Lievin-Le Moal V, Loiseau PM, Djabourov M, et al. Clotrimazole-loaded nanostructured lipid carrier hydrogels: thermal analysis and in vitro studies. Int J Pharm 2013;454:695–702.

[106] Pavelić Ž, Škalko-Basnet N, Filipović-Grčić J, Martinac A, Jalšenjak I. Development and in vitro evaluation of a liposomal vaginal delivery system for acyclovir. J Controlled Release 2005;106:34–43.

[107] Wu SY, Chang H-I, Burgess M, McMillan NAJ. Vaginal delivery of siRNA using a novel PEGylated lipoplex-entrapped alginate scaffold system. J Controlled Release 2011;155:418–26.

[108] Mourtas S, Mao J, Parsy CC, Storer R, Klepetsanis P, Antimisiaris SG. Liposomal gels for vaginal delivery of the microbicide MC-1220: preparation and in vivo vaginal toxicity and pharmacokinetics. Nano Life 2010;1:195–205.

[109] Caron M, Besson G, Etenna SL-D, Mintsa-Ndong A, Mourtas S, Radaelli A, et al. Protective properties of non-nucleoside reverse transcriptase inhibitor (MC1220)

incorporated into liposome against intravaginal challenge of *Rhesus macaques* with RT-SHIV. Virology 2010;405:225–33.

[110] Wang L, Sassi AB, Patton D, Isaacs C, Moncla BJ, Gupta P, et al. Development of a liposome microbicide formulation for vaginal delivery of octylgycerol for HIV prevention. Drug Dev Ind Pharm 2012;38:995–1007.

[111] Malavia NK, Zurakowski D, Schroeder A, Princiotto AM, Laury AR, Barash HE, et al. Liposomes for HIV prophylaxis. Biomaterials 2011;32:8663–8.

[112] Gupta PN, Pattani A, Curran RM, Kett VL, Andrews GP, Morrow RJ, et al. Development of liposome gel based formulations for intravaginal delivery of the recombinant HIV-1 envelope protein CN54gp140. Eur J Pharm Sci 2012;46:315–22.

[113] Alukda D, Sturgis T, Youan BB. Formulation of tenofovir-loaded functionalized solid lipid nanoparticles intended for HIV prevention. J Pharm Sci 2011;100:3345–56.

[114] Zhou QT, Shui Yee Leung S, Tang P, Parumasivam T, Loh ZH, Chan H-K. Inhaled formulations and pulmonary drug delivery systems for respiratory infections. Adv Drug Delivery Rev 2015;85:83–99.

[115] Ong HX, Benaouda F, Traini D, Cipolla D, Gonda I, Bebawy M, et al. In vitro and ex vivo methods predict the enhanced lung residence time of liposomal ciprofloxacin formulations for nebulization. Eur J Pharm Biopharm 2014;86:81–9.

[116] Kelly C, Jefferies C, Cryan S-A. Targeted liposomal drug delivery to monocytes and macrophages. J Drug Delivery 2011;2011 Article ID 727241.

[117] Chono S, Tanino T, Seki T, Morimoto K. Efficient drug targeting to rat alveolar macrophages by pulmonary administration of ciprofloxacin incorporated into mannosylated liposomes for treatment of respiratory intracellular parasitic infections. J Controlled Release 2008;127:50–8.

[118] Hadinoto K, Cheow WS. Nano-antibiotics in chronic lung infection therapy against *Pseudomonas aeruginosa*. Colloids Surfaces B 2014;116:772–85.

[119] LiPuma JL. The changing microbial epidemiology in cystic fibrosis. Clin Microbiol Rev 2010;23:299–323.

[120] Lyczak JB, Cannon CL, Peir GB. Lung infections associated with cystic fibrosis. Clin Microbiol Rev 2002;15:194–222.

[121] Solleti VS, Alhariri M, Halwani M, Omri A. Antimicrobial properties of liposomal azitromycin for *Pseudomonas* infections in cystic fibrosis patients. J Antimicrob Ther 2015;70:784–96.

[122] Pastor M, Moreno-Sastre M, Esquisabel A, Sand E, Vinas M, Bachiller D, et al. Sodium colistimethate loaded lipid nanocarriers for the treatment of *Pseudomonas aeruginosa* infections associated with cystic fibrosis. Int J Pharm 2014;477:485–94.

[123] Hunt BE, Weber A, Berger A, Ramsey B, Smith AL. Macromolecular mechanisms of sputum inhibition of tobramycin activity. Antimicrob Agents Chemother 1995;39:34–9.

[124] Prayle A, Smyth AR. Aminoglycoside use in cystic fibrosis: therapeutic strategies and toxicity. Curr Opin Pulmunary Med 2010;16:604–10.

[125] Lagacé J, Dubreuil M, Montplaisir S. Liposome-encapsulated antibiotics: preparation, drug release and antimicrobial activity against *Pseudomonas aeruginosa*. J Microencapsulation 1991;8:53–61.

[126] Omri A, Beaulac C, Bouhajib M, Montplasir S, Sharkawai M, Lagace J. Pulmonary retention of free and liposome-encapsulated tobramycuin after

intratracheal administration in uninfected rats and rats infected with *Pseudomonas aeruginosa*. Antimicrob Agents Chemother 1994;38:1090–5.

[127] Bruinenberg P, Blanchard JD, Cipolla DC, Dayton F, Mudumba S, Gonda I. Inhlaed liposomal ciprofloxacin; once a day management of respiratory infections. Respir Drug Delivery 2010;1:73–82.

[128] Serisier DJ, Bilton D, De Soya A, Thompson PJ, Kolbe J, Greville HW, the ORBIT-2 investigators Inhaled, dual release liposomal ciprofloxacin in non-cystic fibrosis bronchestasis (ORBIT-2): a randomized, double-blind, placebo-controlled trial. Thorax 2013;68:812–7.

[129] Clancy JP, Dupont L, Konstan MW, Billings J, Fustik S, Goss CH, for Arikace Study Group Phase II studies of nebulized Arikace in CF patients with *Pseudomonas aeruginosa* infection. Thorax 2013;68:818–25.

[130] Alhajlan M, Alhariri A, Omri A. Efficacy and safety of liposomal claritomycin and its effect on *Pseudomonas aeruginosa* virulence factors. Antimicrob Agents Chemother 2013;57:2694–704.

[131] Drulis-Kawa Z, Gubernator J, Dorotkiewics-Jach A, Doroszkiewics W, Kozuber A. A comparison of the in vitro antimicrobial activity of liposome containing meropenem and gentamicin. Cell Mol Biol Lett 2006;11:360–75.

[132] Beaulac C, Clement-Major S, Hawari J, Lagace J. In vitro kinetics of drug release and pulmonary retention of microencapsulated antibiotic in liposomal formulations in relation to the lipid composition. J Microencapsulation 1997;14:335–48.

[133] Ma Y, Wang Z, Zhao W, Lu T, Wang R, Mei Q, et al. Enhanced bactericidal potency of nanoliposomes by modification of the fusion activity between liposomes and bacterium. Int J Nanomed 2013;8:2351–60.

[134] Sachetelli S, Beaulac C, Riffon R, Lagacé J. Evaluation of the pulmonary and systemic immunogenicity of fluidosomes, a fluid liposomal-tobramycicn formulation for the treatment of chronic infections in lungs. Biochim Biophys Acta 1999;1428:334–40.

[135] Chattopadhyay S, Ehrman SH, Bellare J, Venkataraman C. Morphology and bilayer integrity of small lipsoomes during aerosol generation by air-jet nebulization. J Nanoparticle Res 2012;14:779.

[136] Beaulac C, Sachatelli S, Lagace J. Aerosolization of low phase transition temperature liposomal tobramycin as a dry powder in an animal model of chronic pulmonary infection caused by *Pseudomonas aeruginosa*. J Drug Targeting 1999;7:33–41.

[137] Meers P, Neville M, Malinin V, Scotto AW, Sardaryan G, Kurumunda R, et al. Biofilm penetration, triggered release and in vivo activity of inhaled liposomal amikacin in chronic *Pseudomonas aeruginosa* lung infections. J Antimicrob Chemother 2008;61:859–68.

[138] Okusanya OO, Bhavnani SM, Hammel J, Minic P, Dupont LJ, Forrest A, et al. Pharmacokinetic and pharmacodynamics evaluation of liposomal amikacin for inhalation of cystic fibrosis patients with chronic pseudomonal infection. Antimicrob Agents Chemother 2009;53:3847–54.

[139] Halwani M, Yebio B, Suntres ZE, Alipour M, Azghani AO, Omri A. Co-encapsulation of gallium with gentamicin in liposomes enhances antimicrobial activity of gentamicin against *Pseudomonas aeruginosa*. J Antimicrob Chemother 2008;62:1291–7.

[140] Rojanarat W, Changsan N, Tawithong E, Pinsuwan S, Chan H-K, Srichana T. Isoniazid proliposome powders for inhalation – preparation, characterization and cell culture studies. Int J Mol Sci 2011;12:4414–34.

[141] Kuiper L, Ruijgrok EJ. A review on the clinical use of inhaled amphotericin B. J Aerosol Med Pulmonary Drug Delivery 2009;22:213–27.

[142] Fauvel M, Farrugia C, Tsapis N, Gueutin C, Cabart O, Bories C, et al. Aerosolized liposomal amphotericin B: prediction of lung deposition, in vitro uptake and cytotoxicity. Int J Pharm 2012;436:106–10.

[143] Pandey R, Khuller GK. Solid lipid particle-based inhalable sustained drug delivery system against experimental tuberculosis. Tuberculosis 2005;85:227–34.

[144] Gan L, Wang J, Jiang M, Bartlett H, Ouyang D, Eperjesi F, et al. Recent advances in topical ophthalmic drug delivery with lipid-based nanocarriers. Drug Discovery Today 2013;8:290–7.

[145] Mishra GP, Bagai M, Tamboli V, Mitra AK. Recent applications of liposomes in ophthalmic drug delivery. J Drug Delivery 2011 article ID 863734, http://dx.doi.org/10.1155/2011/863734.

[146] Taha EI, El-Anazi MH, El-Bagory IM, Bayomi MA. Design of liposomal colloidal systems for ocular delivery of ciprofloxacin. Saudi Pharm J 2014;22:231–9.

[147] Cortesi R, Argnan IR, Esposito E, Dalpiaz A, Scatturin A, Bortolotti F, et al. Cationic liposomes as potential carriers for ocular administration of peptides with anti-herpetic activity. Int J Pharm 2006;317:90–100.

[148] Fresta M, Panico AM, Bucolo D, Giannavola C, Puglisi G. Characterization and in vivo ocular absorption of liposome-encapsulated acyclovir. J Pharm Pharmacol 1999;51:565–76.

[149] Law SL, Huang KJ. Properties of acyclovir-containing liposomes for potential ocular delivery. Int J Pharm 1998;161:253–9.

[150] Budai L, Budai M, Grof P, Beni S, Noszal B, Klebovich I, et al. Gels and liposomes in optimized ocular drug delivery: studies on ciprofloxacin formulations. Int J Pharm 2007;343:34–40.

[151] Law SL, Huang KJ, Chiang CH. Acyclovir-containing liposomes for potential ocular delivery: corneal penetration and absorption. J Controlled Release 2000;63:135–40.

[152] Chetoni P, Rossi S, Burgalassi S, Monti D, Mariotti S, Saettone MF. Comparison of liposome-encapsulated acyclovir with acyclovir ointment: ocular pharmacokinetics in rabbits. J Ocular Pharmacol Ther 2004;20:169–77.

[153] Seyfoddin A, Al-Kassas R. Development of solid lipid nanoparticles and nanostructured lipid carriers for improving ocular delivery of acyclovir. Drug Dev Ind Pharm 2013;39:508–19.

[154] Mehanna MM, Elmaradny HA, Samaha MW. Mucoadhesive liposomes as ocular delivery system: physical, microbiological, and in vivo assessment. Drug Dev Ind Pharm 2010;36:108–18.

[155] Kaur IP, Rana C, Singh H. Development of effective ocular preparations of antifungal agents. J Ocular Pharmacol Ther 2008;24:481–93.

[156] Kaur IP, Kakkar S. Topical delivery of antifungal agents. Exp Opin Drug Delivery 2010;7:1303–27.

[157] Habib FS, Fouad EA, Abdel-Rhaman MS, Fathalla D. Liposomes as an ocular delivery system of fluconazole: in-vitro studies. Acta Ophthalmol 2010;88:901–4.

[158] Kakkar S, Karuppayil SM, Raut JS, Giansanti F, Papucci L, Schiavone N, et al. Lipid-polyethylene glycol based nano-ocular formulation of ketoconazole. Int J Pharm 2015;495:276–89.

[159] Kalam MA, Sultana Y, Ali A, Aqil M, Mishra AK, Chuttani K. Preparation, characterization, and evaluation of gatifloxacin loaded solid lipid nanoparticles as colloidal ocular drug delivery system. J Drug Targeting 2010;18:191–204.

[160] Philstrom BL, Michalowicz BS, Johnson NW. Periodontal diseases. Lancet 2005;366:1809–20.

[161] Zupančič Š, Kocbek P, Baumgartner S, Kristl J. Contribution of nanotechnology to improved treatment of periodontal disease. Curr Pharm Des 2015;21:3257–71.

[162] Haffajee AD, Bogren A, Hasturk H, Feres M, Lopez NJ, Socransky SS. Subgingival microbiota of chronic periodontitis subjects from different geographic locations. J Clin Periodontol 2004;31:996–1002.

[163] Di Turi G, Riggio C, Marconcini S, Briguglio F, Funel N, Capmani D, et al. Submicrometric liposomes as drug delivery systems in the treatment of periodontitis. Int J Immunopathol Pharmacol. 2012;25:657–70.

[164] Nguyen S, Hiorth M, Rykke M, Smistad G. The potential of liposomes as dental drug delivery system. Eur J Pharm Biopharm 2011;77:75–83.

[165] Longo JP, Leal SC, Simioni AR, DE Fatima Menezes Almeida-Santos M, Tedesco AC, Azevedo RB. Photodynamic therapy disinfection of carious tissue mediate by aluminum-chloride-phthalocyanine entrapped in cationic liposomes; an in vitro clinical study. Lasers Med Sci 2012;27:575–84.

[166] Petelin M, Šentjurc M, Stolič Z, Skalerič U. EPR study of mucoadhesive ointments for delivery of liposomes into the oral mucosa. Int J Pharm 1998;173:193–202.

[167] Sherman MR, Williams LD, Sobczyk MA, Michaels SJ, Saifer GP. Role of the methoxy group in immune responses to mPEG-protein conjugates. Bioconjugate Chem 2012;23:485–99.

[168] Moghimi SM, Andersen AJ, Hashemi SH, Lettiero B, Ahmadvand D, Hunter AC, et al. Complement activation cascade triggered by PEG-PL engineered nanomedicine and carbon nanotubes: the challenges ahead. J Controlled Release 2010;146:175–81.

[169] Colzi I, Troyan AN, Perito B, Casalone E, Romoli R, Piaraccini G, et al. Antibiotic delivery by liposomes from prokaryotic microorganisms: similia cum similis works better. Eur J Pharm Biopharm 2015;94:411–8.

Chapter 6

Organic Polymeric Nanomaterials as Advanced Tools in the Fight Against Antibiotic-Resistant Infections

P.C. Balaure, D. Gudovan and I. Gudovan

University Politehnica of Bucharest, Bucharest, Romania

CHAPTER OUTLINE

Functionalized Nanomaterials for the Management of Microbial Infection.

6.1 INTRODUCTION: ANTIBIOTIC RESISTANCE—A GLOBAL THREAT TO PUBLIC HEALTH

At the beginning of the 20th century diseases created by bacteria were the leading cause of morbidity and mortality worldwide. Two breakthroughs in infectious diseases treatment, the discovery of salvarsan by Paul Ehrlich in 1909 [1] and the somewhat serendipitous discovery of penicillin by Alexander Fleming in 1928 [2], marked the birth of the antimicrobial era. Penicillin discovery was followed by the discovery, large-scale production, and intensive use of many other antibiotics. It was thought for a while that these new medicines could control and eventually eradicate bacterial diseases. However, it was soon noticed that bacteria are able to adapt and protect themselves by the development of antibiotic-resistant mechanisms. Scientists have identified so far four main types of mechanisms through which bacteria can acquire drug resistance. First, the interaction of the antibiotic with its specific biomolecular target can be prevented either by mutational alteration of the target protein [3], or by binding some other proteins to the target thereby changing target conformation and impeding the association of the antibiotic [4,5]. Second, the antibiotic itself can be enzymatically modified or degraded [6]. Third, overexpression of bacterial drug efflux pumps combined with reduced porin expression causing decreased drug permeability was also found to be associated with antibiotic resistance [7–9]. Fourth, acquired resistance genes may cause bacteria to develop alternative pathways for their metabolic/growth requirements that bypass the reaction inhibited by the antibiotic [10]. These self-defense mechanisms enable initially antibiotic-sensitive bacteria to evolve into drug resistant strains among which we mention methicillin-resistant *Staphylococcus aureus* (MRSA) [11], vancomycin-resistant *Enterococcus* (VRE) [12], vancomycin-resistant *S. aureus* (VRSA) [13], and penicillin-resistant *Streptococcus pneumonia* (PRSP) [14a]. Recently, bacteria resistant to the drug of the last resort—colistin—have been identified in patients and livestock in China [14b]. Spreading of these life-threatening pathogens became a major health problem and the crisis was further aggravated by pandrug-resistant and multidrug-resistant organisms [15–17]. According to a recent report from the World Health Organization (WHO) in 2013 there were an estimated 480,000 new cases of multidrug-resistant tuberculosis (MDR-TB) in the world [16]. The Centers for Disease Control and Prevention (CDC) in the United States estimate that more than 2 million people get antibiotic-resistant infections each year and at least 23,000 die because current drugs no longer stop their infections [18]. Obviously, the situation is even more dramatic in the developing countries.

Another challenge that modern medicine faces is the development of antibiotic-resistant biofilms on the surface of implanted prostheses and devices [19,20]. Soon after implantation, a biofouling process begins which results in the formation of a so-called conditioning film on the implant surface. This film consists of small organic compounds and macromolecules like proteins and polysaccharides. Next, bacterial cells adhesion occurs. The latter is a two-step process: phase (I) transient reversible attachment through weak noncovalent intermolecular forces and phase (II) irreversible attachment due to covalent bridging with the substrate mediated by surface adhesin compounds [21]. Beyond phase (II), formation of microcolonies takes place and biofilm maturation begins. A cell-to-cell signaling mechanism known as quorum sensing is involved in the further formation of a mature pathogenic bacterial biofilm with a 3-dimensional structure containing cells packed in clusters with channels between the clusters that allow transport of water and nutrients and waste removal [22–24]. Biofilms are enclosed within an exopolymer matrix also called glycocalyx composed mainly of fibrillar polysaccharides and globular glycoproteins that are secreted by bacteria during biofilm formation. The glycocalyx provides protection for biofilm cells against the immune response of the host and may also restrict the diffusion of substances and bind some antimicrobials [25]. Quorum sensing is a regulatory mechanism that enables bacteria to control the expression of specific genes in a cell-density dependent manner [23]. Therefore, pathogens in biofilms are phenotypically very different from planktonic, free-floating bacteria [26,27] and as the antimicrobial resistance of biofilms is largely phenotypic this could explain along with other factors the extraordinary resistance of biofilms to antimicrobial agents. A 1000-fold decrease in antibiotic susceptibility was reported for bacterial cells in biofilms as compared to cells in a suspension [28]. After biofilm maturation, formation of new biofilms can be initiated through detachment and dispersion of cells from the previously formed biofilm. Pathogenic biofilms including *Enterococcus faecalis, S. aureus, Staphylococcus epidermidis, Streptococcus viridans, Escherichia coli, Klebsiella pneumoniae, Proteus mirabilis*, and *Pseudomonas aeruginosa* can develop on practically any of the indwelling medical devices [19,20,29] like intravenous catheters [30], vascular prosthesis [31], cerebrospinal fluid shunts [32], prosthetic heart valves [29], urinary catheters [29], joint prostheses and orthopedic fixation devices [33], cardiac pacemakers [34], peritoneal dialysis catheters [35], intrauterine devices [36], biliary tract stents [37], dentures [38], breast implants [39], and contact lenses [40]. For instance, *S. aureus* and *S. epidermidis* readily colonize central venous catheters (CVCs) and subsequently enter into the bloodstream, thereby causing about 50–70% of the catheter-related bloodstream infections [41]. It was estimated that about 250,000–500,000 intravascular devices-associated bloodstream infections occur each year in the United States causing prolonged hospital stay and the costs associated with the treatment of these infections range from $4000 to $56,000 per episode [42]. Furthermore, nosocomial (hospital acquired) infections are the fourth leading cause of death in the United States with 99,000 deaths annually, the majority of them (60–70%) being associated with some type of implanted medical device [43]. The costs for the treatment of these infections rose to more than $5 billion in 2007 [44] and in many cases the only solution to save patients' lives was the surgical removal of the infected implant which in turn contributed to increased morbidity in patients and generated supplemental costs. The European Centre for Disease Prevention and Control (ECDC) estimates that about 4,100,000 patients acquire a nosocomial infection in the European Union each year and these infections contribute to an additional 110,000 deaths each year [45]. But pathogenic bacteria can adhere to and colonize not only the surface of medical devices; they also form biofilms on food and food contact surfaces thereby causing food safety concerns [22].

For the reasons outlined above we can conclude that antibiotic resistance became one of the greatest global health threats of the 21st century and as a global problem it requires global solutions. Nanotechnology joined the efforts of finding solutions to address this complex problem. Since nanotechnology enables manipulation of materials structure and properties at the nanoscale, it is a revolutionary approach that opens the door toward innovative and more effective treatments of infectious diseases.

6.2 NANOANTIBIOTICS—A NEW PARADIGM TO FIGHT AGAINST ANTIBIOTIC RESISTANCE BACTERIA

There are two main ways to overcome antibiotic resistance.

One obvious way is to develop new antibiotics. However, most of the molecular targets of antibiotics ensuring selective toxicity have been already discovered [46] and therefore the development of new antibiotics became more complex, more expensive, and more time consuming taking 10–12 years. It is rather hard to believe that the development of new antibiotics could catch up to the fast and frequent development of drug resistance by microbial pathogens [47] and it seems more likely that antibiotic resistance is inevitable [48]. On the other hand, taking into account that antibiotics are administered for a limited period of time, from an economical point of view they look less profitable for pharmaceutical companies than the medication addressing chronic conditions such as cancer, neurological, cardiovascular, musculoskeletal, or psychological disorders, and this could be another explanation for the decline in the introduction of new antibiotics on the pharmaceutical market [49]. Over the past three decades there has been a steady decrease in the number of new antibiotics developed and approved in the United States (see Fig. 6.1B). On the other hand, increasing bacterial resistance trends were apparent during the same period of time as illustrated in Fig. 6.1A for the case of *E. coli* [50a-c]. In a report released in 2009 and entitled "The

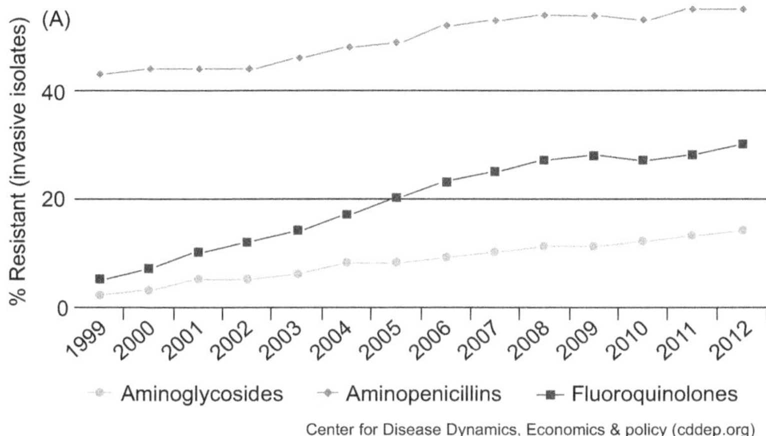

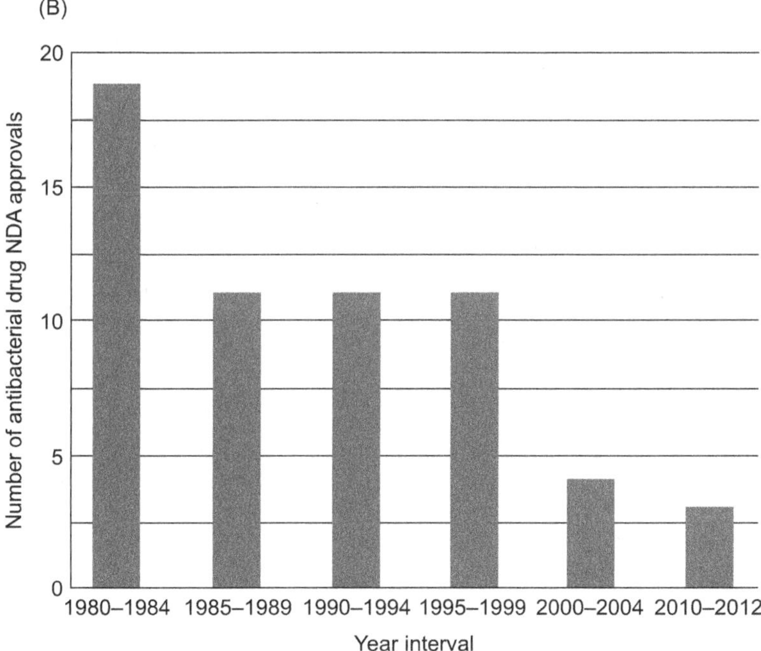

■ **FIGURE 6.1** Trends in the development of antibiotic resistance and in the number of new antibacterial drugs entering the market place in the United States. (A) Antibiotic resistance of *E. coli* in the United States. (B) Number of antibacterial new drug application (NDA) approvals versus year intervals in the United States.

bacterial challenge: time to react," ECDC and The European Medicines Agency (EMEA) identified a gap between multidrug-resistant bacteria in the European Union (EU) and the development of new antibacterial agents [50d]. The report predicted that particular problems caused by Gram-negative bacteria would arise in the future since there is a lack of new agents to treat these infections. A more recent report confirmed these concerns as illustrated by the trends of the percentage of infections with third-generation cephalosporins resistant *E. coli* in the EU countries. As revealed by data collected from the 29 reporting countries, the EU/European economic area (EEA) population-weighted mean percentage for third-generation cephalosporins resistance in *E. coli* increased significantly from 2010 to 2013 (see Table 6.1). Moreover, in 17 European countries substantial increases in the percentages of infections with this cephalosporin-resistant bacterium were observed [50e]. Furthermore, currently used antibiotics have also a series of disadvantages such as:

1. Promote development of resistant bacterial strains
2. Some, e.g. aminoglycosides, tetracyclines, chloramphenicol, and vancomycin, have a low therapeutic index and require patient monitoring due to severe toxicity and side effects
3. May mediate enhancement of bacterial virulence [51]
4. When presented as conventional dosage forms, antibiotics suffer from poor absorption, nonspecific body distribution associated with systemic side-effects and they also may be degraded and inactivated before reaching their targets

Table 6.1 *Escherichia coli*. Total number of invasive isolated tested (N) and percentage resistant to third-generation cephalosporins (%R), including 95% confidence intervals (95% CI) for EU/EEA population weighted mean and in Romania

| Country | 2010 | | | 2011 | | | 2012 | | | 2013 | | | | Trend 2010–2013 | Comment** |
	N	%R	(95% CI)	N	%R	(95% CI)	N	%R	(95% CI)	N	%R	(95% CI)			
EU/EEA (population-weighted mean)		9.5	(9–10)		9.6	(9–11)		11.9	(11–13)		12.6	(12–14)			>
Romania	34	20.6	(9–38)	95	21.1	(13–31)	191	25.1	(19–32)	298	22.8	(18–28)			

Reproduction of data from European Centre for Disease Prevention and Control (ECDC). Antimicrobial resistance surveillance in Europe. Annual report of the European Antimicrobial Resistance Surveillance Network (EARS-Net) 2013. ISNB 978-92-9193-603-8, Stockholm: ECDC, 2014, <http://dx.doi.org/10.2900/3977> is acknowledged.
The symbol > indicates a significant increasing trend.

The other way to fight antibiotic resistance is the development of new modified-release dosage forms that enable controlled and site-specific delivery of existing antibiotics. Nanosized drug delivery systems (nanoDDS) provide multiple possibilities to enhance the efficiency of anti-infectious therapy since they are able to overcome both the drawbacks associated with conventional dosage forms and the drug resistance mechanisms of bacteria. As compared to unmodified dosage forms of antibiotics, nanoDDS display several advantages such as: (1) increase the bioavailability of poorly water soluble antibiotics by increasing their solubility; (2) prevent effective opsonization by the reticuloendothelial system (RES) through pegylation of antibiotic loaded nanoparticles; (3) protect the antibiotic payload against premature degradation and inactivation [52]. The above three features contribute to much improved pharmacokinetic and pharmacodynamic properties of nanoDDS encapsulated antibiotics. With respect to the ability to overcome drug resistance mechanisms some remarkable features of nanoDDS need to be highlighted. First, nanoDDS are able to codeliver simultaneously several antimicrobials acting by various mechanisms. This makes resistance development unlikely since multiple simultaneous gene mutations in the same bacterial cell would be required [53–56,47]. Second, some nanoparticle platforms like liposomes and miosomes are able to fuse with the membrane of the bacterial cell and to release into the cytoplasm a high concentration of drug all at once thereby saturating transmembrane efflux pumps and bypassing the resistance mechanism of increased drug efflux and decreased drug uptake [53]. Third, nanoDDS provide multiple possibilities of functionalization by covalent attachment of several types of targeting ligands such as antibodies, antibody segments (nanobodies), aptamers, peptides, and small molecules that selectively bind receptors at the site of infection. Thereby, higher doses of antibiotic can be selectively delivered at the infected site resulting in the decrease of adverse effects concomitant with a significant increase of the therapeutic efficacy. In addition to this active targeting mechanism, nanoDDS can be also passively targeted to the infected sites taking advantage of the so-called enhanced permeability and retention (EPR) effect that is a consequence of the increased blood vessel permeability at the sites of inflammation [57,58]. Biofilm development can be prevented by using nanoDDS as antibiotic-releasing coatings on catheters and other implanted medical devices [59,60]. Due to their nanoscale dimensions, nanoDDS can be readily internalized by phagocytosis and this ability underlies their capacity to overcome another drug resistance mechanism, i.e., intracellular bacteria. Intracellular bacteria are those bacteria that have acquired the capacity to survive and replicate inside mononuclear phagocytes and also some other types of host cells. Such bacteria have developed complex mechanisms to evade the harsh conditions of the degradative intracellular compartments either by avoiding acidification and subsequent phagosome-lysosome fusion or by the lysis of the phagosomal membrane thereby escaping into the cytoplasm [61]. Once internalized antibiotic-loaded nanoDDS can release their content inside the infected host cells to combat intracellular bacteria. The advantage of this mode of action is the possibility of achieving high local (intracellular) concentrations of the antibiotic without the need to use high overall doses, thereby decreasing the probability of generating resistant mutants [53,62,63].

The ultimate goal of any drug delivery system is the release of its therapeutic payload. An ideal drug delivery system should provide both spatial and temporal control over the drug release process. NanoDDS can be engineered so that the drug release process is triggered in sharp response to some endogenous or exogenous stimuli [64,65]. Intrinsic triggers are specific to the pathological area such as changes in pH, temperature, redox condition, and activity of certain enzymes. Extrinsic triggers like magnetic field, ultrasound, various irradiations are locally applied at the disease site and initiate drug release. Such nanosized programmable drug delivery systems are also named "smart" since their most striking characteristic is their ability to mimic the stimuli-responsiveness feature of natural supramolecular polymers. Obviously smart nanoDDS should be built up from nontoxic, nonimmunogenic, biocompatible, and, as much as possible, biodegradable materials. Smart delivery brings a range of important benefits when compared to systemic delivery, which are summarized and schematically illustrated in Fig. 6.2.

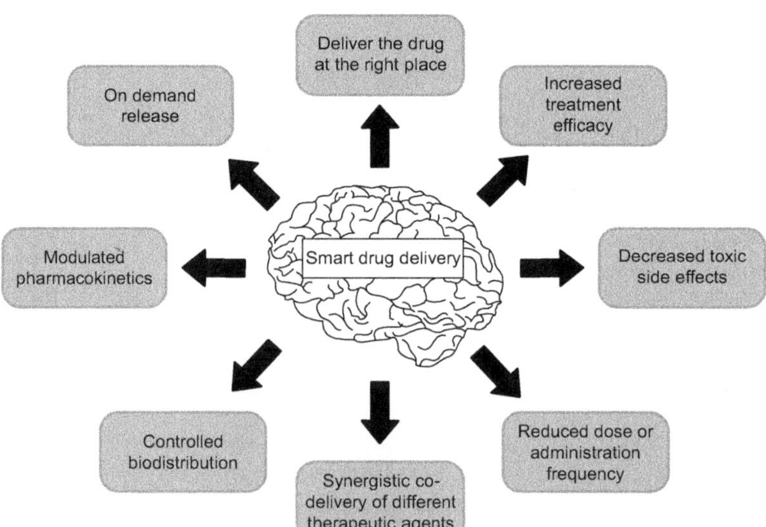

■ **FIGURE 6.2** Advantages of a smart nano drug delivery system.

Both inorganic and organic materials as well as inorganic/organic (core/shell) hybrids have been used as nanoplatforms for drug delivery. Inorganic nanoplatforms can be based on quantum dots, metals, mesoporous silica or hybrids. Organic nanoplatforms are built up of carbon nanotubes, graphenes, lipid-based materials (liposomes, niosomes, solid lipid nanoparticles, nanostructured lipid carriers), and polymer-based materials including biocompatible natural and synthetic homopolymers, amphiphilic block copolymers, polymer-drug conjugates, polymeric micelles, polymersomes, polymeric nanospheres and nanocapsules, as well as dendritic polymers [66].

Compared to other delivery platforms, polymer-based nanocarriers present a series of important advantages conferred by the special properties of polymers. We mention here: (1) structural stability in biological fluids; (2) precisely tunable properties such as size, zeta potential, and release profile that can be adjusted to meet the requirements by manipulating a series of polymer structural factors such as molecular weight, chemical composition, monomer sequence, chain conformation, degree of branching, cross-linking density, or by changing the surfactants and solvents used in the preparation of polymeric nanoparticles; (3) facile and versatile surface functionalization for covalent attachment of drug molecules and targeting ligands [47,52,67].

The term "nanoantibiotics" was introduced in the literature to designate either nanomaterials that are antimicrobials by themselves [68], or nanoplatforms, which are used for antibiotics delivery [47].

The present chapter is devoted to the presentation of nanoantibiotics based on polymeric materials. The complex molecular architecture of smart polymers, their supramolecular morphologies, the main methods by which polymeric nanoparticles can be prepared, and their most important medical applications in antiinfectious therapy will be highlighted. Antimicrobial polymers and polymer coatings used to protect medical devices against biofilm development will also be presented.

6.3 POLYMERIC ARCHITECTURES RELEVANT TO DRUG DELIVERY APPLICATIONS

Precisely localized delivery of the drug at the diseased site is one of the most challenging tasks a nanocarrier may face. Drug release from smart polymer nanocarriers is triggered in sharp response to certain stimuli such as local physiological changes in pH, redox potential, enzyme activity, temperature, or to remote physical stimuli (local application of external electric or magnetic fields, light irradiation, ultrasounds, and so on) [69–71]. In the case of internal triggering stimuli, effective drug targeting can be achieved only if the sensitivity range of the smart polymer to that particular stimulus matches the range of the stimulus physiological values [72]. The problem is that sometimes the latter range is very narrow or in other words the smart polymer must sense very small changes in its environment to target drug release [73]. For instance, if drug delivery to the interstitial space of solid tumors is intended, than the polymer must respond to changes in pH in the range of tenths of the pH unit [74]. Perhaps the most remarkable property of smart polymers is their tunable sensitivity and this is also a great advantage over the other types of nanoDDS. By applying various polymer engineering techniques the sensitivity of the polymer toward a given stimulus can be tuned up within a very narrow range [67,75]. For instance, the lower critical solution temperature (LCST), i.e., the temperature above which a thermosensitive water-soluble polymer undergoes a phase transition in water from a soluble hydrated form to an insoluble (dehydrated) state, can be modulated by modifying molecular weight, chemical composition and molecular architecture of the polymer [76,77].

The molecular architectures of the macromolecular constituents of a polymeric nanocarrier have a determining influence on its properties such as: stability in biological fluids, morphology, size and size distribution, circulation time, drug release profile, biodistribution, and endocytosis mechanism [78]. As previously mentioned, the characteristic feature of smart polymers used as drug carriers is their ability to mimic the response capabilities of biological polymers such as proteins. It is well known that proteins are able to perform their complex biological functions properly only if their amino acids chains adopt a particular spatial organization, a native 3D conformation that is specific for each protein. Similarly, the molecular architecture of synthetic polymers which describes the shape of the individual macromolecules is mainly responsible for their properties.

Four main types of polymer architectures are encountered among polymeric drug nanocarriers: linear, branched, crosslinked, and dendritic polymers (see Fig. 6.3). With respect to the chemical composition, polymers may be built up from molecules of the same monomer (homopolymers), or from molecules of different monomers (copolymers).

Therefore, linear polymers can be homopolymers (**1A** in Fig. 6.3) or copolymers. Taking into account the monomer sequence along the copolymer chain, one can distinguish several types of linear copolymers: alternating copolymers (**1E** in Fig. 6.3), gradient copolymers (**1F** in Fig. 6.3), random copolymers (**1G** in Fig. 6.3), and block copolymers (**1B–1D** in Fig. 6.3). Block copolymers are composed of several homopolymer segments joined together by covalent bonds. Depending on the number of

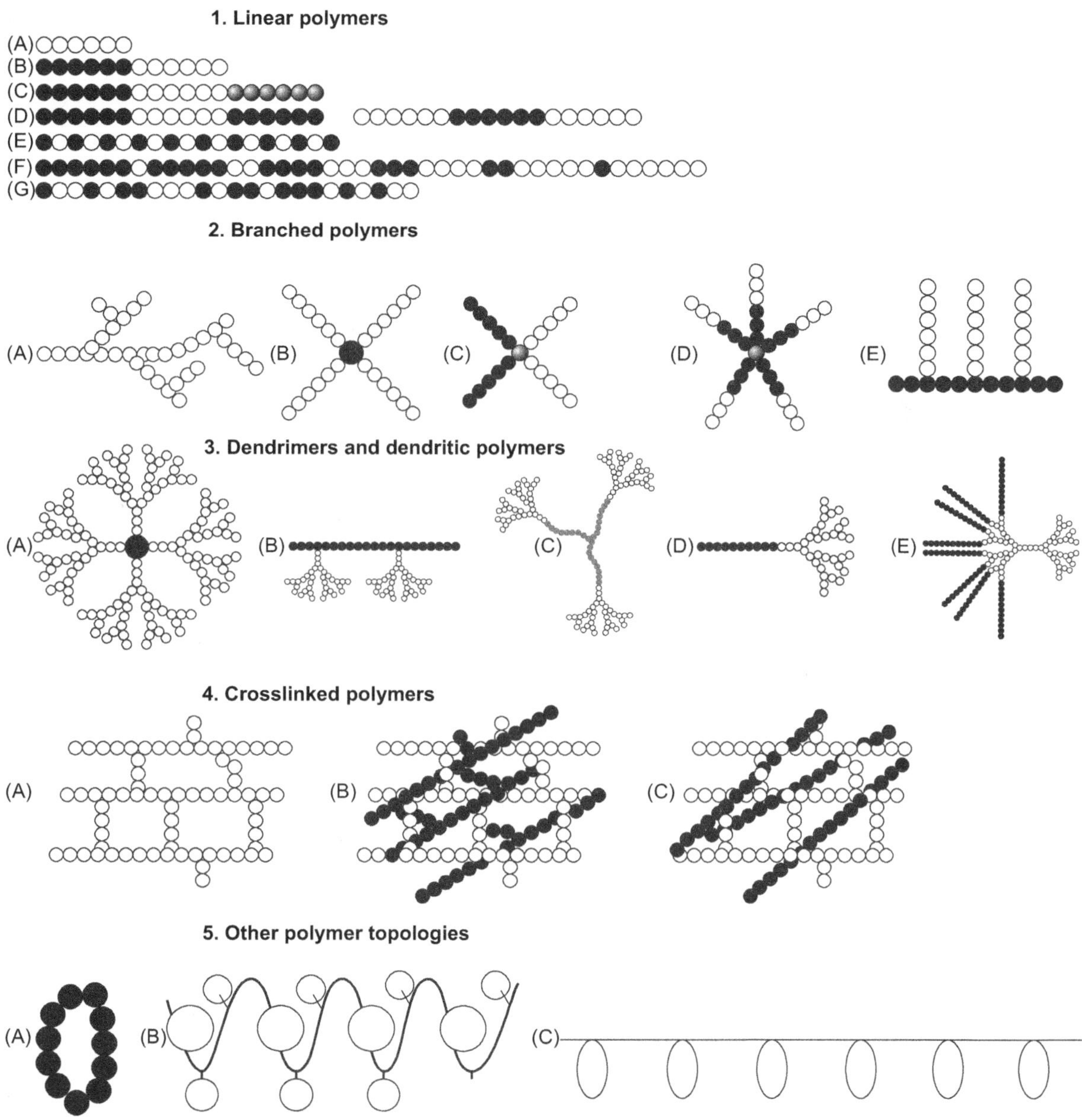

■ **FIGURE 6.3**　Classification of polymers by chemical composition and topology.

homopolymer segments, diblock, triblock, and multiblock copolymers are possible. To illustrate these possibilities a diblock copolymer (**1B**) and three types of triblock copolymers, ABC (**1C**), ABA, and, BAB respectively (**1D**), are depicted in Fig. 6.3.

Fig. 6.3 also presents a series of nonlinear polymer topologies (**2–5**). Several examples of branched polymers are given: hyperbranched polymers (**2A**), regular star polymers having the same arm segments that can be either homopolymer segments (**2B**) or diblock copolymer segments (**2D**), miktoarm star polymers in which the arm segments differ from one another (**2C**), and a particular type of comb polymers i.e., graft copolymers (**2E**). A comb polymer is composed of macromolecules consisting of a linear main chain from which a large number of oligomeric side chains emanate, whilst a graft copolymer is a comb-like macromolecule in which the chemical composition of the side chains is different from that of the main chain. A particular type of branched polymers are the dendrimers. Dendrimers have a peculiar highly organized globular tree-like architecture consisting entirely of repetitive branched units emanating from a central core (**3A** in Fig. 6.3). The combination of dendritic and linear topologies allows for many more variations of polymer architectures such as: linear polymers with one or both ends linked to the focal point of a dendron (**3D** in Fig. 6.3), dendri-graft copolymers (**3B** in Fig. 6.3), or linear polymer bow tie dendrimer hybrids (**3E** in Fig. 6.3). Similarly, if one combines the dendritic and star topologies in one macromolecule the result is a dendronized star polymer (**3C** in Fig. 6.3).

Cyclic (**5A** in Fig. 6.3) and helical (**5B** in Fig. 6.3) topologies are encountered in some naturally occurring biopolymers (peptides and proteins). A cardo polymer is composed of macromolecules consisting of a linear main chain and many ring units, each having one atom in common with the main chain (**5C** in Fig. 6.3).

In a crosslinked polymer (**4A** in Fig. 6.3), different adjacent chains are bound to one another by means of covalent bridges forming a network. An interpenetrating polymer network (**4B** in Fig. 6.3) is an intimate mixture of at least two different crosslinked polymers without any covalent binding between the networks of different types. In semiinterpenetrating polymer networks (**4C** in Fig. 6.3) only one of the two intimately mixed polymers is crosslinked while the other is not.

Fabrication of smart nanoDDS based on supramolecular polymeric materials was not possible prior to the emergence and continuous development within the last two decades of appropriate macromolecular engineering techniques such as controlled polymerization methods, postpolymerization functionalization methods or routes to control the macromolecular self-assembly. These techniques allowed the synthesis of polymers with low polydispersity indexes and with precisely controlled macromolecular architectures to become possible and opened new routes toward intricate structures such as block copolymers, star polymers, graft polymers, and dendrimers. The most successfully used techniques of controlled/living polymerization are:

1. Atom transfer radical polymerization (ATRP) [79]
2. Nitroxide-mediated polymerization (NMP) [80,81]
3. Reversible addition-fragmentation chain transfer polymerization (RAFT) [82]
4. Ring-opening polymerization (ROP) which according to the reaction mechanism can be anionic (AROP) [83], cationic (CROP) [84], radicalic (RROP) [85], metal catalyzed ("coordination-insertion" mechanism [86]), organocatalyzed, or may involve metathesis (ring opening metathesis polymerization (ROMP) [87])

The peculiar dendritic architecture also requires specific synthetic methods and there are two different approaches for the synthesis of dendrimers and dendrons—the divergent route [88] and the convergent one [89], respectively. Each of these controlled polymerization techniques is a powerful synthetic tool by itself but the combination of two or more methods exponentially increases their synthetic value and allows access to more and more complex macromolecular architectures. The detailed description of these techniques is beyond the scope of the present chapter but we do mention here a series of excellent reviews [90–93], covering the subject. In addition, Table 6.2 summarizes some of the newest and most illustrative achievements in the field; the monomers, the polymerization methods, the polymeric architectures, the properties of the polymers relevant to drug delivery applications, and the bibliographic sources are mentioned.

6.4 TYPES, STRUCTURES, AND SUPRAMOLECULAR MORPHOLOGIES OF POLYMERIC NANOCARRIERS

There are two main types of polymeric nanovehicles used for drug delivery.

One type of nanovehicles named polymer-drug conjugates consist of a single macromolecule, the drug being either covalently inserted within the main polymer chain or attached to the polymer backbone as pendant groups [78,115]. If the drug retains its whole pharmacological activity while being covalently bound to the polymer, than the polymer-drug conjugate is referred to as a polymer drug. If the drug regains its physiological activity only after it has been cleaved from the polymer, than the polymer-drug conjugate is called a macromolecular prodrug. The drug can

Table 6.2 Controlled Polymerization Methods and Polymer Architectures for Drug Delivery Applications

Monomer(s), Initiator(s), Activators, and Catalyst(s)	Polymerization Method(s)	Polymer Architecture	Essential Properties for Drug Delivery Applications/ References
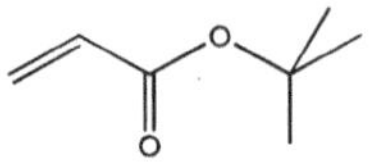 PEG initiator	PEG initiated ROP of caprolactone (CL) by coordination-insertion mechanism, esterification of terminal hydroxyl groups with 2-bromoisobutyryl bromide (Br-iBuBr), and eventually ATRP of *tert*-butyl acrylate and hydrolysis	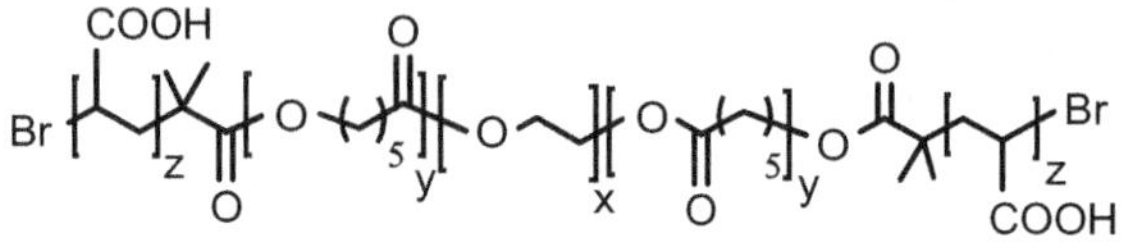 Amphiphilic pentablock copolymer	Self-assemblies in water into pH-sensitive spherical nanosized micelles/[94]
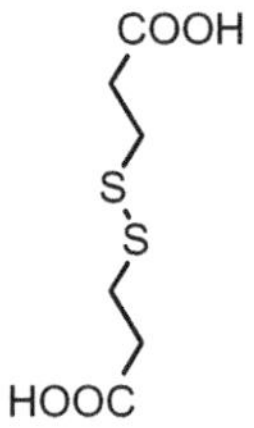 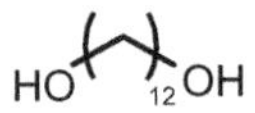 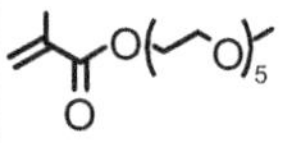	Carbodiimide mediated polycondensation of 3,3′-dithiodipropionic acid and 1,12-dodecyl diol, esterification with Br-iBuBr, and ATRP of oligo(ethylene glycol) monomethyl ether methacrylate	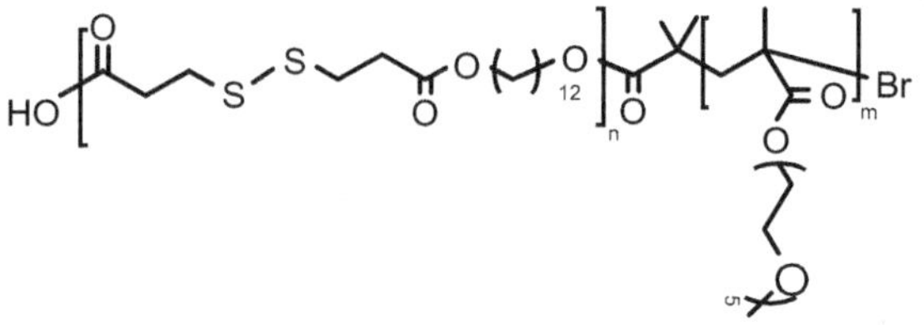 Amphiphilic block copolymer	Self-assemblies in water forming redox responsive core/shell micelles consisting of a hydrophobic polyester core surrounded with polymethacrylate corona/[95]
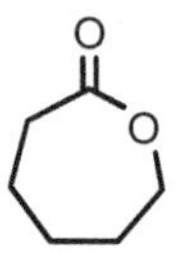 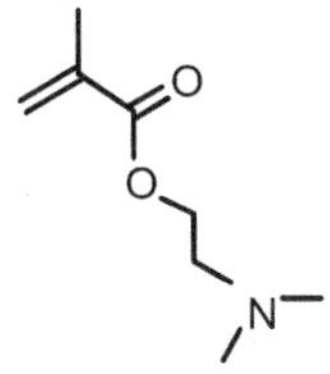 MPEG initiator	MPEG initiated ROP of caprolactone, esterification of the terminal hydroxyl group with Br-iBuBr, and ATRP of 2-(dimethylamino) ethyl methacrylate (DMAEMA)	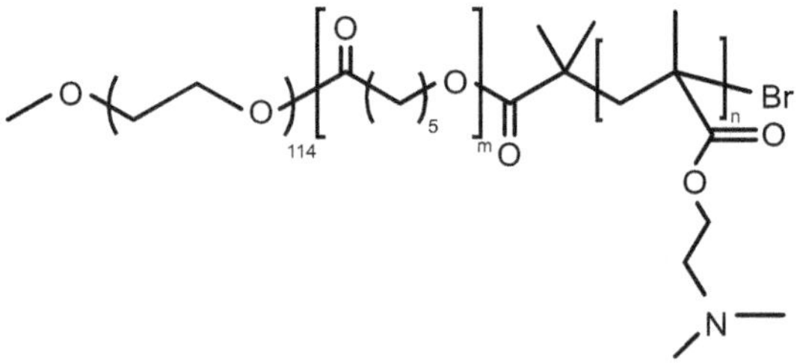 Amphiphilic triblock copolymer MPEG-*b*-PCL-*b*-PDMAEMA	Biocompatible; pH-responsive; pH-dependent critical aggregation concentrations; depending on the length of the poly[2-(dimethylamino) ethyl methacrylate segment forms in water either micelles or vesicles/[96]

(*Continued*)

Table 6.2 Controlled Polymerization Methods and Polymer Architectures for Drug Delivery Applications (Continued)

Monomer(s), Initiator(s), Activators, and Catalyst(s)	Polymerization Method(s)	Polymer Architecture	Essential Properties for Drug Delivery Applications/References
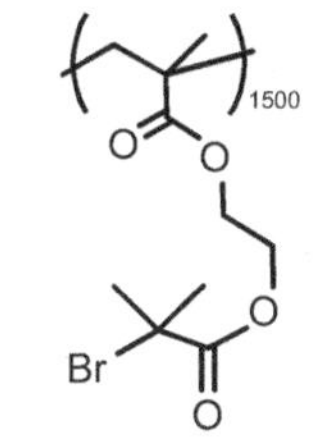 as macroinitiator for ATRRP of 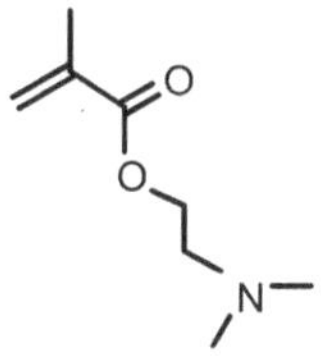	"Grafting from" polymerization of 2-dimethylaminoethyl methacrylate by ATRP	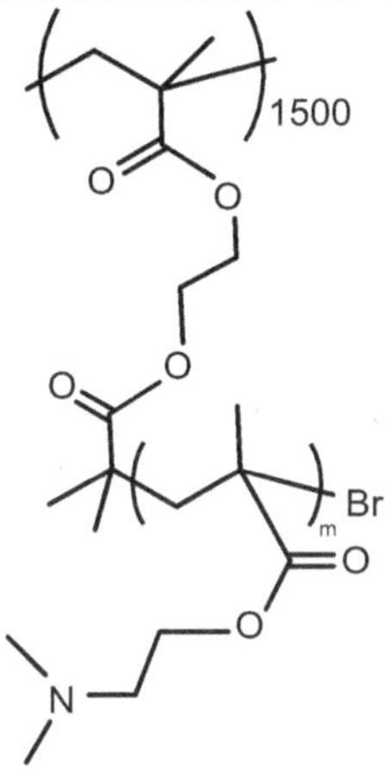 Cylindrical polymer brushes	pH and salt responsive/[97,98]
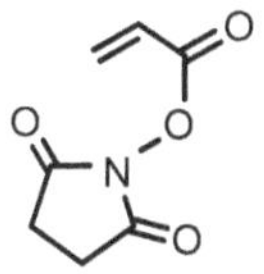 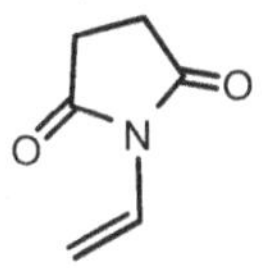 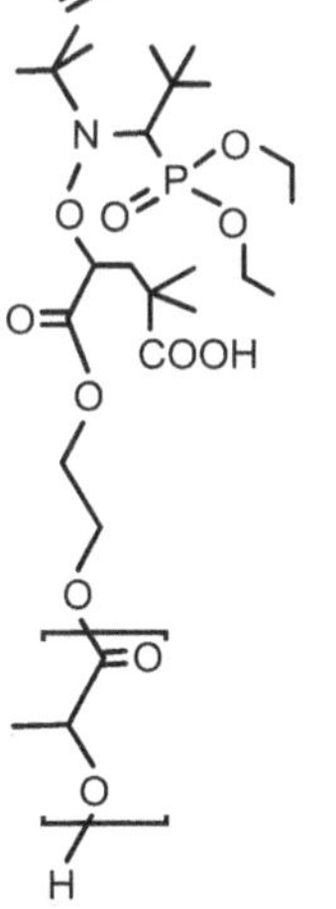 NMP macroinitiator	NMP	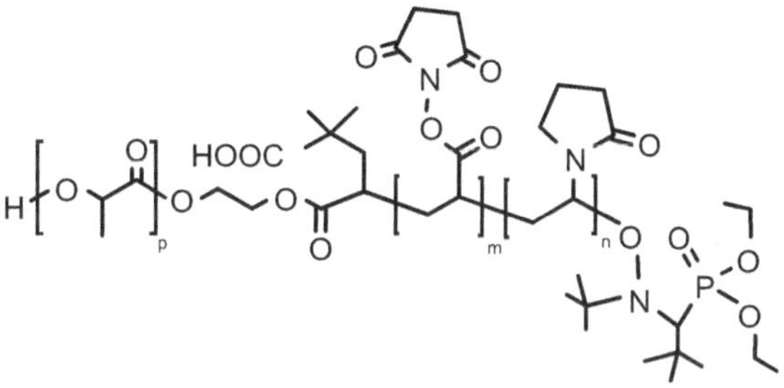 Amphiphilic polylactide (PLA)-block-(N-acryloxysuccinimide, N-vinylpyrrolidone copolymer) PLA-b-poly(NAS-co-NVP); Polylactide nanoparticles with a hydrophilic and highly functionalized surface	Possibilities of conjugation with important biomolecules such as immunostimulating molecules for vaccine delivery applications/[99]

Table 6.2 Controlled Polymerization Methods and Polymer Architectures for Drug Delivery Applications (Continued)

Monomer(s), Initiator(s), Activators, and Catalyst(s)	Polymerization Method(s)	Polymer Architecture	Essential Properties for Drug Delivery Applications/ References
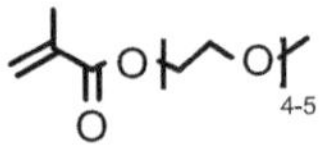 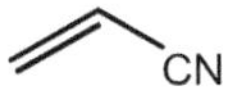 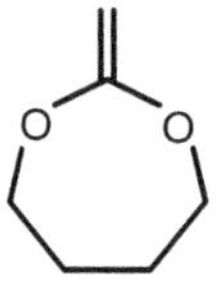 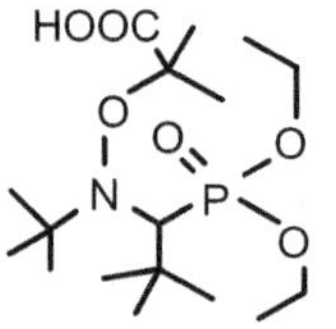 NMP initiator NMP initiator	Nitroxide-mediated radical ring-opening co-polymerization of 2-methylene-1,3-dioxepane (MDO) with oligo(ethylene glycol) methyl ether methacrylate (MeOEGMA) and acrylonitrile (AN)	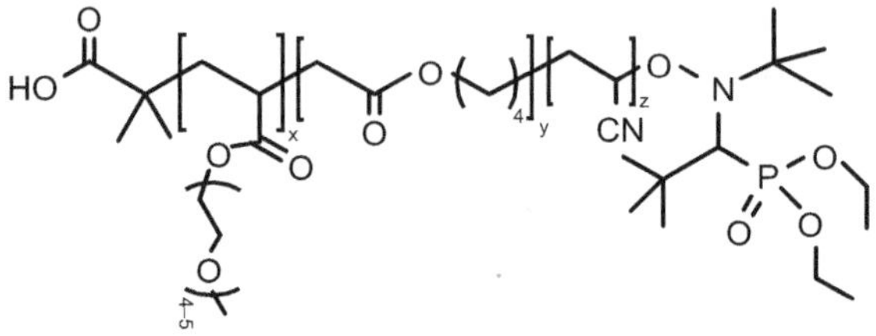 Comb-like PEG-based copolymer poly(MeOEGMA-*co*-MDO-*co*-AN)	Hydrolytically degradable; demonstrated high cell viability in cytotoxicity assays/[100,101]
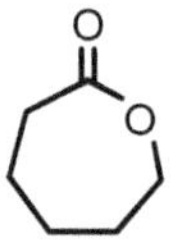 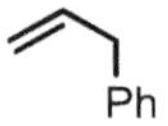 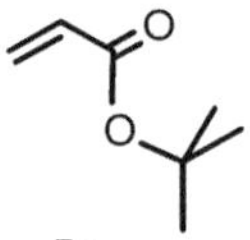 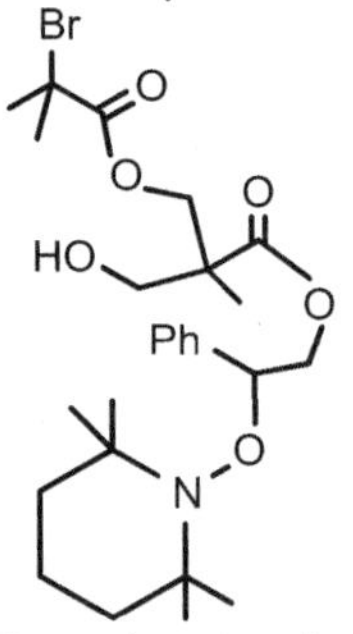 Heterofunctional initiator	Combination of ROP of caprolactone, NMP of styrene, and ATRP of *t*-butyl acrylate	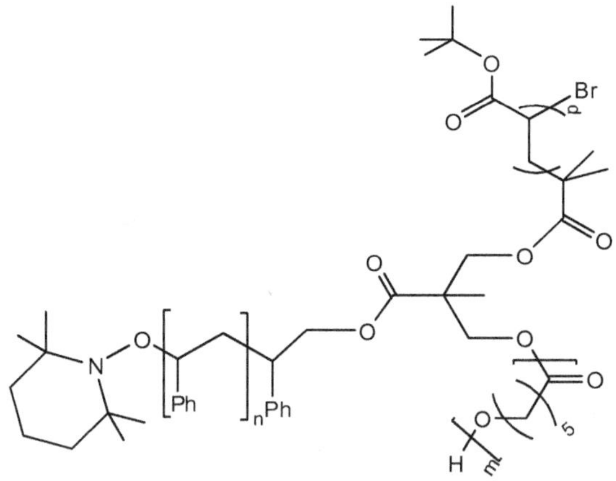 ABC-type miktoarm star copolymer	/[102,103]

(*Continued*)

Table 6.2 Controlled Polymerization Methods and Polymer Architectures for Drug Delivery Applications   (Continued)

Monomer(s), Initiator(s), Activators, and Catalyst(s)	Polymerization Method(s)	Polymer Architecture	Essential Properties for Drug Delivery Applications/ References
Initiator (AIBN) RAFT chain transfer agent	Sequential RAFT polymerization of 2-(4-oxopentanoate) ethyl methacrylate (OEMA), pyridyl disulfide ethyl acrylate (PDEA), and poly(ethylene glycol) acrylate (PEGA) using a pyrene terminated RAFT agent for subsequent attachment through π-π stacking to graphene	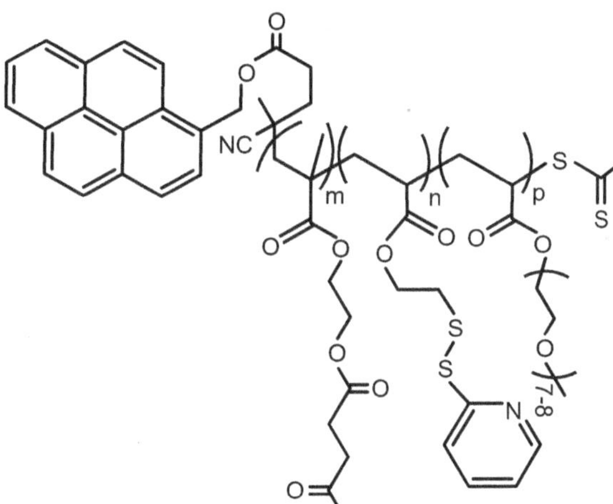 ABC-type triblock copolymer polyOEMA-*b*-polyPDEA-*b*-polyPEGA 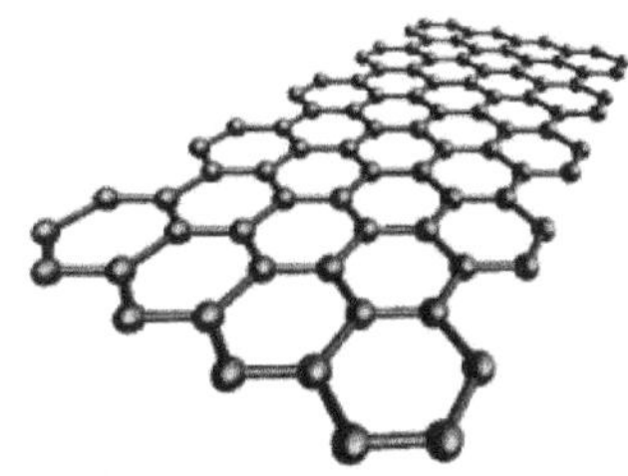 Graphene	Water soluble, biocompatible, biodegradable, contains active keto groups for drug conjugation through pH labile oxime bond, selective drug release through pH manipulation and redox processes, little toxicity to human cells/[104]

Table 6.2 Controlled Polymerization Methods and Polymer Architectures for Drug Delivery Applications   (Continued)

Monomer(s), Initiator(s), Activators, and Catalyst(s)	Polymerization Method(s)	Polymer Architecture	Essential Properties for Drug Delivery Applications/References
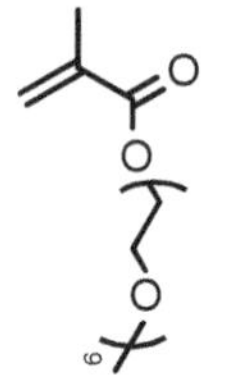 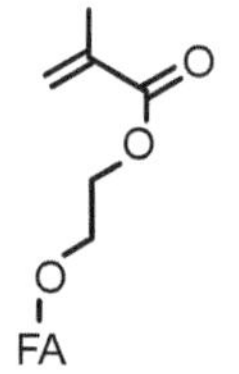 FA Initiator AIBN 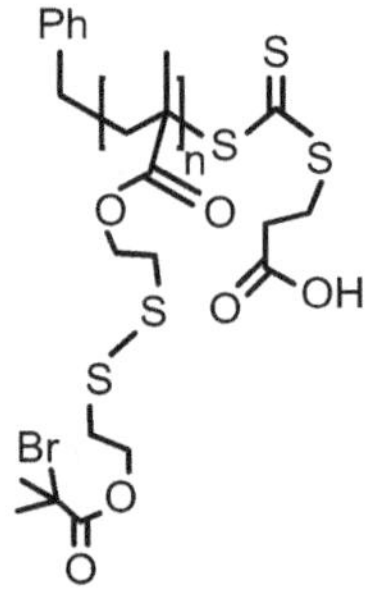 macroRAFT chain transfer agent 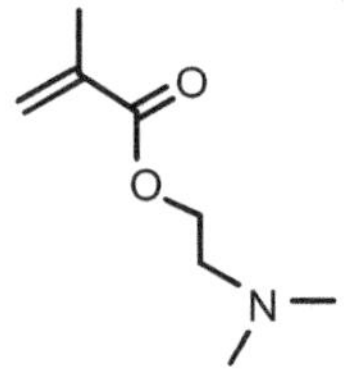	Co-polymerization of oliogoethylene glycol methacrylate (OEGMA) and folic acid (FA)-conjugated methacrylate (FAMA) monomers using a bifunctional macroRAFT chain transfer agent based on poly[2-((2-(2-bromo-2-methylpropanoyloxy)ethyl)disulfanyl) ethyl methacrylate] (PBSSMA) followed by grafting from polymerization of 2-dimethylaminoethyl methacrylate (DMAEMA) by ATRP	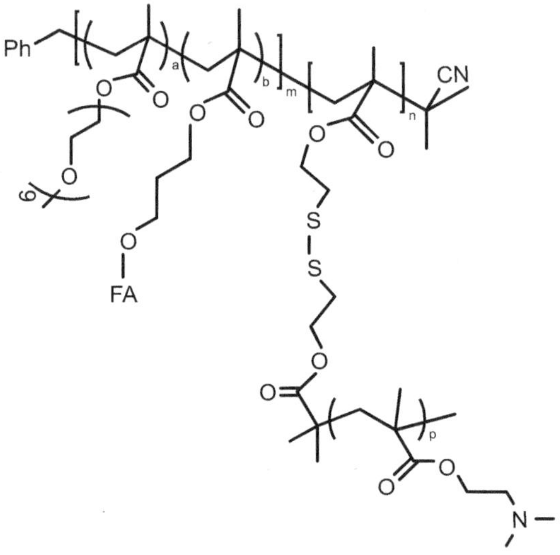 Cationic brush block copolymer poly(OEGMA-*co*-FAMA)-*b*-poly(BSSMA-*g*-PDMAEMA)	Highly functionalized polycationic nonviral intracellular delivery vector able to form complex micelles with negatively charged macromolecules such as nucleic acids or with anionic drugs. PEG chains impart high stability in the bloodstream by sterically hindering interaction with plasma proteins. Further functionality is brought to these long circulating nanocarriers by the conjugation to the targeting ligand folate. Eventually, the disulfide bridges are redox responsive and the nanocarrier is prone to biodegradation in the cytoplasm due to much higher concentration of glutathione as compared to the extracellular milieu. The MTT assay showed low cytotoxicity for the brush block copolymer/[105]

(Continued)

Table 6.2 Controlled Polymerization Methods and Polymer Architectures for Drug Delivery Applications (Continued)

Monomer(s), Initiator(s), Activators, and Catalyst(s)	Polymerization Method(s)	Polymer Architecture	Essential Properties for Drug Delivery Applications/ References
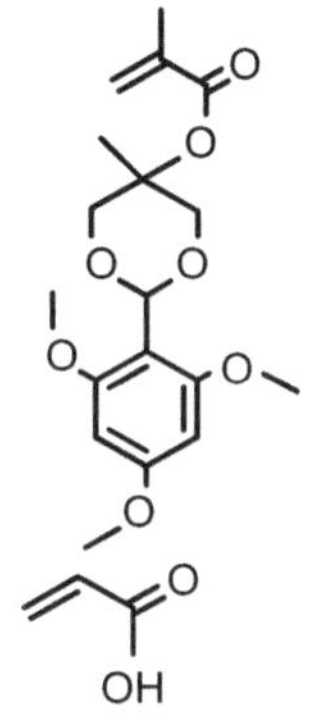 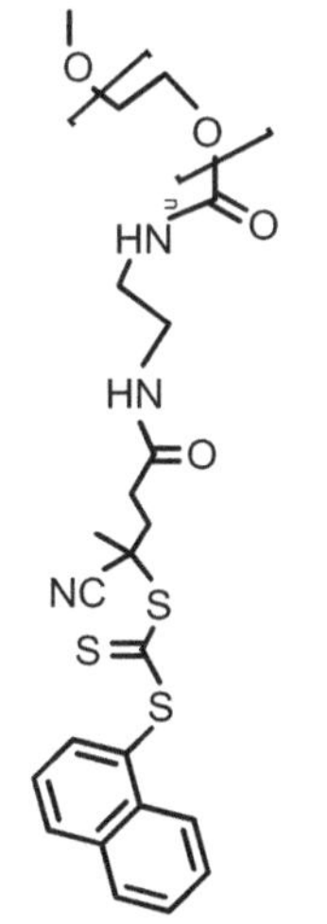 (see structures) Initiator AIBN (see structure) MPEG-CPADN macroRAFT chain transfer agent	Sequential RAFT polymerization of 2,4,6-trimethoxybenzy lidene-1,1,1-tris(hydroxymethyl) ethanemethacrylate (TTMA) and acrylic acid (AA) using a methoxy poly(ethylene glycol) (MPEG) functionalized macroRAFT chain transfer agent based on 4-cyanopentanoic acid dithionaphthalenoate (CPADN), MPEG-CPADN	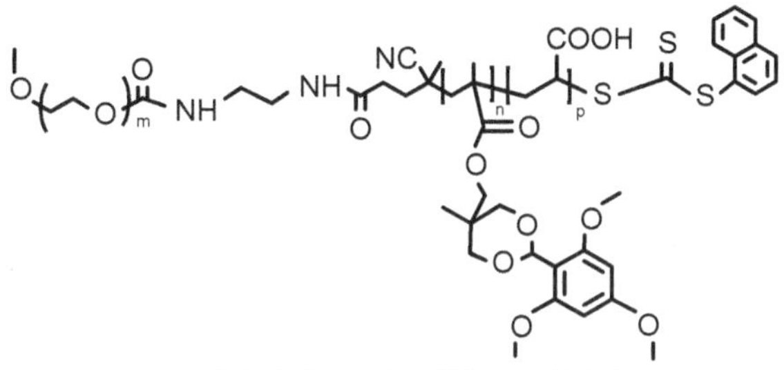 (see structure) Asymmetric poly(ethylene glycol)-*b*-poly(2,4,6-trimethoxybenzylidene-1,1,1-tris(hydroxymethyl)ethane methacrylate)-*b*-poly(-acrylic acid) (MPEG-*b*-PTTMA-*b*-PAA) triblock copolymers	Self-assemblies forming pH-sensitive and hydrolytically degradable chimeric polymersomes; excellent biocompatibility due to the "stealth" properties of the MPEG chains exposed on the outer surface of the polymersomes; high drug loading efficiency; targeted intracellular delivery due to endosomal acid-triggered hydrolysis of acetal groups in the membrane of the polymersomes; lack of toxicity/[106]

Monomer(s), Initiator(s), Activators, and Catalyst(s)	Polymerization Method(s)	Polymer Architecture	Essential Properties for Drug Delivery Applications/ References
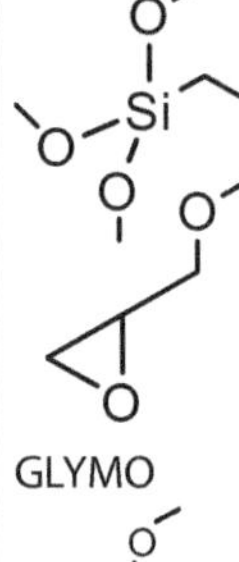 GLYMO 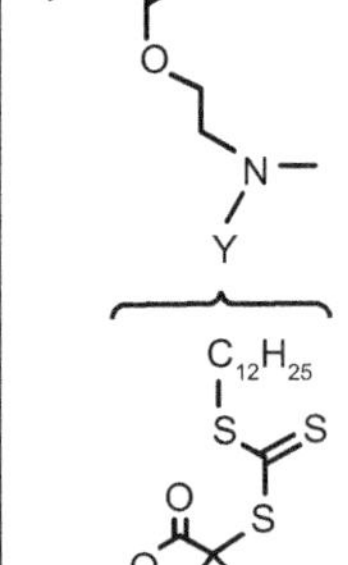 RAFT chain transfer agent 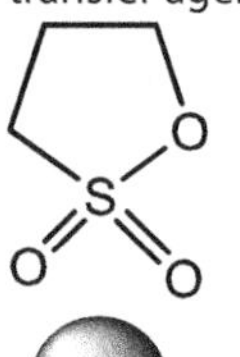 AIBN initiator Mesoporous silica nanoparticle (MSN)	Zwitterionic sulfobetaine copolymer coated MSNs were prepared by "grafting from strategy." The RAFT chain transfer agent used to grow PDMAEMA chains from the surface of the mesoporous silica nanoparticle was synthesized in three steps: (1) surface epoxidation by reaction of the silane coupling agent (3-glycidyloxypropyl) trimethoxysilane (GLYMO) with the silanol groups of MSNs; (2) epoxide ring opening on treatment with a methanol solution of hydrochloric acid; and (3) carbodiimide mediated coupling to the RAFT agent S-dodecyl-S '-(α, α' -dimethyl-α ''-acetic acid) trithiocarbonate. Eventually some pendant tertiary amino groups were converted to sulfobetaine groups by treatment with 1,3-propanesultone.	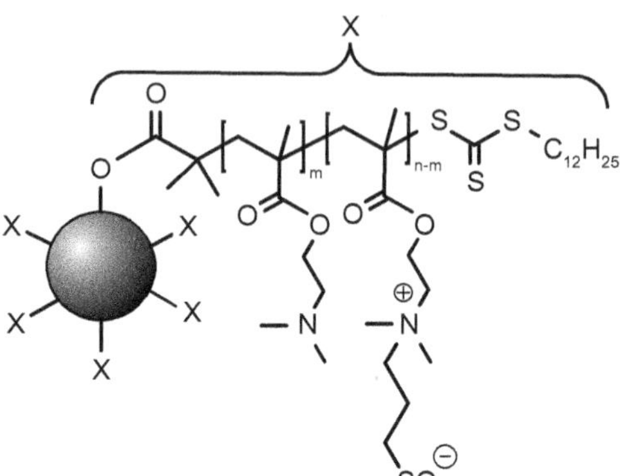	Biocompatible; temperature-induced conformational modifications of the polymer chains in the copolymer shell control the opening and closing of the MSNs pores thereby controlling the drug release process; increased loading capacity as compared to naked MSNs since the polymeric coating provides by itself an additional compartment for drug loading; low cytotoxicity revealed by MTT assay/[107]

Table 6.2 Controlled Polymerization Methods and Polymer Architectures for Drug Delivery Applications (Continued)

Monomer(s), Initiator(s), Activators, and Catalyst(s)	Polymerization Method(s)	Polymer Architecture	Essential Properties for Drug Delivery Applications/ References
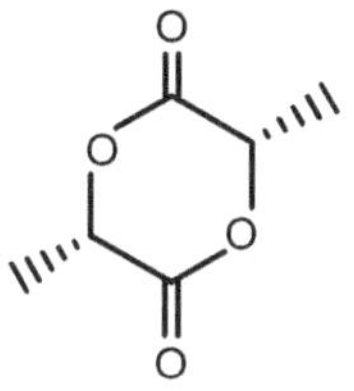 L-lactide (LA) 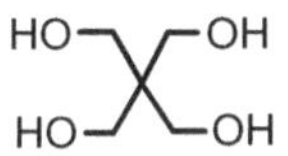 Pentaerythritol initiatorZn prolinate as biocompatible catalyst	Pentaerythritol initiated metal catalyzed ROP of L-lactide, conversion of the star PLA-OH to carboxyl-terminated polylactide PLA-COOH, and eventual conjugation to the antibiotic ciprofloxacin.	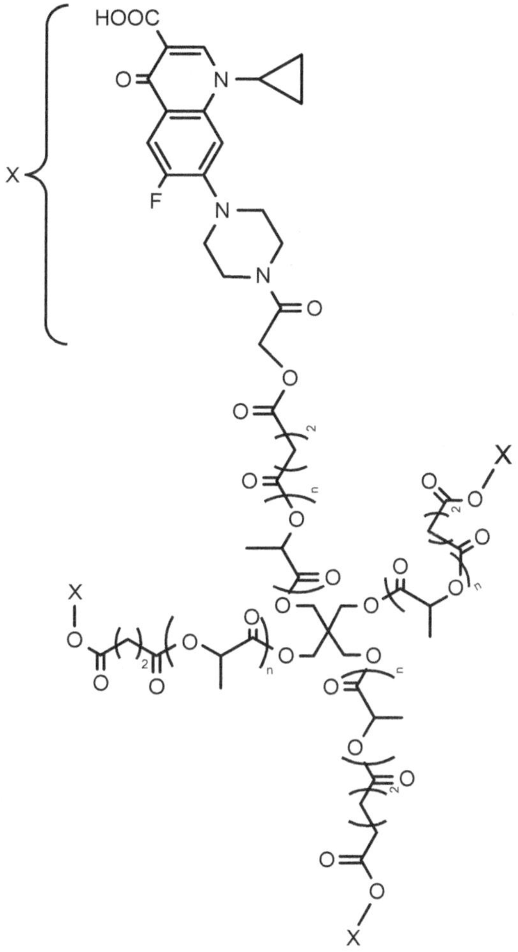 Electrospun fabricated nonwoven nanofibers of star polymer-ciprofloxacin conjugates	Controlled release of ciprofloxacin; release rate is strongly dependent on the pH-medium and on the length of the PLA segment; effective inhibition of *S. aureus* and *E. coli* development both in static (agar) and dynamic (liquid) environment/[108]

(Continued)

Monomer(s), Initiator(s), Activators, and Catalyst(s)	Polymerization Method(s)	Polymer Architecture	Essential Properties for Drug Delivery Applications/ References
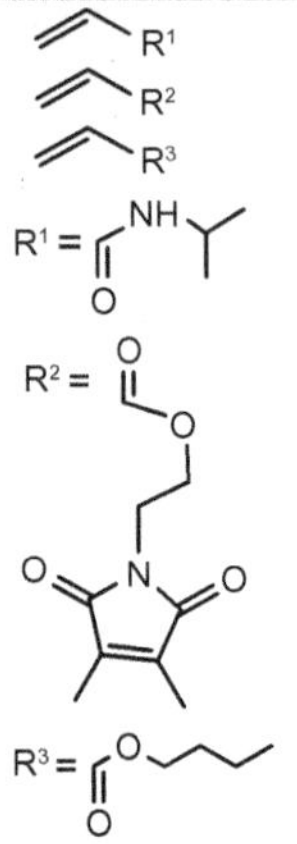 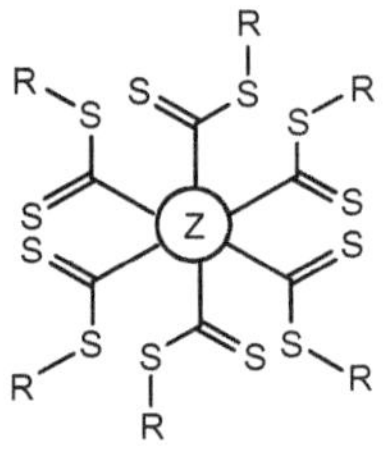 $R^1 = $ $R^2 = $ $R^3 = $ Multifunctional star-shaped RAFT agent $R = $ 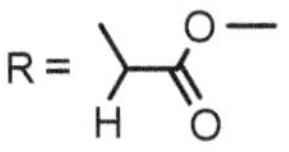	Synthesis of six-armed star amphiphilic diblock copolymers *via* the "core first" approach using a multifunctional star-shaped RAFT agent. Sequential Z-RAFT co-polymerization of a hydrophilic monomer together with a UV cross-linkable monomer and polymerization of a hydrophobic monomer leads to a globular core-corona structure. The outer hydrophilic shell is cross-linked on UV irradiation. Eventually, the RAFT-core is removed *via* aminolysis to give hallow spheric nanocontainers with a hydrophobic cavity. By a similar procedure, but reversing the order of monomers addition hallow sphered nanoparticles for encapsulation of hydrophilic drugs can be obtained.	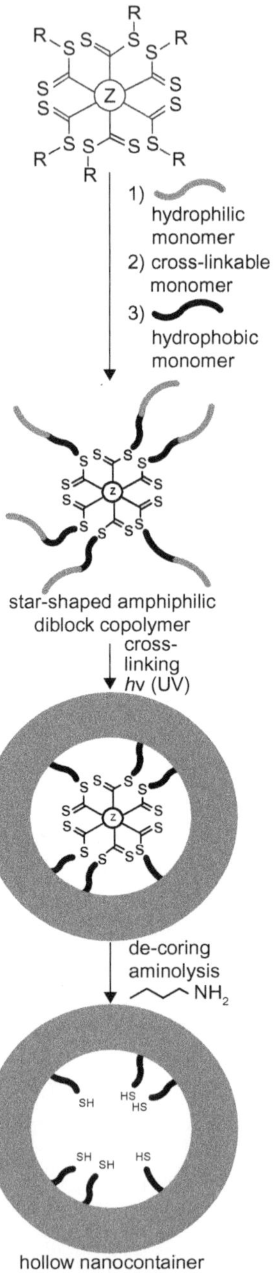	/[109,110]

Monomer(s), Initiator(s), Activators, and Catalyst(s)	Polymerization Method(s)	Polymer Architecture	Essential Properties for Drug Delivery Applications/References
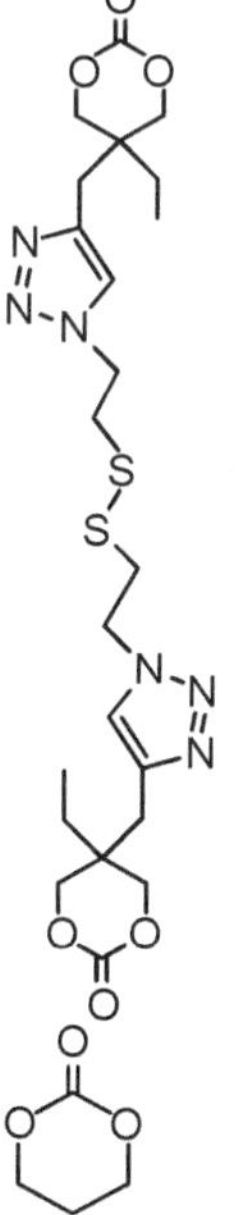 Grubb third-generation catalyst[(H$_2$Imes)(3-Br-py)$_2$(Cl)$_2$Ru=CHPh]	Ring opening metathesis polymerization of a Boc-protected diamine derived from oxanorbornene dicarboxylic anhydride, using a third-generation Grubb catalyst. Eventual removal of the protecting groups leads to a predominantly hydrophilic polycationic homopolymer which mimics the biochemical activity of antimicrobial peptides.	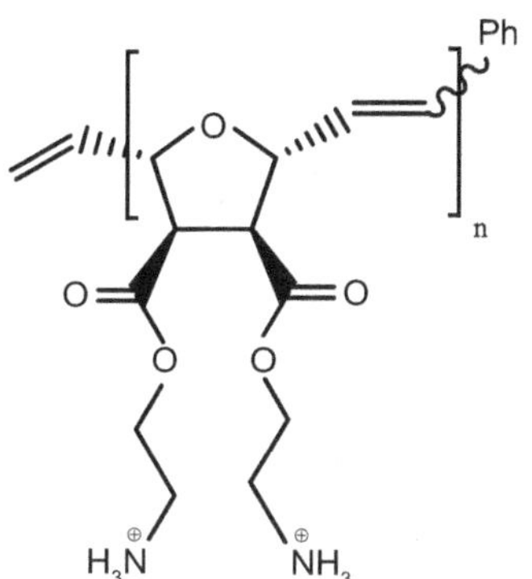	Antimicrobial polymer/[111]
MPEG initiator	Ring-opening co-polymerization (ROP) of 5,5′-(((disulfanediyl bis(ethane-2,1-diyl)) bis(1H-1,2,3-triazole-1,4-diyl) bis(methylene)) bis(5-ethyl-1,3-dioxan-2-one) (TDCSS) and trimethylene carbonate (TMC) initiated by mono-methoxy poly(ethylene glycol) (MPEG).	Crosslinked amphiphilic polycarbonate diblock copolymer MPEG-*b*-poly(TMC-*co*-TDCSS)	Biodegradable; self assembles in water into nanosized micelles; redox-responsive drug release [112]

(Continued)

Table 6.2 Controlled Polymerization Methods and Polymer Architectures for Drug Delivery Applications (Continued)

Monomer(s), Initiator(s), Activators, and Catalyst(s)	Polymerization Method(s)	Polymer Architecture	Essential Properties for Drug Delivery Applications/ References
CL Malic acid—initiator for ROP of CL Monomer used in the coupling step of the divergent growth of dendrons Activation reagent used in the divergent synthesis of dendrons MPEG-adipoyl chloride Magnetic ferrite nanoparticles stabilized with an oleic acid shell	Malic acid initiated ROP of CL followed by coupling of the terminal hydroxyl group of PCL to the acryloyl chloride monomer; Michael addition of diethanolamine to the resulted acryloyl ester activates the latter for the next coupling step; iterative application of the coupling and activation steps results in the construction of successive dendron generations. After the last activation step, linear MPEG segments are attached to the linear-dendritic hybrid by reaction with MPEG-adipoyl chloride. Eventually, dendronized magnetic ferrite nanoparticles were prepared by the ligand exchange method from the above-obtained linear-dendritic-linear copolymers and oleic acid-stabilized ferrite nanoparticles.	Amphiphilic hybrid linear-dendritic-linear block copolymers consisting of PCL linear block, poly(amino-ester) dendritic block (G2) and MPEG linear block 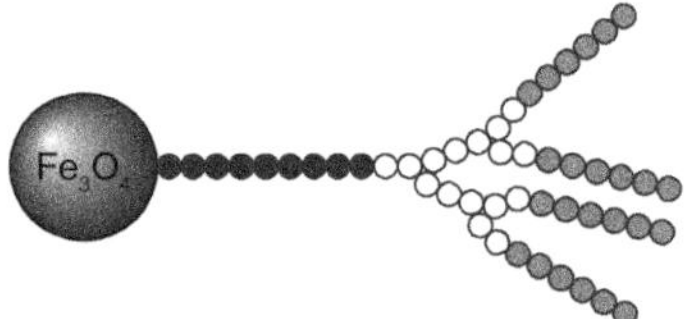Dendronized magnetic ferrite nanoparticles	pH-sensitive superparamagnetic amphiphilic dendrimers; low polydispersity; high encapsulation efficiency; hydrolytically degradable and reasonable release profile; high magnetic properties/[113,114]

be attached to the polymer *via* a spacer (see Fig. 6.4A) and the polymer nanocarrier can gain further functionalities by the attachment of targeting ligands and solubilizing groups [116].

The other type of polymer-based nanocarriers consists in various polymeric supramolecular systems formed through the association of several individual macromolecules that are held together within the supramolecular assembly by weak noncovalent intermolecular forces. Loading drugs into such polymeric supramolecular assemblies broadly named nanoparticles can be achieved either during the preparation of nanocarriers (incorporation) or after the formation of nanocarriers (incubation) [117]. The drug

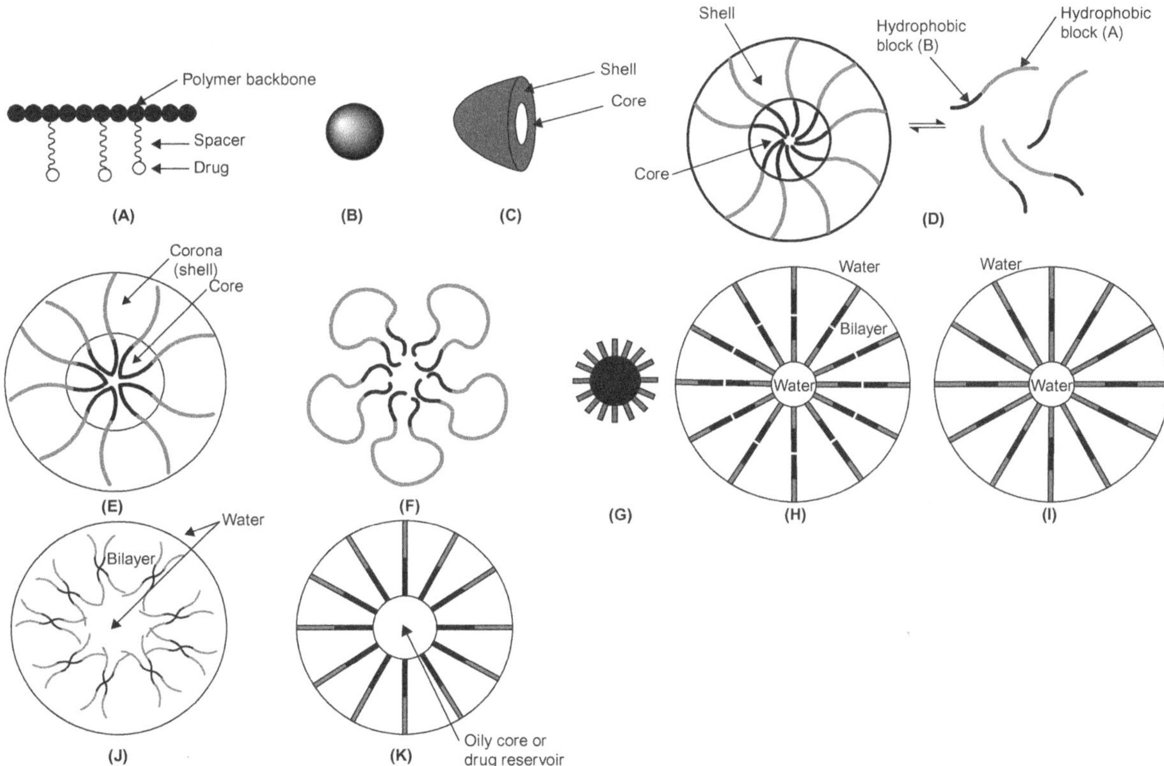

■ **FIGURE 6.4** Different types of polymeric nanocarriers: (A) polymeric prodrug; (B) nanosphere; (C) nanocapsule; (D) micelle in dynamic equilibrium with unimers (single polymer molecules) in solution; (E) normal spherical micelle built up from amphiphilic ABA triblock copolymers; (F) flower-like micelle formed by amphiphilic triblock copolymers of BAB type; (G) nanosphere of amphiphilic block copolymers; (H) polymersome composed of a bilayer membrane of amphiphilic AB type diblock copolymers; (I) polymersome composed of a monolayer membrane of amphiphilic symmetric ABA type triblock copolymers; (J) polymersome composed of a bilayer membrane of ABA type amphiphilic triblock copolymers; (K) nanocapsule composed of a bilayer membrane of amphiphilic AB type diblock copolymer molecules surrounding an oily core or drug reservoir.

can be either physically entrapped within the nanoparticle or adsorbed on its surface. Alternatively, the drug can be covalently bound to the polymeric matrix or conjugated to the outside of the particle. The weak and reversible nature of supramolecular associations can be exploited in drug delivery applications since in response to some internal, physiological, or external stimuli the supramolecular nanocarrier can be destabilized and collapses with concomitant release of its drug payload at a specific site or with a preprogrammed timing [118]. Since several supramolecular morphologies are possible such as hollow spheres, solid spheres, and rods, controlling and programming the supramolecular assembly process by engineering the polymeric building blocks is a formidable task.

The term "nanoparticle" designates a particulate structure of colloidal dimensions i.e., of less than 1000 nm in diameter, and encompasses both nanospheres and nanocapsules. Nanospheres (see Fig. 6.4B) belong to the monolithic (homogenous) systems meaning that they are composed of an entirely solid polymeric matrix. The drug is dispersed throughout the whole mass of the nanoparticle. Nanocapsules (see Fig. 6.4C) belong to the core-shell morphologies or reservoir-type systems and are heterogeneous systems comprising an inner core surrounded by a continuous polymer shell or membrane. [119,120]. The inner core that confines the active substance is usually liquid but it can also be semisolid or even solid [121]. Depending on what types of raw materials and preparation methods are used, the inner cavity entrapping the drug can be either lipophilic or hydrophilic.

Among the polymeric materials used in the construction of nanoDDS, amphiphilic block copolymers received a special attention since they are able to self-assemble when placed in contact with an aqueous milieu forming several types of colloidal particles such as polymeric micelles, nanospheres or vesicular structures like polymersomes and nanocapsules [122].

Polymeric micelles are colloidal particles formed by the self-assembly of amphiphilic block copolymers in water. They have a core-shell morphology. The core of the micellar structure is built up from the hydrophobic blocks of the amphiphilic block copolymers and is enveloped with a hydrophilic corona (see Fig. 6.4D). The driving force directing the self-assembly process is provided by the hydrophobic interactions developed on the formation of the micellar core. Lipophilic drugs can be encapsulated within the hydrophobic core. However, micellar aggregation of individual amphiphilic macromolecules (unimers) occurs only after the amphiphile concentration in aqueous solution exceeds a specific value. Aggregation is evidenced by abrupt changes of many physicochemical properties of the aqueous solution of amphiphiles, and the narrow concentrations range over which such changes occur is called the critical micellar concentration (CMC). If amphiphile concentration drops below CMC then micelles will disassemble. Hence, micelles are dynamic structures in the sense that their assembly and disassembly is concentration dependent. From this point of view, amphiphilic block copolymers exhibit an important advantage over other surfactants such as Cremophor EL or polysorbates since they exhibit lower CMCs [123]. Therefore, nanoformulations based on amphiphilic block copolymer micelles will remain stable to dilution within body fluids thereby avoiding a premature release of their drug payload. The stability of micellar associations can be further increased by reinforcement of the weak intermolecular forces holding together the unimers through covalent cross-linking of either the hydrophobic core or the hydrophilic corona [124]. Amphiphilic diblock copolymers and amphiphilic triblock copolymers of ABA type, where A stands for the hydrophilic block, self-assemble into normal spherical micelles in water (see Fig. 6.4D and E, respectively). By contrary BAB amphiphilic triblock copolymers tend to form "flowerlike" micelles as depicted in Fig. 6.4F [125]. This is an example of how molecular architecture influences the morphologies of the resulted aggregates. Being flanked on both ends by the core forming hydrophobic blocks B, the middle hydrophilic corona forming block A is forced to form a loop in order to bring the hydrophobic blocks into the core of the micelle [126]. Such constraints disfavor aggregation by additionally decreasing entropy. Therefore, some copolymer chains in the micelle might have one of the hydrophobic blocks B extended into solution. A different alternative to circumvent the entropic penalty is the formation of a bridged structure within which several hydrophobic cores (droplets) are linked by the hydrophilic segments [127].

As previously mentioned, the term "nanosphere" generally designates a solid matrix particle. However, nanospheres prepared from amphiphilic block copolymers have a particular structure since in this case there is no cleancut division between nanospheres and micelles. That means that nanospheres of amphiphilic block copolymers have a core shell-type structure (see Fig. 6.4G) resembling micelles but at the same time the unimers are in a "frozen" state forming a phase-separated solid matrix core [123,127]. The structure of nanospheres prepared from amphiphilic diblock copolymers can be tuned up by manipulating the ratio between the molecular weights of the two blocks. For instance, the central core of nanoparticles prepared from the amphiphilic macromolecules of methoxy poly(ethylene glycol) (MPEG)-block poly(D,L-lactic acid) (PDLLA) copolymer (MPEG-*b*-PDLLA) become more dense resembling nanospheres as the length of the hydrophobic PDLLA block increases, while shorter PDLLA blocks results in formation of micelle-like assemblies [128,129]. As compared to polymeric micelles, polymeric nanospheres are much more prone to opsonization and clearance by the RES. Therefore, in order to increase the circulation time of polymeric nanospheres, one has to prevent binding of opsonins to their outer surface; otherwise recognition by the phagocytic cells of the RES cannot be avoided. To this purpose, shielding groups must be grafted to the surface of nanospheres with the role of a steric barrier that impedes opsonins access. Such shielding groups impart "stealth" properties to nanoparticles making them "invisible" to the RES. Hydrophilic poly(ethylene glycol) (PEG) chains are among the most effective shielding groups [130]. From this point of view, nanospheres prepared from amphiphilic diblock copolymers like the above mentioned MPEG-*b*-PDLLA exhibit improved stability in the blood plasma as compared to nanospheres without MPEG coating.

Another type of polymeric nanocarrier is those with a vesicular structure like polymersomes and nanocapsules. With a supramolecular morphology resembling liposomes, polymersomes are reservoir-type structures with a central aqueous cavity surrounded by a polymeric membrane. The polymeric membrane may consist of a monolayer of amphiphilic ABA type triblock copolymer molecules [131] (see Fig. 6.4I). However, ABA amphiphilic triblock copolymers can also self-assemble to form a bilayer polymeric envelope around the aqueous core as depicted in Fig. 6.4J [132]. Obviously, diblock copolymers form vesicles with bilayer walls (see Fig. 6.4H). Polymersomes are of special interest as nanoDDS since they can transport both hydrophilic and hydrophobic drugs. Hydrophilic drugs are solubilized in the internal aqueous cavity while hydrophobic drugs are confined within the polymeric membrane through hydrophobic interactions. This feature of polymersomes can be exploited for synergistic codelivery of hydrophobic antitumor antibiotics and hydrophilic therapeutic peptides in cancer treatment [133]. If a polymeric membrane comprising a bilayer of amphiphilic diblock copolymer molecules encloses an oily cavity (see Fig. 6.4K), then the whole supramolecular assembly is called a nanocapsule. Nanocapsules are used to deliver hydrophobic drugs like the antibacterial agent triclosan [134] encapsulated within the oily cavity.

6.5 METHODS FOR PREPARATION OF DRUG-LOADED POLYMERIC NANOPARTICLES

There are two main ways to prepare polymeric nanoparticles:

1. Polymeric nanoparticles synthesized during the polymerization of monomers
2. Polymeric nanoparticles synthesized from preformed polymers

6.5.1 **Preparation of Polymeric Nanoparticles Starting With Monomers**

This paragraph outlines the methods used to prepare drug-loaded polymeric nanocarriers through the polymerization of monomers.

6.5.1.1 Conventional Emulsion Polymerization

Depending on the nature of the continuous phase, emulsion polymerization can be carried out either in organic solvents or in water.

Toxicological concerns associated with the use of large amounts of organic solvents which needs to be eventually removed severely restrict application of the continuous organic phase methodology for drug encapsulation.

The alternative of the aqueous continuous phase method involves dispersion of a relative hydrophobic monomer in water using an oil-in-water emulsifier followed by initiation of the polymerization reaction using either a water soluble initiator like sodium persulfate or an oily soluble initiator like 2–2′-azoisobutyronitrile (AIBN) [135]. So, the key components are water, the monomer, the emulsifier, and the initiator. Polymerization occurs within a very large number of monomer droplets dispersed in the continuous aqueous phase although monomer-swollen micelles may also exit in the reaction system. Application of this method is extremely profitable in the case of anionic polymerization of alkylcyanoacrylate monomers since water initiates the polymerization reaction by itself. Antibiotic-loaded nanospheres of poly(alkylcyanoacrylate), e.g., poly(isohexylcyanoacrylate), poly(*n*-butylcyanoacrylate), poly(isobutylcyanoacrylate) have been prepared by emulsion polymerization in aqueous media following the synthetic procedure briefly described next [136–139]. The monomer is dispersed in acidified water containing the antibiotic (ampicillin [136], moxifloxacin [140], ciprofloxacin [141]) and a suitable surfactant that is approved for biomedical applications like poloxamer 188, polysorbate 80, dextran. The polymerization is allowed to proceed for a few hours under vigorous stirring and eventually the reaction mixture is neutralized and the nanoparticle suspension is filtered and freeze-dried. The so prepared antibiotic-loaded nanospheres were tested against intracellular pathogens. Encapsulation of ampicillin in poly(isohexylcyanoacrylate) resulted in 120-fold increase of the antibiotic efficacy in treatment of experimental salmonellosis [136] while moxifloxacin encapsulated poly(*n*-butylcyanoacrylate) nanoparticles caused a ten-fold decrease in the antibiotic concentration required for inhibition of intracellular *Mycobacterium tuberculosis* growth [140]. Emulsion polymerization has been also applied to produce nanospheres with modified surface properties [142].

6.5.1.2 Mini-Emulsion Polymerization

This method differs from conventional emulsion polymerization by the generation of nanoemulsions using high-energy methods, for instance high shear rate technique, which is required for the system to reach a state with an interfacial tension much greater than zero. In order to retard Ostwald ripening in the nanosized monomer droplets, mini-emulsion polymerization also requires the use of a costabilizer prior initiating polymerization. Hexadecane is a good choice since it fulfills all the requisite conditions for a costabilizer that are high monomer solubility, poor water solubility, and low molecular weight [143]. Hence, a typical formulation consists in water as the continuous phase, monomer, co-stabilizer, surfactant, and initiator. A standard procedure to perform mini-emulsion polymerization of n-butyl-cyanoacrylate consists in (1) generation of a nanoemulsion by ultrasonication a mixture of a solution of the surfactant, e.g., sodium dodecyl sulfate, in aqueous hydrochloric acid and a monomer solution of the costabilizer followed by (2) polymerization initiation on addition of the nucleophilic reagent for instance a solution of sodium hydroxide or ammonium hydroxide [144]. The mechanism of the anionic polymerization of alkylcyanoacrylates ensures that the initiating nucleophilic group will be exposed on the outer surface of the growing polymer particle thereby providing also a way for surface functionalization of the prepared polymer nanoparticles [144].

6.5.1.3 Micro-Emulsion Polymerization

In contrast to the two above discussed polymerization techniques in the case of micro-emulsion polymerization, a thermodynamically stable two phase system is generated spontaneously with the help of large amounts of surfactants, the interfacial tension at the water-oil interface being close to zero. Consequently, no high energy input is required. The polymer particles synthesized by this technique are of smaller dimensions (typically less than 80 nm) than those produced by mini-emulsion polymerization. Polymerization is initiated by a water soluble initiator which diffuses within the monomer swollen micelles [145].

6.5.1.4 Interfacial Polymerization

While emulsion polymerization produces drug-loaded nanospheres, interfacial polymerization leads to drug encapsulation within nanocapsules. If polymerization occurs in an oil-in-water emulsion, then nanocapsules with an oily core able to entrap hydrophobic drugs are obtained. Conversely, if the polymerization reaction is conducted in a water-in-oil emulsion, than nanocapsules with an aqueous reservoir able to encapsulate hydrophilic drugs will be produced. In a typical procedure producing oil-containing nanocapsules, the oil and the alkylcyanoacrylate monomer, e.g., isobutyl-2-cyanoacrylate, are dissolved in a water-miscible organic solvent such ethanol or acetone along with the drug, and next this organic phase is slowly injected into a stirred aqueous solution that contains a hydrophilic surfactant like Pluronic F68. As a consequence of the diffusion of the water miscible organic solvent into the aqueous phase, the oil is dispersed as tiny droplets and polymerization of the monomer occurs at the oil-water interface on contact with the nucleophilic hydroxyl groups in the water molecule [137,138,146]. The resulted milky suspension is eventually concentrated. To obtain nanocapsules containing an aqueous core that encapsulate water soluble drugs, an aqueous-ethanol solution of the hydrophilic drug is added under vigorous stirring to an oily (miglyol 812) organic phase that contains a suitable surfactant having a low hydrophilic-lipophilic balance (HLB) value, for instance sorbitan monooleate. Then, isobutylcyanoakrylate monomer is added slowly to the microemulsion under continuous stirring and interfacial polymerization is allowed to proceed for a few hours [147].

Polymeric nanoparticles of copolymers built up from a lipophilic monomer such as phthaloyldichloride and a hydrophilic monomer like diethylenetriamine can be synthesized through interfacial polycondensation with or without a surfactant [146]. The polycondensation reaction takes place at the oil-water interface within an aqueous emulsion spontaneously formed on the dispersion of the oily organic phase into water. A typical composition of the organic phase is: an oil such as miglyol 812, the lipophilic monomer, a lipophilic surfactant such as lipoid S75, and a water miscible solvent like acetone. The aqueous phase contains the hydrophilic monomer and a hydrophilic surfactant like Pluronic F68 [148].

6.5.2 Synthesis of Polymeric Nanoparticles From Preformed Polymers

There are several drawbacks associated with the fabrication of drug-loaded polymeric nanoparticles by drug encapsulation during the in situ polymerization of monomers. First of all, the drug should not be altered under the conditions required for polymerization of monomers and should be inert toward monomers, initiators, solvents, stabilizers, and surfactants [149]. Only few monomers can be polymerized in situ in the presence of a drug to produce drug-loaded nanoparticles suited for in vivo applications. As shown in the previous paragraph alkylcyanoacrylates are among the most exploited monomers for this purpose [150]. Another example are acrylamides [151]. Polyamides-based nanocapsules have been prepared by in situ interfacial polycondensation of phthaloyl chloride and

1,6-hexamethylenediamine [119,148]. Chemical inertia of the drug toward the monomer is not always required. In some few cases the drug itself acts as an initiator of the monomer polymerization to eventually get a drug-polymer conjugate. As an example of this situation we mention here the paclitaxel-initiated controlled ROP of lactide [152]. Other drawbacks intrinsically associated with the involvement of a chemical reaction in the preparation procedure are incomplete monomer conversion, formation of byproducts, production of unwanted oligomers, and so on. A solution to overcoming these drawbacks is the use of the other preparation variant i.e., synthesis of nanoparticles from preformed polymers. Below is a survey of the main synthetic procedures to fabricate polymeric nanoparticles starting with polymers instead of monomers.

6.5.2.1 Nanoprecipitation Method

This method, also called solvent displacement method, relies on the spontaneous emulsification that occurs when a solution of a polymer in a water miscible organic solvent is injected into the stirred external aqueous phase which is an antisolvent for the polymer and may optionally contain a surfactant with a high HLB value like polyvinyl alcohol (PVA) [153,154] or Pluronic F68 [155]. Fast diffusion of the organic solvent into the aqueous phase causes precipitation of the polymer at the interface between the two phases. An illustrative example is given here. A general procedure for the synthesis of antibiotic-loaded nanoparticles by the nanoprecipitation method is presented in Fig. 6.5. Azithromycin (AZI)-encapsulated nanospheres of poly(D,L-lactide-*co*-glycolide) (PLGA) were prepared as follows: AZI and PLGA powders were solubilized in acetone and the resulted solution was injected in water containing PVA at a constant rate. The resulted nanosuspension was kept under stirring for 12h in order to allow acetone to evaporate at room temperature and nanoparticles were harvested through eventual centrifugation [153]. The minimum inhibitory concentration (MIC) of so prepared AZI-encapsulated PLGA nanospheres showed an eight-fold decrease as compared to the free drug when tested against indicator bacteria including *E. coli* (PTCC 1330), *Haemophilus influenzae* (PTCC 1623), *Streptococcus pneumoniae* (PTCC 1240) [154] as well as against the intracellular bacterium *Salmonella typhi* [153]. The release profile of AZI from the PLGA nanospheres was characterized by burst release within the first 2h followed by sustained release for up to one day and the encapsulation efficiency was about 60% [156]. Levofloxacin (LEV)-loaded PLGA nanoparticles have been prepared by the same procedure with the minor difference of using Pluronic F68 as hydrophilic surfactant instead of PVA [155]. The nanoparticle suspension was readily transformed into inhalable dry-powder nanoparticle aggregates by spray drying and the

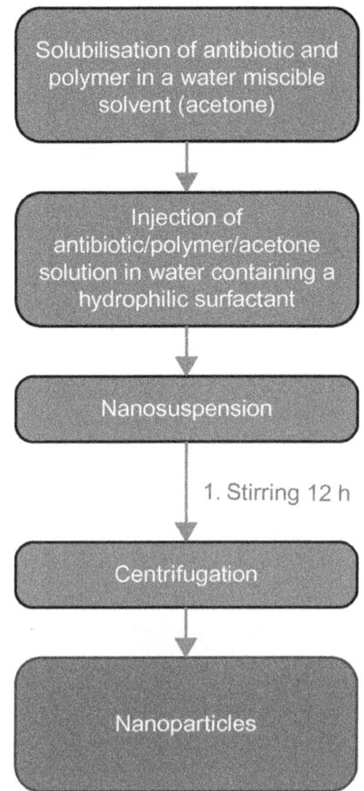

■ **FIGURE 6.5** Synthesis of antibiotic-loaded nanoparticles by the nanoprecipitation method.

latter formulation was tested against *P. aeruginosa* biofilm cells. Both dried powder and suspension nanoformulations exhibited similar antibacterial activities eradicating 99.9% of the biofilm cells after 24 h [155].

6.5.2.2 Emulsion-Solvent Diffusion Method

This method involves two steps. In the first step an oil-in-water emulsion is prepared using two immiscible mutually saturated organic and aqueous phases. The organic phase is composed of a partly water soluble organic solvent such as ethyl acetate. Mutual saturation of the above two phases is obtained by mixing equal volumes of each phase when the mass transfer process between phases results in the establishment of a thermodynamic interphase equilibrium state. In other words, the organic phase will be saturated with water, while the aqueous phase will be saturated in the organic solvent. The polymer poly(ε-caprolactone) (PCL) or PLGA, an oil, for instance miglyol 812, and the drug are dissolved in the organic phase. A hydrophilic stabilizer like PVA is dissolved in the aqueous phase. Then, the organic phase is dispersed in the aqueous phase with the aid of a high

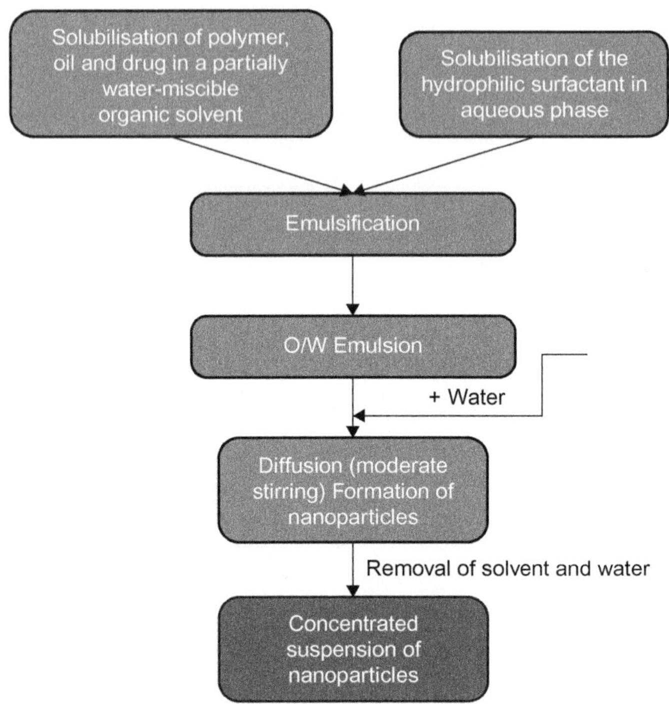

■ **FIGURE 6.6**   General procedure for the synthesis of drug-loaded nanoparticles by the emulsion-solvent diffusion method.

shear mixer when an oil-in-water emulsion is produced. A volume of water four times greater than the volume of the emulsion is added slowly to the emulsion in the second step. Solvent (ethyl acetate) diffusion to the external phase causes precipitation of the polymer and formation of nanoparticles [157,158]. Complete removal of the organic solvent and partial removal of water is carried out under reduced pressure leading to a concentrated suspension of nanoparticles. In Fig. 6.6 a scheme of preparation of nanoparticles by the emulsion-solvent diffusion method is depicted. Depending on the oil-to-polymer ratio either nanospheres or nanocapsules can be obtained by this method [146]. In a slightly modified variant of this method, the drug and the polymer (PLGA) are dissolved in a mixture of methylene chloride and acetone. So, instead of using a partly water-miscible organic solvent, one uses a water-immiscible solvent (methylene chloride) and a freely water-miscible solvent (acetone). This organic phase is gently poured into the aqueous phase that contains PVA as emulsifier. Nanosized emulsion droplets spontaneously form within the aqueous phase due to the rapid diffusion of acetone from each emulsion droplet out to the external aqueous phase, a process that drastically reduces the dimension of the droplets. If necessary a high speed homogenizer may be used in order to obtain a fine emulsion. Eventually, the remaining methylene chloride is evaporated at room temperature under stirring and the produced nanoparticles are separated by ultracentrifugation, washed with distilled water and freeze-dried [159]. Rifampicin (RIF)-encapsulated PLGA nanoparticles have been prepared by this method and they showed a significant increase in their antibacterial activity as compared to the free antibiotic when tested against gram-positive bacteria [160]. Similarly, pegylated PLGA nanoparticles (PEG-PLGA) loaded with roxithromycin (RXN) have been prepared by the emulsification-solvent diffusion method, and their in vitro activity against *S. aureus*, *Bacillus subtilis*, and *S. epidermidis* was estimated in terms of MIC values. A nine-fold decrease of the MIC value on *S. aureus* and a 4.5-fold decrease on MIC values on *B. subtilis* and *S. epidermidis* were revealed compared to RXN solution [161].

6.5.2.3 Emulsion-Solvent Evaporation Method

This method resembles much the emulsification-solvent diffusion method. Like in the emulsification-solvent diffusion method an oil-in-water nanoemulsion is prepared in the first step by dissolving the polymer and the drug in a volatile water-immiscible solvent like methylene chloride or chloroform and subsequent dispersing this organic solution in the external aqueous phase that contains a hydrophilic stabilizer like PVA. The difference between the two methods lies in how the polymer solvent is removed from the dispersed organic phase droplets thereby producing precipitation of polymer nanoparticles. In the emulsion-solvent evaporation method this purpose is achieved by slow evaporation of the volatile organic solvent, which transforms the nanoemulsion into a nanosuspension. Levofloxacin (LEV)-loaded PLGA nanoparticles have been prepared following this procedure. Better encapsulation efficacies were obtained through increasing water-miscibility level of the oil phase by adding lecithin in the aqueous phase [155]. AZI- and RIF-loaded pegylated polylactic acid (PEG-PLA) copolymer nanoparticles have been prepared by emulsification-solvent evaporation technique using chloroform as the organic solvent which was eventually removed under reduced pressure [63]. These nanoparticles have been tested against the intracellular bacteria *Chlamydia trachomatis* and *Chlamydia pneumoniae*. The co-delivery of both antibiotics enhanced treatment efficiency in reducing microbial burden [63].

Double emulsification is also useful for fabricating drug-encapsulated nanoparticles. Double and multiple emulsions are complex colloidal systems in which the droplets dispersed in the continuous phase are emulsions themselves; hence one can say that double emulsions are "emulsions of emulsions" [120,162]. There can be (water-in-oil)-in-water ((W/O)/W) and (oil-in-water)-in-oil ((O/W)/O) emulsions. Obviously, two emulsification processes are required to produce such nanoformulations. For instance, preparation of (W/O)/W emulsions primarily involves formation of a water-in-oil (W/O) emulsion by dispersing water as small droplets within an oily organic phase, which in turn is further dispersed as larger droplets into another aqueous phase. Taking into account the above mode of preparation, it is clear that two different types of stabilizers are required: a hydrophobic surfactant is needed to stabilize the primary W/O simple emulsion and then a hydrophilic surfactant has to be used to stabilize the final (W/O)/W double emulsion. Ciprofloxacin (CIP)-encapsulated PLGA nanoparticles were prepared by the above method as follows: an aqueous solution of CIP was added dropwise to a methylene chloride solution of PLGA and the mixture was sonicated to produce the first W/O emulsion. Then, this emulsion was poured in an aqueous phase containing PVA and homogenized using a high-pressure homogenizer to obtain the (W/O)/W emulsion. Separation of nanoparticles was carried out by ultracentrifugation, the nanoparticles were washed with distilled water, again subjected to centrifugation and eventually lyophilized [163]. The release profile was characterized by a burst effect lasting 12 h, followed by continuous release for two weeks. This sustained release profile enables CIP-loaded PLGA to effectively inhibit growth of *E. coli* bacteria when tested in vivo on ICR mice [163]. Eudragit RS100 and RL100 are other polymeric materials used to encapsulate CIP by the (W/O)/W emulsification and solvent evaporation method [164]. The same method was used to encapsulate LEV within PLGA nanoparticles [155]. Vancomycin (VCM)-loaded PLGA nanoparticles with significantly increased intestinal permeability as compared with VCM solution have been similarly prepared [165].

6.5.2.4 Salting-Out Method

This method derives from the emulsification-solvent diffusion method too. Also in the salting-out method, the first step consists of preparing an oil-in-water emulsion. However, the polymer solvent is a completely water-miscible solvent like acetone and emulsification is achieved with the aid of a salting-out agent that is added in high amounts to the aqueous phase and dramatically decreases the solubility of the organic solvent in water. Therefore, when the aqueous phase is added to the organic phase rather than the diffusion, phase separation takes place, the organic phase being squeezed out from the aqueous phase. The salting-out agent is usually an electrolyte such as magnesium chloride or calcium chloride but also non-electrolytes like sucrose can be used. Along with the salting-out agent the aqueous phase contains a colloidal stabilizer like polyvinylpyrrolidone, hydroxyethylcellulose or sodium carboxymethylcellulose, which acts as a viscosity-increasing agent too [146]. In the second step, polymer solvent removal provoking precipitation of nanoparticles is achieved by addition of a large amount of water which causes a large drop in the concentration of the salting-out agent while the organic solvent regains its water solubility and diffuses out of the emulsion droplets entering the external aqueous phase. Following this procedure, isoniazid (INH), which is the first line antimicrobial in treatment of tuberculosis, was incorporated in to the pH-sensitive methacrylic acid-ethyl acrylate copolymer (MAEA). The key components of the organic phase were: MAEA as preformed polymer, acetone as polymer solvent, and INH. The aqueous phase was composed of $ZnSO_4$ heptahydrate as salting-out agent and sodium carboxymethylcellulose (NaCMC) as viscosity-enhancing agent. Next a brief description of the synthesis of pH-sensitive INH-loaded MAEA nanoparticles is given. An oil-in-water emulsion is prepared by addition of the aqueous phase to the organic phase under vigorous stirring. Thereafter, water is added to allow the mass transfer between the two phases resulting in formation of INH-loaded MAEA nanoparticles [166].

6.5.2.5 Dialysis Method

Dialysis is in fact a modified solvent displacement method. It involves a dialysis tube within which is inserted a semi-permeable membrane with a suitable molecular weight cutoff so that it acts as a physical barrier for the polymer. The polymer, the drug, and optionally a surfactant are dissolved in the polymer solvent and dialyzed against a polymer nonsolvent miscible with the polymer solvent. Gradual displacement of the polymer solvent by the polymer nonsolvent through the membrane results in polymer precipitation and formation of nanoparticles [167,168]. Lactose-conjugated PLGA nanoparticles were prepared by dialysis and tested for pulmonary delivery of RIF. By fluorescence studies it was proved that conjugation to lactose is beneficial for an enhanced lung uptake of the antibiotic [169].

6.5.2.6 Methods Based on the Application of an Electric Field

In the following paragraphs the principle underlying two related methods, electrospinning and elctrospraying, will be presented.

Fabrication of Polymer Nanofibers Through Electrospinning

By electrospinning, polymer materials in solution or melt are formed into continuous nanosized diameter fibers. A typical electrospinning setup consists of three major components: a syringe pump containing the polymer solution or melt, a high-voltage power supply, and an electrically conductive collector that may be either stationary or rapidly rotating. By applying a high voltage electric field, the surface of the tiny liquid drop emerging from the tip of the syringe pump needle becomes electrically charged and the electric field distorts the spherical drop into a conical object known as a Taylor cone [170,171]. A polymer jet moving toward the collector is formed from the tip of the pendant drop when the electric force overcomes the surface tension of the polymer solution. During the movement of the jet toward the collector, the polymer solvent is evaporated and the polymer material is eventually deposited on the collector. If the collector is stationary, randomly oriented fiber mats are deposited on it, while the use of a rapidly rotating collector results in aligned nanofibers that are very useful in tissue engineering applications. Using two-stream electrospinning technique, nanofiber composites of biodegradable poly(ester urethane) urea (PEUU) and tetracycline hydrochloride-loaded PLGA have been prepared and their applicability to use as material in abdominal wall closure in surgery was demonstrated [172]. Active wound dressing nanofiber mats loaded with tetracycline hydrochloride antibiotic were fabricated through electrospinning from 50:50 blends of poly(lactic acid) (PLA) and PCL [173]. Next, a nice application of the electrospinning method in producing biocompatible and biodegradable polymeric materials for dental tissue engineering is presented. Bone regeneration after dental surgery is compromised mainly due to infiltration of fibroblasts from gingiva at the place of the periodontal defect, thereby interfering with new bone formation. Therefore, prior to performing implants, the damaged alveolar site has to be protected using a so-called guided tissue regeneration/guided bone regeneration (GBR/GTR) membrane which has multiple functionalities: (1) it acts as an obstructive barrier against infiltration of fibroblasts; (2) it sustains the appropriate cells to regenerate the lost tissue; (3) it provides a scaffold for bone regeneration [174]. A hybrid inorganic-organic anti-infective GBR/GTR microfiber implant membrane was fabricated using the electrospinning technique. Briefly, PCL and gelatin were dissolved in trifluoroethanol and to this polymer solution (PG) metronidazole-loaded halloysite nanotubes (MNA-HTN) were added in a weight ratio (PG)/(MNA-HTN) of 5:1 to get the electrospinning solution. The membranes electrospun from the latter solution showed very promising properties for dental tissue engineering applications, such as: (1) excellent mechanical strength and flexibility; (2) biocompatibility and biodegradability due to polyester and polyamide components; and (3) sustained release profile of the embedded metronidazole (MNA) antibiotic [175].

Fabrication of Polymer Nanoparticles Through Electrospraying

Electrospraying and electrospinning share the same working principle and the apparatus setup is similar for both methods. The main difference between the two techniques is the concentration of the polymer solution. The viscosity and the surface tension of the polymer solution depend on its concentration. When the surface tension is balanced by the electrostatic force, a Taylor cone will be formed at the tip of the metallic needle. In electrospinning, further increase in the strength of the electrostatic field results in ejection of a charged stream of polymer solution moving from the tip of the Taylor cone toward the collector. However, a stable jet can be obtained only if the concentration of the polymer solution exceeds a certain value also called the critical entanglement concentration that is specific for each given polymer. Below this concentration, instead of a fiber jet, nanoparticles will be produced. To conclude, electrospraying involves more diluted polymer solutions and much lower voltages than electrospinning. Metronidazole (MNA)-loaded PLGA nanoparticles were fabricated by electrospraying. The electropraying solution was obtained by dissolving the polymer and the drug in trifluoroethanol. The prepared nanoparticles exhibited sustained drug release for at least 41 days [176]. Table 6.3 summarizes the main methods used to prepare drug-loaded polymeric nanomaterials.

6.6 POLYMER-BASED NANOMATERIALS FOR ANTIINFECTIOUS THERAPY

This paragraph describes the principal types of polymeric materials used for combating antibiotic-resistant infections.

6.6.1 Antimicrobial Polymers

Antimicrobial polymers are polymers possessing intrinsic antibacterial activity. The bacterial cell walls are negatively charged and therefore it is not surprising that polymers containing positively charged quaternary ammonium groups have been extensively studied for use as potential biocides [182–184]. The membrane-disrupting activity of antimicrobial polymers is strongly affected by their hydrophobic/hydrophilic balance. It is believed that upon interaction with a cell membrane, the antimicrobial polymer undergoes a conformational change adopting an amphiphilic structure with the hydrophobic parts being in contact with the lipid domains of the membrane while the hydrophilic positively charged moieties are oriented toward the negatively charged surface of the membrane. Therefore,

Table 6.3 Methods to Prepare Drug-Loaded Polymeric Nanomaterials

Synthetic Strategy	Method	Polymer	Loaded Antibiotic	Reference(s)
By in situ polymerization of the corresponding monomers	Emulsion polymerization	Poly(isohexylcyanoacrylate)	Ampicillin	[136]
		Poly(n-butylcyanoacrylate)	Moxifloxacin	[140]
		Poly(isobutylcyanoacrylate)	Ciprofloxacin	[141]
	Mini-emulsion polymerization	Poly(n-butylcyanoacrylate)	–	[144]
	Interfacial polymerization	Poly(isobutylcyanoacrylate)	–	[137,147]
	Interfacial polycondensation	Polyaminoamide which is a copolymer of phthaloyl chloride and diethylenetriamine	–	[148]
From preformed polymers	Nanoprecipitation	Poly(D, L-lactic-co-glycolic) acid (PLGA)	Azythromycin	[153,154]
		PLGA	Levofloxacin	[155]
	Emulsion-solvent diffusion	Poly(ε-caprolactone) (PCL)	–	[157]
		PLGA	–	[158]
		PLGA	Rifampicin	[160]
		Pegylated PLGA (PEG-b-PLGA) diblock copolymer	Roxithromycin	[161]
		PLGA	Tobramycin	[177]
		PEG-b-PLGA	Amphotericin B	[178]
		Eudargit S100	Amoxicillin	[179]
	Single emulsification-solvent evaporation	PLGA	Levofloxacin	[155]
		PEG-b-PLGA	Azithromycin+ Rifampicin	[63]
		PEG-b-PLGA	Gentamicin	[180]
		PLGA high glycolic acid content (10:90)	Meropenem	[181]
	Double emulsification-solvent evaporation	PLGA	Ciprofloxacin	[163]
		Eudargit RS100 and RL100	Ciprofloxacin	[164]
		PLGA	Levofloxacin	[155]
		PLGA	Vancomycin	[165]
		PEG-b-PLGA	Gentamicin	[180]
	Salting-out	Methacrylic acid-co-ethylacrylate (MAEA) copolymer	Isoniazid	[166]
	Dialysis	Lactose-conjugated PLGA	Rifampicin	[169]
	Electrospinning	Poly(Lactic acid) (PLA)/PCL blends	Tetracycline	[173]
		Poly(ester urethane) urea (PEUU) based on poly(caprolactone)diol and putrescine/ PLGA blends	Tetracycline	[172]
		PCL/gelatin/metronidazole-loaded halloysite nanotubes mixtures	Metronidazole	[175]
	Electrospraying	PLGA	Metronidazole	[176]

manipulation of the hydrophobic/hydrophilic balance greatly impacts the selective toxicity of the antimicrobial polymer which is intended to damage only the microbial membranes and should not alter mammalian cell membranes. But the overall balance between hydrophilic and hydrophobic residues is not the only thing that matters, the spatial distribution of these moieties being equally important. For instance, highly organized micelles of amphiphilic triblock polycarbonate copolymers having the hydrophobic region localized at the micellar core and the hydrophilic positively charged moieties exposed on the surface are very active bactericides toward antibiotic-resistant Gram-positive bacteria, but are almost inoffensive against Gram-negative ones. This is due to the different chemical nature of the outermost layer of the cell walls in the two types of bacteria. Exposure of the positively charged moieties on the nanoparticle surface enables the effective interaction with the negatively charged teichoic acid residues inserted within the peptidoglycan outermost layer of the cell wall of Gram-positive bacteria while hiding the hydrophobic moieties in the micelle core is unsuitable for an efficient association with the outermost lipopolysaccharide shell of Gram-negative bacteria [185,186]. Different synthetic strategies have been developed in order to gain control over the spatial distribution of the hydrophobic and charged moieties within the polymer backbone. In one approach called "segregated monomer," a statistical copolymer is synthesized by random copolymerization of a hydrophobic monomer and a cationic monomer, respectively [187]. Another approach is to polymerize "facially amphiphilic monomers" bearing segregated charged and non-polar moieties on each repeating unit so that in the final macromolecule these moieties will be positioned on opposite sides of the homopolymer backbone [187]. A third approach referred to as the "same center" approach involves the synthesis of polymers bearing a pendant non-polar alkyl chain that is directly bound to the positive charge carrying moiety [188]. Some of the most illustrative examples are given here. In the following, the polymers with quaternary nitrogen atoms will be classified according to their molecular architectures and chemical composition.

6.6.1.1 Linear Polymers Based on Diallylammonium Monomers

Timofeeva et al. [189,190] have synthesized new poly(N,N-diallyl, N-methylammonium trifluoroacetate) (poly(DAMATFA)) **1** and poly(N,N-diallylammonium trifluoroacetate) (poly(DAATFA)) **2** by the ammonium persulfate-initiated radical cyclopolymerization of the equimolecular salts with trifluoroacetic acid of the corresponding secondary and tertiary diallylamines, respectively. The radical polymerization reaction depicted in Scheme 6.1 proceeds under mild conditions in aqueous solution. The obtained polycationic electrolytes showed high biocidal activity toward *E. coli*.

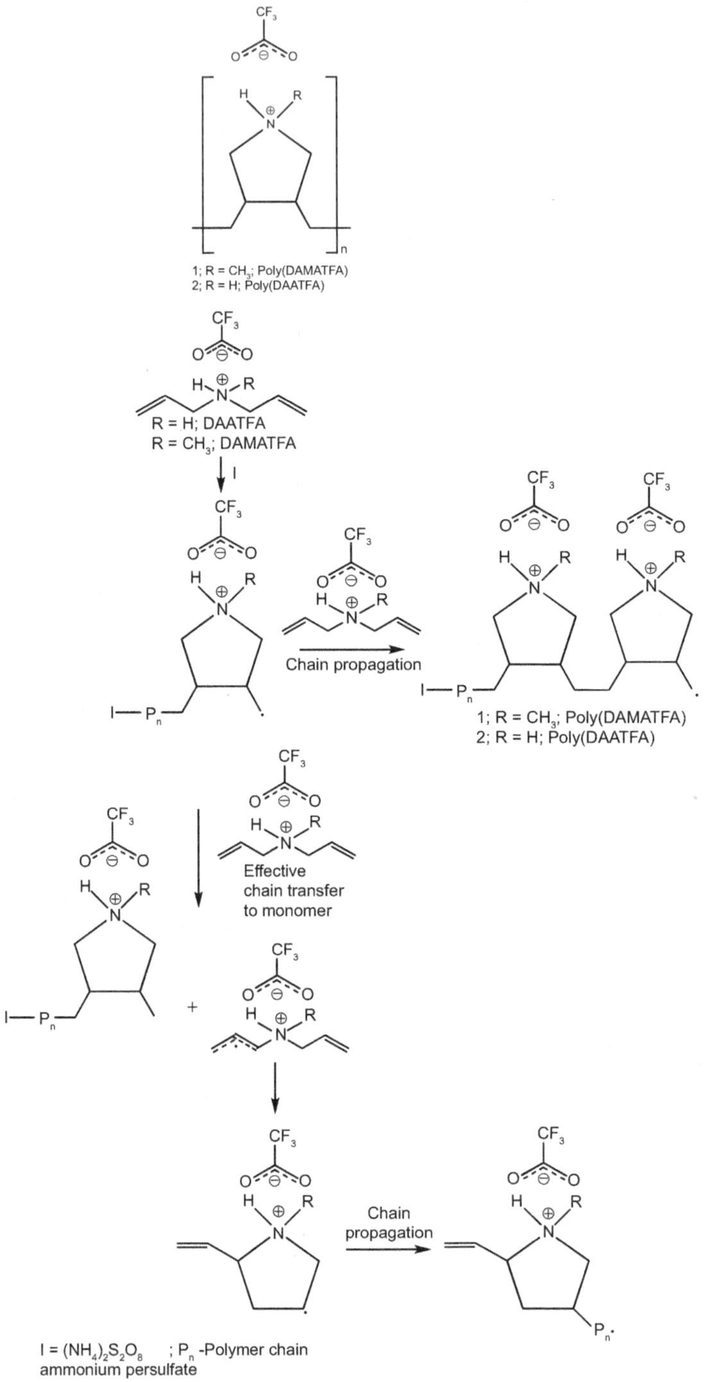

■ **SCHEME 6.1** Synthesis of biocides poly(DAMATFA) and poly(DAATFA) by radical polymerization of the corresponding tertiary and secondary diallylammonium trifluoroacetates.

6.6.1.2 Linear Polymers Based on Acrylic and Methacrylic Monomers

In a "segregated monomer" approach, Kuroda and coworkers [191–193] have synthesized a series of cationic amphiphilic random copolymers containing primary ammonium groups and hydrophobic alkyl side chains. The synthesis depicted in Scheme 6.2 involves the free radical copolymerization of 2-((*tert*-butoxycarbonyl)amino)ethyl methacrylate and various esters of methacrylic acid (C_1, C_2, C_4, and C_6 alkyl esters as well as benzyl ester) using AIBN as a free radical initiator. In order to obtain low molecular weight copolymers, 3-mercaptomethylpropionate (MMP) was used as a chain transfer agent. Eventually deprotection of *tert*-butoxycarbonyl (Boc) groups in trifluoroacetic acid (TFA) led to copolymers **3**.

Increasing the mole fraction of the hydrophobic segments (x) resulted in increased antibacterial activity but the undesirable hemolytic activity increased too [192]. Recently, Punia et al. have synthesized pegylated copolymers using poly(ethylene glycol) methyl ether methacrylate (PEGMA) as a co-monomer [194,195]. Both pegylated terpolymer **4** prepared through the "monomer segregated" approach and random copolymer **5** obtained according to the "same center" strategy were investigated for their antibacterial and

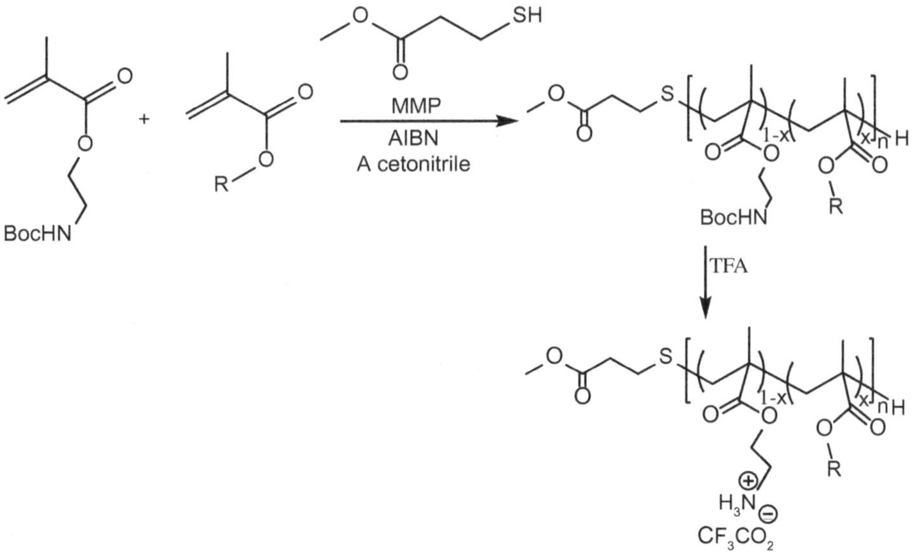

■ **SCHEME 6.2**   Synthesis of biocidal random methacrylate copolymers **3** ("segregated monomer" approach).

hemolytic activities. Copolymer **5** in which the hydrophobic hexyl chain and the cationic ammonium moiety are both on the same repeating unit maintained high antibacterial activity toward *E. coli* while it exhibited 1300-fold less hemolytic activity as compared to the corresponding unpegylated homopolymer [195]. For the other topology with the cationic and lipophilic groups being placed on different monomer units like in terpolymer **4**, pegylation has little or no influence on the hemolytic activity [194].

4

5

Using similar synthetic protocols, the membrane-disrupting methacrylamide random copolymer **6** has been obtained and tested for its antibacterial activity against Gram-negative *E. coli* and Gram-positive *S. aureus* bacteria. Copolymer toxicity was estimated in terms of its hemolytic activity. For mole fraction ranging from 0 to 0.2, the activity against *E. coli* was diminished as the mole fraction of the hydrophobic hexyl chain was increased. Further increasing the mole fraction of the hydrophobic substituent dramatically enhanced the antibacterial activity of the copolymer against Gram-negative *E. coli*. On the other hand, the activity toward Gram-positive *S. aureus* monotonically decreased with increasing hydrophobicity. The hemolytic activity was enhanced continually with increasing the mole fraction of the hydrophobic hexyl side chain [196].

6

6.6.1.3 Linear Polycarbonates

Qiao *et al.* [186] synthesized linear polycarbonates **11** with various molecular weights and charge densities by random metal-free organocatalytic ROP of cyclic carbonates **8** and **9** using benzyl-protected 2,2-bis(methylol) propionic acid (**10**) as initiator. Along with the polymer synthesis, Scheme 6.3 depicts the syntheses of both monomers 5-methyl-5-(3-chloropropyl) oxycabonyl-1,3-dioxan-2-one (**8**) and 5-methyl-5-ethyloxycabonyl-1,3-dioxan-2-one (**9**), respectively, and the synthesis of the initiator **10**. *N*-(3,5-trifluoromethyl)phenyl-*N'*-cyclohexylthiourea (TU) and 1,8-diazabicyclo [5,4,0]undec-7-ene (DBU) were used as organocatalysts for the controlled ROP [197,198]. The random polycarbonate copolymers self-assemble in simulated bacterial growth medium forming highly dynamic micellar aggregates, which readily undergo dissociation at the bacterial membrane. By carefully manipulating the overall hydrophobic/charged moieties composition the authors were able to find the optimal ratios at which the copolymers exhibited increased interaction with bacterial cell membranes being active against both Gram-positive (*S. aureus*) and Gram-negative (*E. coli* and *P. aeruginosa*) bacteria. These random polycarbonate copolymers also showed selective toxicity toward bacteria over mammalian cells as demonstrated by hemolysis tests performed on rat red blood cells [186].

6.6.1.4 Linear Polymers Containing Aromatic Heterocyclic Structures

The most known synthetic amphiphilic polymers in this category belong to poly(vinylpyridine)s. Their synthesis is easy, involving AIBN-initiated radical copolymerization of 4-vinylpyridine with various esters of methacrylic acid followed by quaternization of the pyridine moieties within the polymer with different haloalkanes. Both "segregated monomer" and "same center" approaches, as well as a combination of these two strategies, namely "the both centers" approach, are illustrated in Scheme 6.4 [188]. The highest membrane-disrupting activity evaluated in the terms of MICs was found for copolymers prepared by "segregated monomer" approach. Unfortunately, space separation also favored enhancement of mammalian cell toxicity.

6.6.1.5 Amphiphilic Polynorbornene and Polyoxanorbornene Derivatives

Amphiphilic polymers with tunable antibacterial and hemolytic activities have been prepared by Tew and coworkers from the appropriate norbornene and oxanorbornene derivative monomers [187,199–202] through ROMP using Grubb's third-generation catalysts [203]. Facially

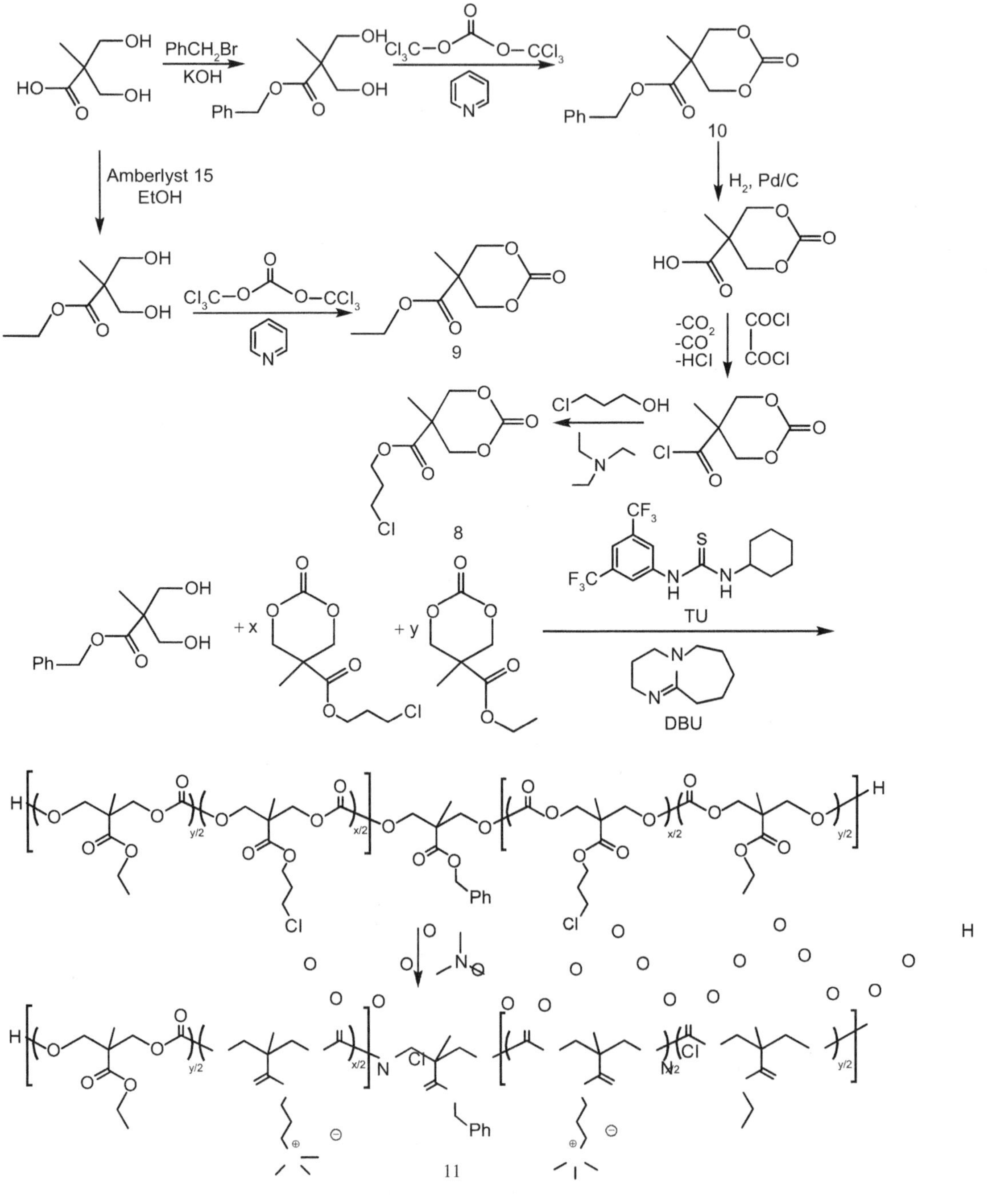

■ **SCHEME 6.3** Synthesis of bactericidal polycarbonates **11**.

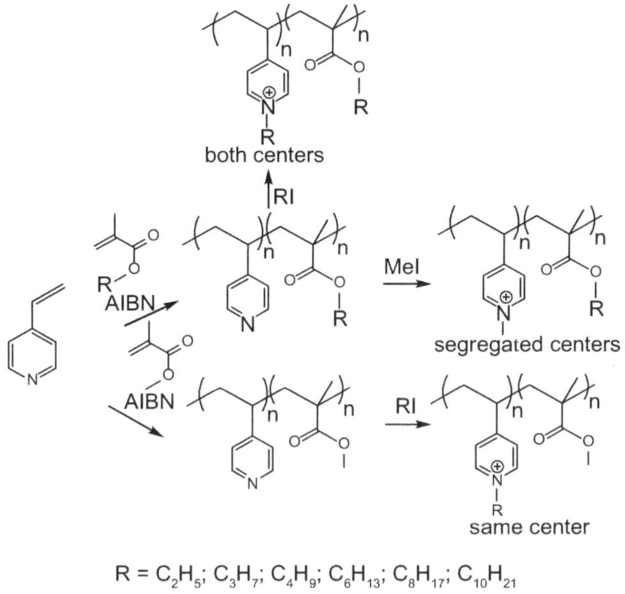

both centers

RI

MeI

segregated centers

AIBN

RI

same center

$R = C_2H_5;\ C_3H_7;\ C_4H_9;\ C_6H_{13};\ C_8H_{17};\ C_{10}H_{21}$

■ **SCHEME 6.4** Synthesis of pyridinium polymers with different topologies with respect to the positive charged moieties and the pendant alkyl chains.

amphiphilic norbornene monomers **12** and **13** were synthesized for further ROMP polymerization and copolymerization as depicted in Scheme 6.5 [199,200]. Similarly, the oxanorbornene derivatives **15**, **16**, and **18** have been prepared and utilized as monomers for the synthesis of the copolymer **17** having a segregated topology with charged and hydrophobic moieties placed on different repeating units, and of the facially amphiphilic homopolymer **19**, respectively, in which the same monomer bears both cationic and hydrophobic groups albeit bound to separate centers (see Scheme 6.6) [187,201].

"Same center" amphiphilic polyoxanorbornenes having the hydrophobic alkyl tails directly bound to the positively charged groups have been also prepared [204] as shown in Scheme 6.7.

Among these polymers differing from each other by the relative topographical positions of the hydrophilic and hydrophobic moieties, some facially amphiphilic copolymers of oxanorbornene derivative monomers **18** showed the most remarkable properties, especially outstanding selective toxicity. Homopolymer **19** (R= CH_3) is inactive and nontoxic, while homopolymer **19** (R= $n\text{-}C_3H_7$) is both highly bactericidal and hemolytic. As expected, copolymerization of methyl and propyl monomers **18**

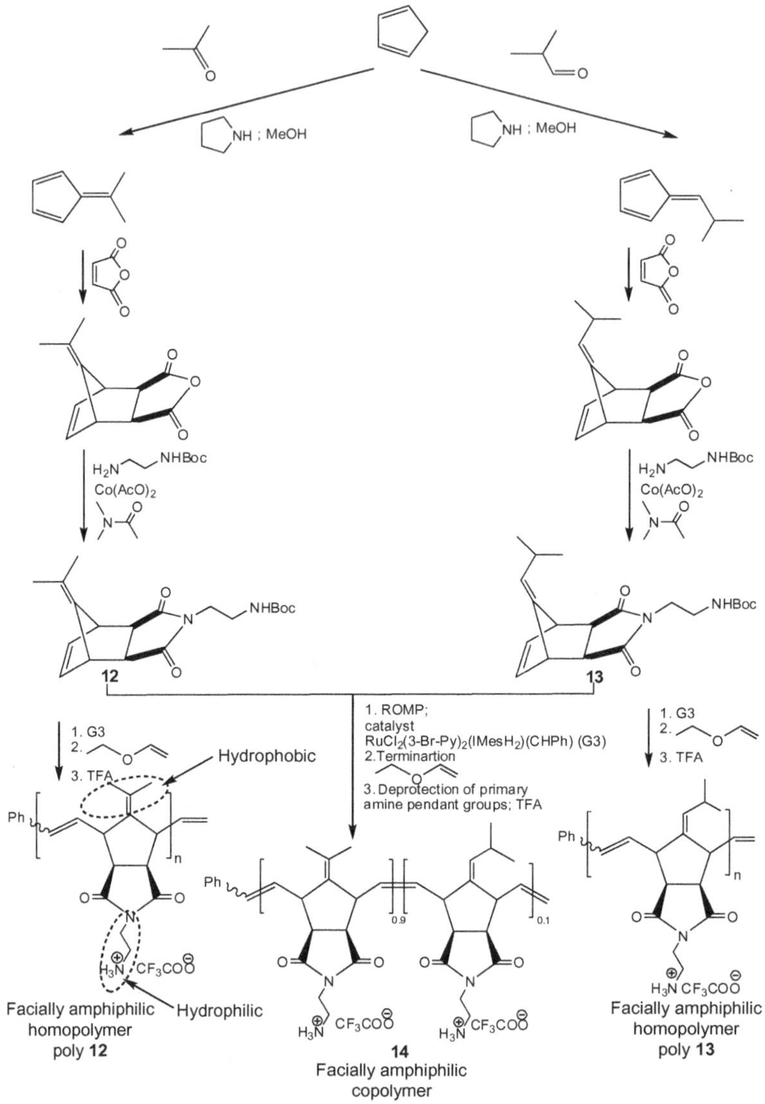

■ **SCHEME 6.5** Facially amphiphilic polynorbornene homo- and co-polymers.

at various molar ratios (methyl monomer/propyl monomer= 1/9; 5/5, and 9/1) resulted in copolymers with low hemolytic activity while the high bactericidal activity was preserved. In other words, the selectivity defined as the ratio HC_{50}/MIC_{90} was tremendously increased (HC_{50}=hemolytic concentration lysing 50% of blood cells; MIC_{90}=minimal inhibitory concentration preventing 90% bacterial growth). The authors [201] were

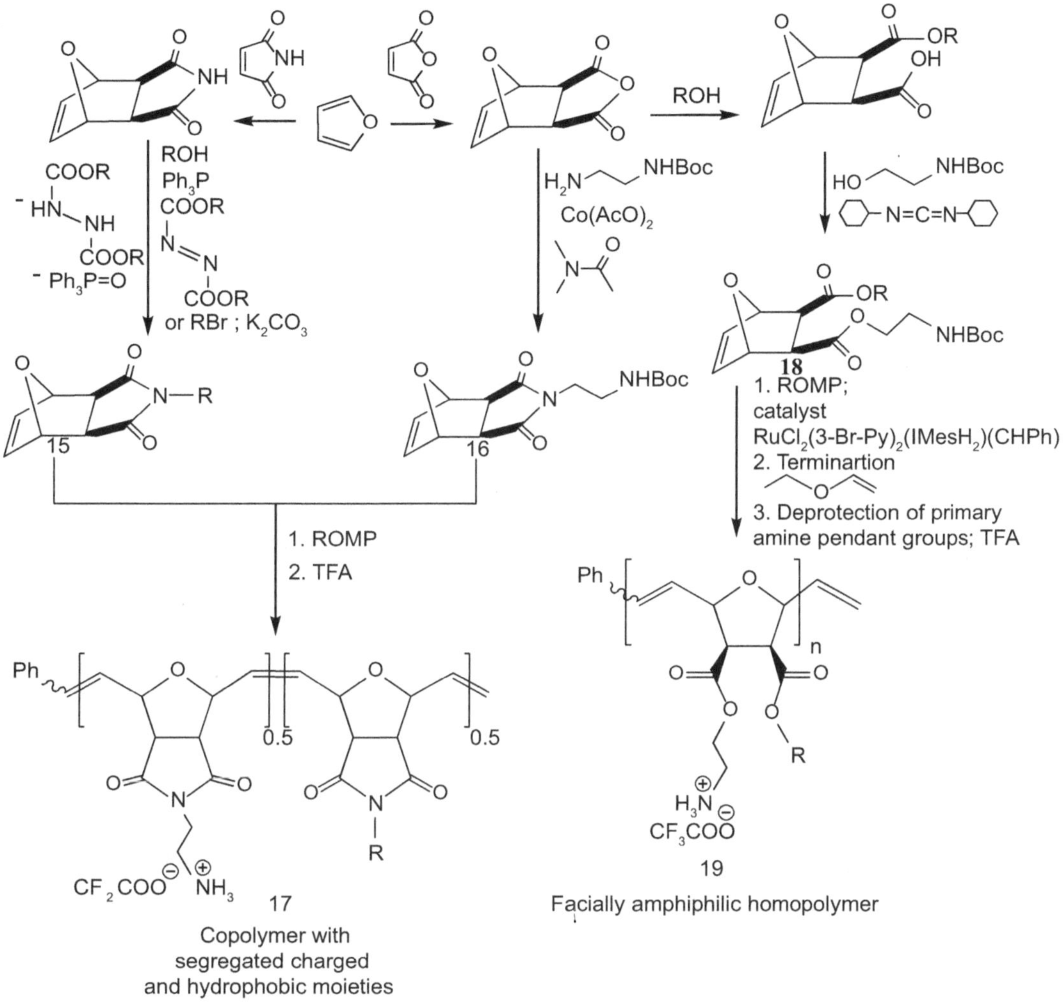

■ SCHEME 6.6 Polyoxanorbornenes prepared by the "segregated monomer" and "facially amphiphilic monomer" approaches, respectively.

able to obtain copolymers with as much as 533-fold higher selectivity for bacteria over mammalian cells. Moreover, these copolymers were double selective, being 53 times more active against *S. aureus* than against *E. coli*. Such an exceptional selectivity against *S. aureus* is especially desirable since MRSA is one of the "superbugs" threatening hospital patients. On the other hand, selectively targeting a particular type of bacteria is also salutary since a problematic request of antibacterial therapy is how to kill pathogenic bacteria while leaving other beneficial bacteria unharmed.

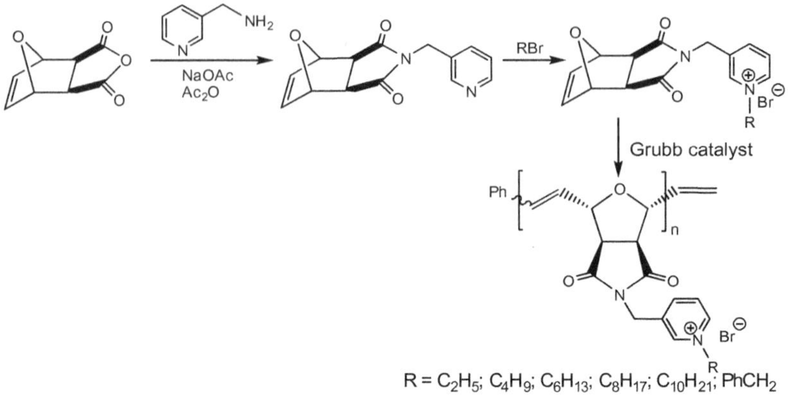
NaOAc
Ac2O
RBr
Grubb catalyst
R = C2H5; C4H9; C6H13; C8H17; C10H21; PhCH2

■ **SCHEME 6.7** Synthesis of polyoxanorbornenes by the "same center" approach.

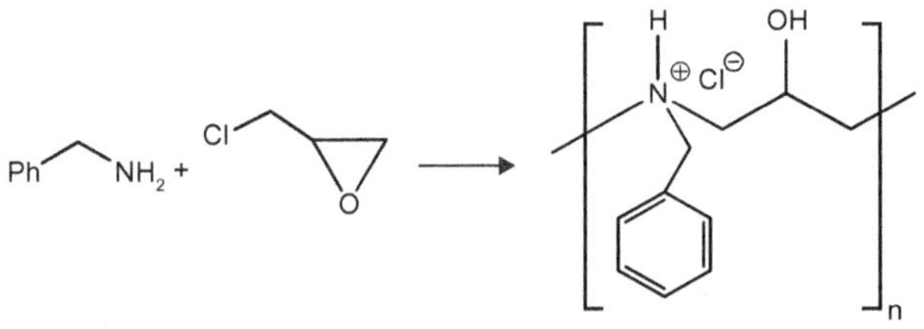

■ **SCHEME 6.8** Synthesis of an ionene possessing antifungal, antibacterial, and antiyeast activities.

6.6.1.6 Ionenes

Ionenes are antimicrobial cationic polyelectrolytes in which the quaternary ammonium groups are part of the polymer backbone. Cakmak et al. [205] have synthesized cationic polyelectrolytes through the polycondensation reaction between epichlorhydrin and benzyl amine (see Scheme 6.8).

6.6.1.7 Block Copolymers

Hedrick and coworkers [185] have reported the synthesis of triblock polycarbonates of BAB type by the sequential ROP of trimethylene carbonate derivative monomers **8** and **9** initiated from the diol **10** followed by quaternization of the central block with trimethylamine. Hence, the molecular construction kit used by the authors includes: (1) two hydrophobic head and tail B blocks based on monomer **9** that direct polymer self-assembly into a nanoparticle; (2) a middle cationic A block based on quaternized monomer **8** that selectively interacts with the bacterial cell membrane; and (3) a biodegradable polycarbonate backbone that is broken down into nontoxic compounds thereby assuring good biocompatibility. When placed in water, these amphiphilic triblock polycarbonates self-assemble into cationic micellar nanoparticles. These biodegradable nanoparticles exhibited selective membrane-disrupting properties being active against Gram-positive bacteria such as *B. subtilis*, *S. aureus*, and MRSA as well as toward fungi. The biological activity estimated in terms of MICs was proved to be similar to that of the antibiotic vancomycin for *S. aureus* and MRSA and that of amphotericin B for the fungus *Cryptococcus neoformans*. Hemolysis tests were performed to evaluate nanoparticles toxicity in vitro. The polycarbonate cationic micelles did not affect red blood cell membranes. In vivo toxicity assays carried out on mice revealed that no significant liver and kidney damages occurred within the first 14 days after intravenous administration of the micelles. Another advantage is the easiness of the preparation methods that uses nontoxic and inexpensive starting materials. Therefore, these block copolymer micelles appear as a valuable alternative to antibiotics.

6.6.1.8 Antimicrobial Dendrimers

Dendrimers are nearly monodisperse macromolecules with a peculiar highly branched globular molecular architecture resembling the shape of a tree crown. The molecular architecture of a dendrimer comprises three topologically distinct regions: (1) a central core or focal moiety; (2) an exact number of concentric layers of dendritic branches also referred to as generations emanating from the core; and (3) an exact number of functional end groups of the outermost layer of repeating units that is mathematically determinable for each dendrimer generation. Dendrimers can be built up using two distinct synthetic strategies namely the convergent and the divergent route, respectively. In the divergent approach, dendrimer growth starts from the initiator core and proceeds toward the periphery [206], while the convergent build up follows exactly the opposite direction: it is initiated from the exterior and progresses inwards terminating at the core [207]. The divergent method illustrated in Scheme 6.9 for poly(amidoamine) (PAMAM) dendrimers leads to dendrimers with symmetrical architecture, while macromolecules having different dendritic segments or mixed structural elements have become available by using the convergent approach. By engineering dendrimer synthesis, their outer surface can be made hydrophilic, while the specific branched architecture provides hydrophobic cavities at their interior.

Kanan and coworkers [208] have studied the effect of the functional end groups on the periphery of G4-PAMAM dendrimers on the antimicrobial

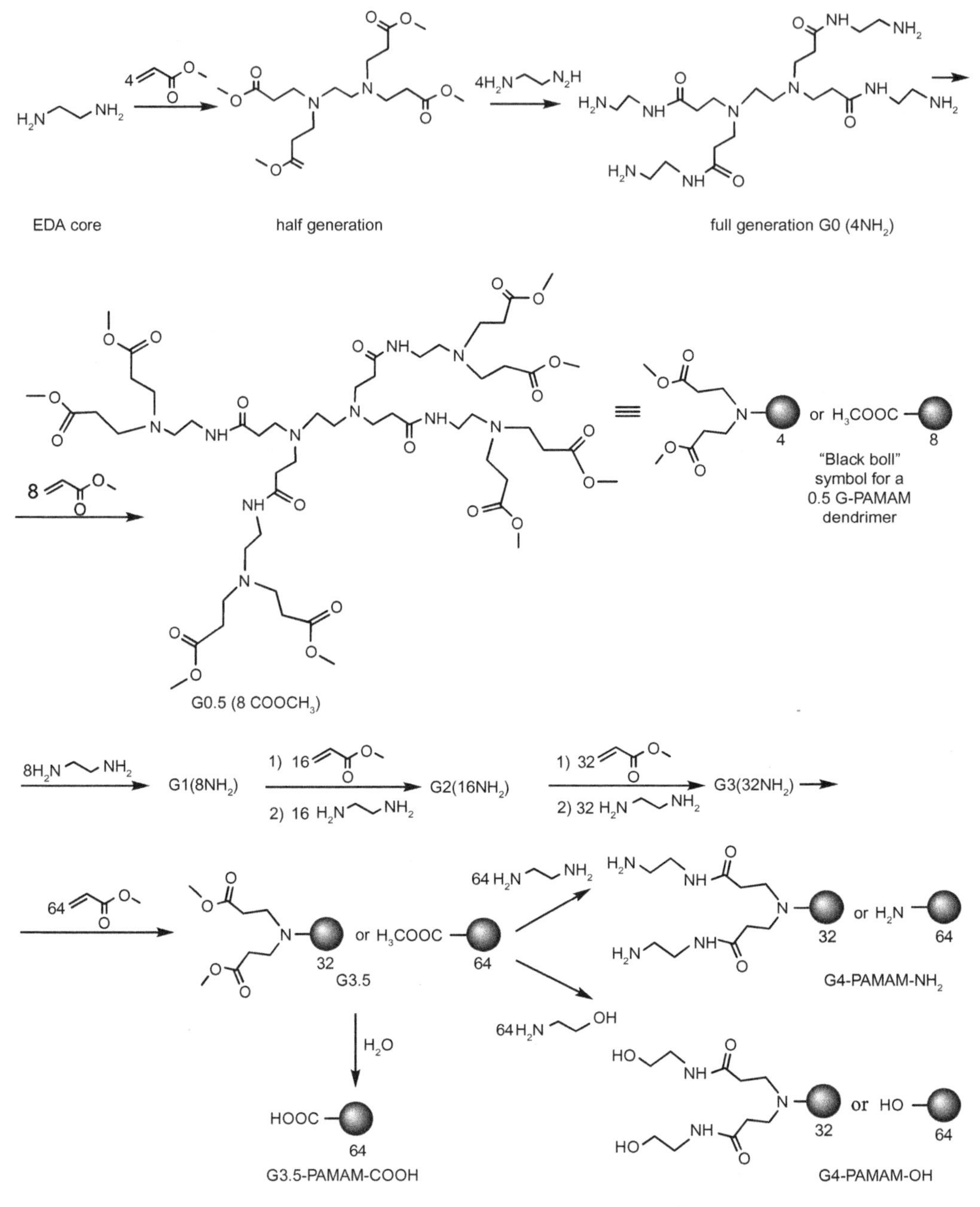

■ **SCHEME 6.9** Iterative divergent synthesis of PAMAM dendrimers with etylenediamine (EDA) core.

activity against *E. coli* in a guinea pig model. The authors have induced chorioamnionitis in pregnant guinea pigs by intra-cervical inoculation with *E. coli*. Five minutes after inoculation, various types of antimicrobial dendrimers, i.e., G4-PAMAM-NH$_2$, G4-PAMAM-OH, and G3.5-PAMAM-COOH were injected into the cervix of animals. Treated animals were compared to a group of animals without treatment (positive control). Another group composed of healthy animals was used as negative control. The levels of interleukins (IL-6 and IL-1β) and of tumor necrosis factor (TNFα) in placenta of the G4-PAMAM-OH treated animals were similar to those of animals in the negative control group and much lower than those in the positive control group. Moreover, the G4-PAMAM-OH has been found to be non-cytotoxic up to 1 mg mL^{-1} concentrations to human cervical epithelial (End1/E6E7) cells and immune cells (BV-2), while G4-PAMAM-NH$_2$ dendrimer albeit highly bactericidal was also highly cytotoxic to above 10 μg mL^{-1}. Different mechanisms seem to underlie the bactericidal activity of the three dendrimers. G3.5-PAMAM-COOH dendrimer acts as a polyanion at the physiological pH and affects the outer membrane of Gram-negative bacterium *E. coli* by chelating the calcium and magnesium ions that stabilize the negatively charged residues on the lipid head groups. G4-PAMAM-NH$_2$ dendrimer acting as a polycation binds to the polyanionic lipopolysaccharide outermost shell of the bacterial cell wall. G4-PAMAM-OH dendrimer alters the permeability of the outer membrane of *E. coli* by hydrogen binding its O-antigens (outer polysaccharide). Cell integrity studies showed that G4-PAMAM-NH$_2$ and G3.5-PAMAM-COOH damaged both outer and inner membranes of *E. coli*, provoking the bacterial lysis. This was the first study to reveal the bactericidal activity of G4-PAMAM-OH dendrimer as well as its low cytotoxicity toward human cells [208].

Following the reactions sequence depicted in Scheme 6.10, Urbanczyk-Lipowska and coworkers [209] have synthesized the antimicrobial dendrimer **22** based on the basic amino acid *L*-lysine. The *tris*-amino acid dendrimer core **20** was prepared in two steps from *L*-lysine through Michael addition to excess acrylonitrile followed by reduction. After protection and deprotection steps the final coupling step using the amino-protected *L*-lysine derivative, 2-Cl-Z-Lys(Boc)-OH **21**, led to the first generation of the dendritic cationic peptide **22** with an electric charge of (+5). The dendritic poly(*L*-lysine) was found to be active against a series of Gram-positive and Gram-negative bacteria including the multi-resistant strains *S. aureus* ATCC 43300 and *E. coli* ATCC BAA-198 and against fungi from the *Candida* genus as well.

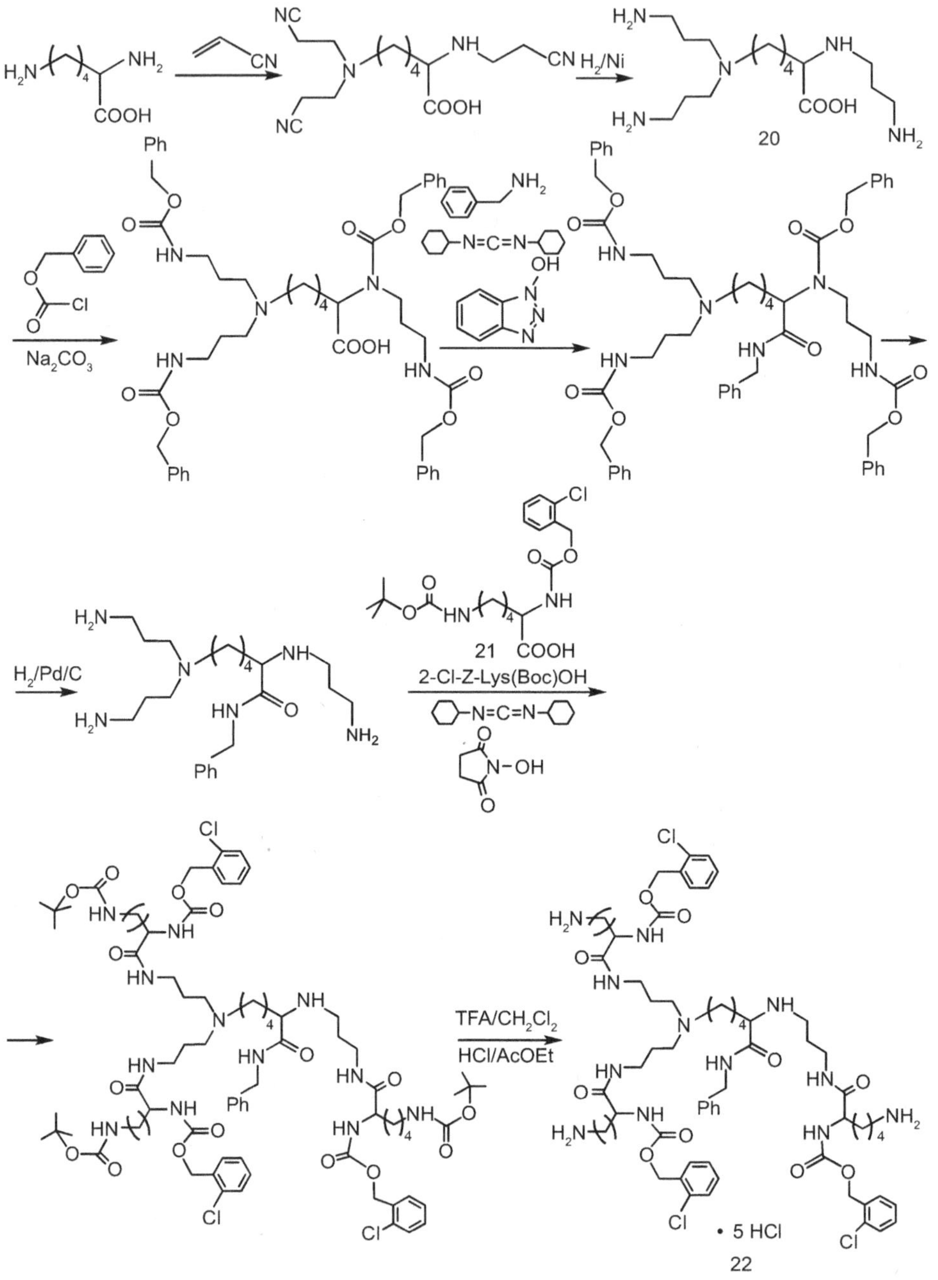

■ **SCHEME 6.10**   Synthesis of the cationic dendritic poly(*L*-lysine) **22**.

6.6.1.9 Polymers Mimicking Natural Cationic Host Defense Peptides

Host defense peptides, e.g., defensins, cathelicidins (LL-37), and magainins, are components of the innate immune system of various species like humans, animals, plants, and invertebrates defending these organisms against unwanted invaders such as pathogenic bacteria, viruses, and fungi. The common structural feature these evolutionarily highly conserved oligopeptides share is the facially amphiphilic conformation i.e., the spatial segregation of cationic and nonpolar amino acid residues onto opposite faces or regions of the folded peptide backbone, thereby conferring them membrane-disrupting properties. For instance, magainin from the African clawed frog has an amphiphilic α-helix secondary structure with pendant cationic and hydrophobic amino acid side chains arranged on opposite faces of the helix [210], while human defensin has an amphiphilic β-sheet secondary structure [211]. Unlike antibiotics that target specifically a single cellular activity for instance DNA synthesis, protein synthesis or cell wall synthesis, cationic host defense peptides have multiple targeting sites varying from the outer membrane to signaling pathways [212–214]. Therefore, it is more difficult for microorganisms to develop resistance to host defense peptides than resistance to antibiotics. Consequently, tremendous efforts have been made to produce synthetic mimics of natural antimicrobial peptides and to establish the structural determinants of antimicrobial activity. The conclusions drawn from multiple studies was that the local amphiphilicity and the ability of the synthetic mimics to self-organize into a conformation with segregated hydrophobic and hydrophilic regions on the interface with the bacterial membrane are the two crucial factors for antimicrobial activity. With respect to these requirements, all the antimicrobial polymers presented so far are in fact synthetic mimics of host defense peptides.

Keeping in mind the above requirements, the groups of DeGrado [215] and Tew [216–218] have designed and synthesized facially amphiphilic arylamide oligomers **25** by the reactions sequence shown in Scheme 6.11. Arylamides **25** are inexpensive and can be obtained in very high yields through polycondensation between the facially amphiphilic diamine monomer **23** and the hydrophobic isophthaloyldichloride **24**. The molecular design of these oligomers stabilizes the facially amphiphilic structure. Indeed, hydrogen bonding between the thioester groups and hydrogens on the amide groups prevents rotation around the sp^2C-N bond, locking the oligomer backbone into a conformation in which the pendant hydrophobic t-butyl groups and the hydrophilic ammonium groups are segregated to opposite sides of the molecule. The arylamide oligomer with n = 8 showed

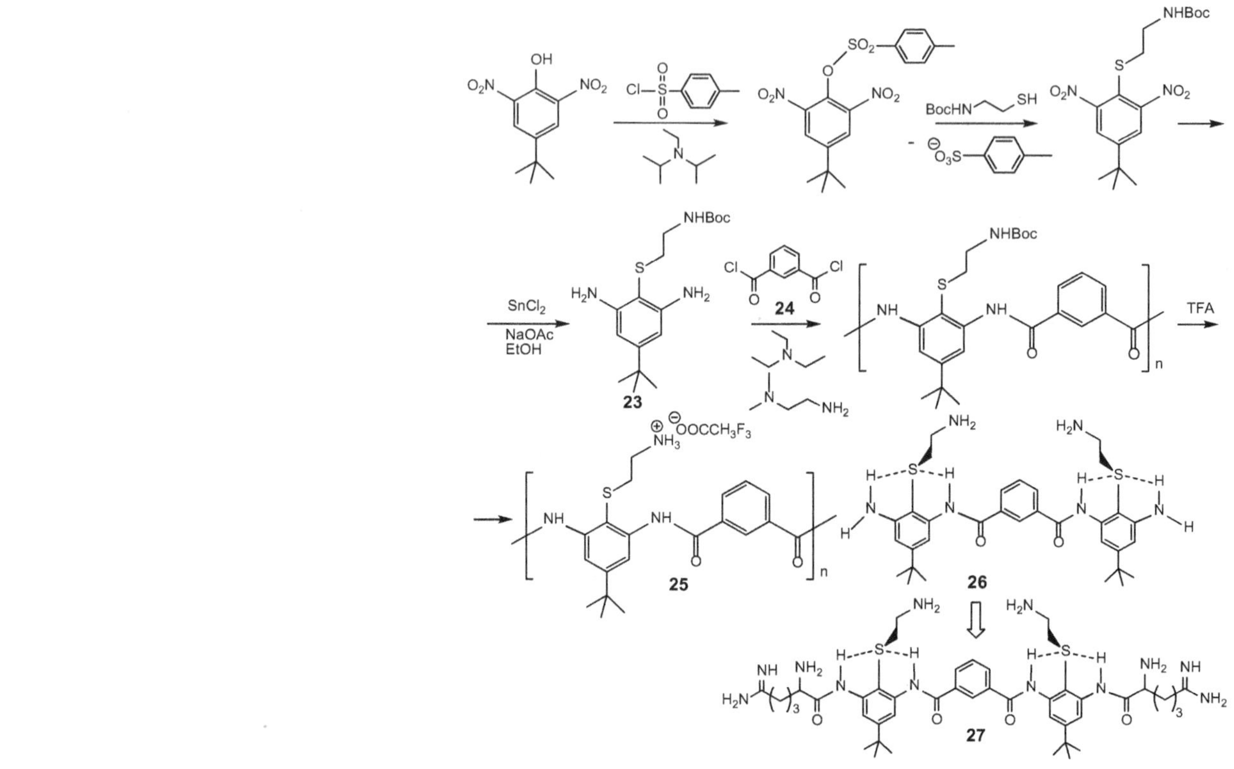

■ **SCHEME 6.11** Synthesis of facially amphiphilic arylamide oligomers and further optimization of the balance between the cationic hydrophilic moieties and the hydrophobic moieties.

high antibacterial activity against both Gram-positive and Gram-negative bacteria. MIC_{90} values against *B. subtilis* and *E. coli* were as low as 16 and $7.5\,\mu g\,mL^{-1}$, respectively [216]. However, toxicity toward human red blood cells was also very high. Therefore, in order to increase selectivity, hydrophobicity had to be reduced and the oligomer **26** with only three aromatic rings was synthesized and tested. The latter exhibited lower hemolytic activity while it maintained reasonable antibacterial activity. Therefore, it served as the starting point for further substituent optimization.

Introduction of the dibasic amino acid arginine with hydrophilic positively charged end groups resulted in a spectacular increase in selectivity. Compound **27** exhibited the following MIC_{90} values: $6.25\,\mu g\,mL^{-1}$ against *E. coli* and $12.5\,\mu g\,mL^{-1}$ against *S. aureus*, while HC_{50} was $715\,\mu g\,mL^{-1}$ resulting in the remarkable selectivity values of 110 and 57, respectively [215].

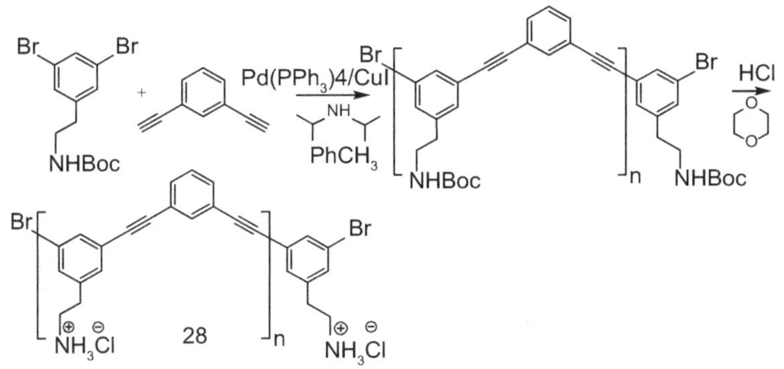

■ **SCHEME 6.12** Synthesis of nonhemolytic antimicrobial phenylene ethynylene oligomers **28**.

Tew and coworkers have also synthesized phenylene ethynylene oligomers through Sonogashira coupling between the appropriate aromatic dihalogeno derivatives and *m*-diethynylbenzene as shown in Scheme 6.12. Contrary to the previously presented arylamide oligomers, amphiphilic phenylene ethynylene oligomers have no intramolecular hydrogen bonds. Hence, rotation of the repeating units around the single bonds of the backbone is allowed. It has been proved that free rotation enables these amphiphilic oligomers to orient their hydrophilic and hydrophobic segments in such a way as to adopt a facially amphiphilic conformation on contact with a hydrophilic-hydrophobic (water-oil) interface. It is presumed that a similar self-organization process would occur upon contact with the cell membrane [218]. The phenylene ethynylene oligomer with three aromatic rings **28** (n = 1) was found to be bactericide and nonhemolytic exhibiting a selectivity of 440 for *S. aureus* over erythrocytes [219,220]. Oligomer **28** is also much more selective than **29** (see Scheme 6.13) which has an additional hydrophobic pentyl chain in one of the aromatic rings in the repeating unit [221,222].

6.6.1.10 Polysaccharides

Chitosan is a linear polysaccharide that is obtained through partial deacetylation of chitin on treatment with aqueous sodium hydroxide. Therefore, chitosan is a random copolymer of 2-acetamido-2-deoxy-*D*-glucose and 2-amino-2-deoxy-*D*-glucose in which the monosaccharide units are joined together by β- 1→4 glycosidic bonds (see Fig. 6.7A). Chitosan exhibits intrinsic antimicrobial activity against fungi, viruses, and bacteria and has also the advantage of being biocompatible, biodegradable and nontoxic.

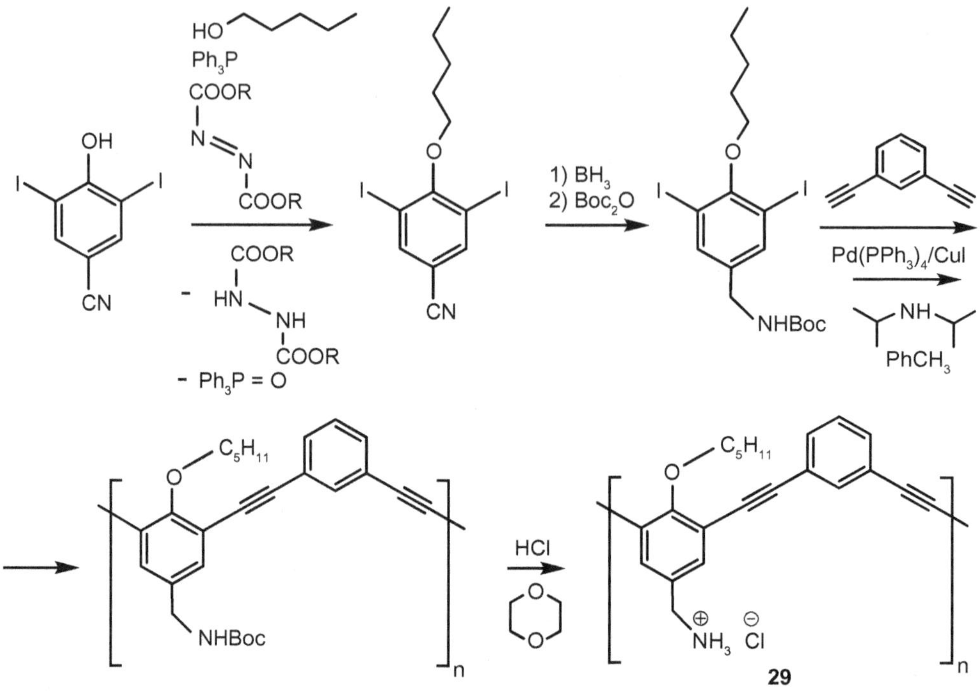

■ SCHEME 6.13   Synthesis of facially amphiphilic phenylene ethynylene oligomers **29**.

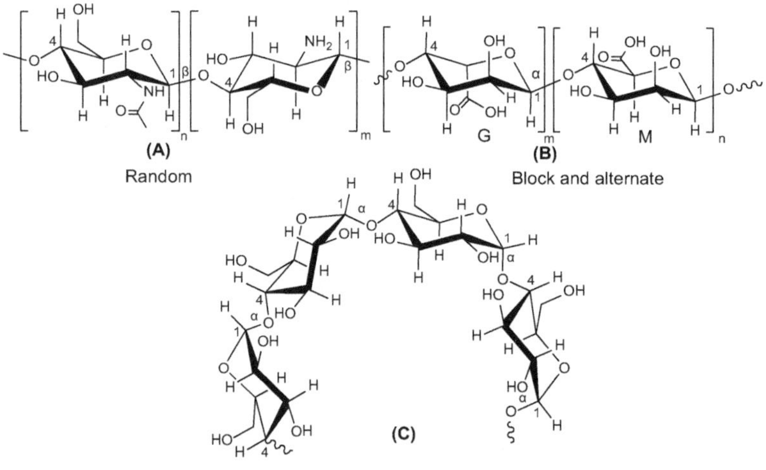

■ FIGURE 6.7   Antimicrobial polysaccharides: (A) chitosan, (B) alginate, (C) amylose.

Chitosan is active against both Gram-positive and Gram-negative bacteria in a molecular weight dependent manner. It was found [223] that low molecular weight chitosan nanoparticles are more active against Gram-negative bacteria and less active toward Gram-positive ones, while the latter can be efficiently combated using nanoparticles containing higher molecular weight chitosan. Chitosan nanoparticles fight against infections produced by microorganisms [47, 53] using the following mechanisms: (1) at pH values lower than 6.5 which is the pK_a of the amino groups, chitosan becomes polycationic and binds to the negatively charged bacterial cell wall causing increased permeability and eventually leakage of intracellular components; (2) chitosan can replace Ca^{2+} ions that stabilize anionic phospholipids causing loss of fluidity and eventually phase separation; thus more and more discrete lipid domains in the membrane undergo a progressive transition to much smaller micelles [224,225]; (3) chitosan binds the negatively charged sugar phosphate backbone of DNA thereby impeding DNA transcription and hence protein synthesis; (4) chitosan might inactivate some metalloproteins through chelation; (5) chitosan helps wound healing by inhibiting the release of inflammatory cytokines and favoring recruitment of fibroblasts and deposition of collagen III. Li et al. have synthesized amphiphilic chitosan nanoparticles by introducing hydrophobic oleoyl chains through acylation with oleoyl chloride of some amino groups in the polysaccharide backbone [226]. The amphiphilic oleoyl-modified chitosan nanoparticles have been prepared through an oil-in-water emulsification method and were tested for the bacterial activity against *E. coli* and *S. aureus* using various experimental methods such as: (1) counting viable organisms at different incubation times; (2) measuring integrity of cell membranes through spectrophotometric quantification of the amount of nucleic acids released from the cytoplasm; (3) estimating outer membrane permeabilization by fluorescence-based *N*-phenyl-1-naphthylamine (NPN) uptake assay; (4) estimating inner membrane permeabilization through detection of leakage of cytoplasmic β-galactosidases by measurement of the enzymatic activity using *o*-nitrophenyl-β-*D*-galactoside as a substrate; (5) analyzing the protein content in the cell-free supernatant by sodium dodecyl sulfate-polyacrylamide gel electrophoresis (SDS-PAGE); (6) imaging using transmission electron microscopy (TEM) to visualize the damages produced to the bacterial cells [227,228]. These studies revealed that oleoyl-modified chitosan nanoparticles caused extensive damages to both *E. coli* and *S. aureus* cell membranes as evidenced by rapid release of intracellular components. Another study [229] showed that chitosan synergistically increases the antibacterial activity of sulfamethoxazole against *P. aeruginosa*.

Alginate hydrogels coated with chitosan hydrochloride exhibited antibacterial activity against *E. coli* causing complete killing of bacteria after 24 h of contact. Cytotoxicity tests performed on mesenchymal stromal cells (MSC) revealed neither substantial morphological changes nor growth inhibition. Moreover, release studies carried out using encapsulated rhodamine B as a model drug evidenced a sustained release profile. All these properties, i.e., antibacterial activity, lack of cytotoxicity, and sustained release profile, make these chitosan coated alginate hydrogels promising candidates for novel medicated wound dressings [230]. Alginate hydrogels were prepared as follows: calcium carbonate was finely dispersed in distilled water by sonication and sodium alginate was dissolved in the suspension under stirring. Then, glucono delta lactone (GDL) was added and slowly hydrolyzed into gluconic acid thereby causing dissolution of calcium carbonate particles and gelation of alginate. Coating with chitosan was achieved by simple immersion into chitosan hydrochloride solution [230]. Alginates are block copolymers consisting in homopolymer segments of (1,4)-linked β-D-mannuronate (M) and α-L-guluronate (G) residues as depicted in Fig. 6.7B.

Potato starch films incorporated with chitosan have been found to have antibacterial activity against *S. aureus* and *E. coli* [231]. Starch is a mixture of amylose (20–25% by weight) and amylopectin (75–80%). Amylose is a polymer which depending on source consists of 500–20,000 α(1→4)-*D*-glucose units. The α (1→4) bonds promote the formation of a helix structure (see Fig. 6.7C). Amylopectin differs from amylose in being highly branched. Short side chains of about 30 glucose units are attached with α (1→6) linkages approximately every twenty to thirty glucose units along the chain. Amylopectin molecules may contain up to 2 million glucose units.

Table 6.4 summarizes the main classes of antimicrobial polymers by molecular architecture and chemical composition.

6.6.2 Polymer-Antibiotic Conjugates

There are two main ways to synthesize antibiotic-conjugated polymers. In one approach the antibiotic is covalently attached to a preformed polymer. An alternative strategy is based on the synthesis of a prodrug monomer which is subsequently polymerized. A prodrug monomer consists of a polymerizable moiety that is covalently attached to the drug through a cleavable linker [232]. As compared to the synthesis of polymer-drug conjugates starting with preformed polymers, polymerization of prodrug monomers allows higher drug loadings to be achieved, but at the same time its

Table 6.4 Main Types of Antimicrobial Polymers

Type of Polymer by Molecular Architecture	Type of Polymer by Chemical Composition	Antimicrobial Activity	Reference(s)
Linear homopolymers	Poly(N,N-diallyl, N-methylammonium trifluoroacetate) (poly(DAMATFA)) and poly(N,N-diallylammonium trifluoroacetate) (poly(DAATFA)) 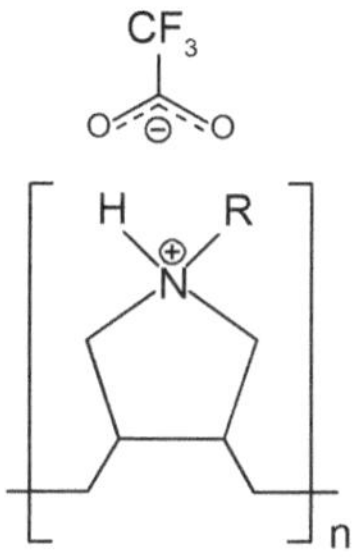 1; R = CH_3; Poly(DAMATFA) 2; R = H; Poly(DAATFA)	*E. coli*	[189,190]
Linear random copolymers	Poly[(2-amino)ethyl methacrylate-*co*-alkyl methacrylate] 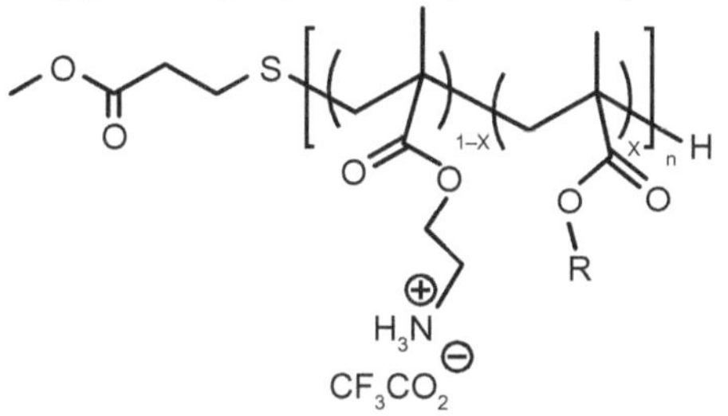 R = CH_3, C_2H_5, n-C_4H_9, n-C_6H_{13}, $C_6H_5CH_2$	*S. aureus; E. coli*	[191–193]
	Poly[(2-amino)ethyl acrylate-*co*-hexylacrylate-*co*-poly(ethylene glycol) methyl ether methacrylate] 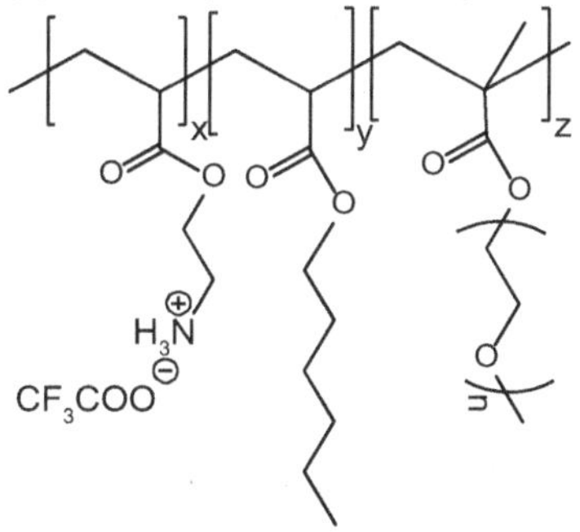 "monomer segregated" approach		[194,195]

Table 6.4 Main Types of Antimicrobial Polymers (Continued)

Type of Polymer by Molecular Architecture	Type of Polymer by Chemical Composition	Antimicrobial Activity	Reference(s)
	Poly[(6-aminohexyl)acrylate-*co*-poly(ethylene glycol) methyl ether methacrylate] 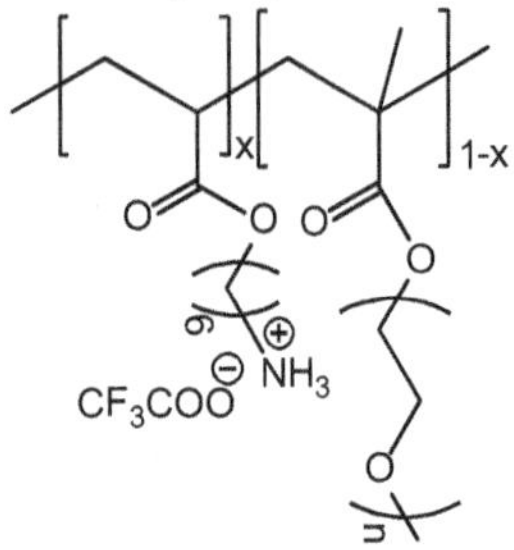 "same center" approach		[194,195]
	Poly[(3-aminopropyl) methacrylamide-*co*-hexyl methacrylamide] 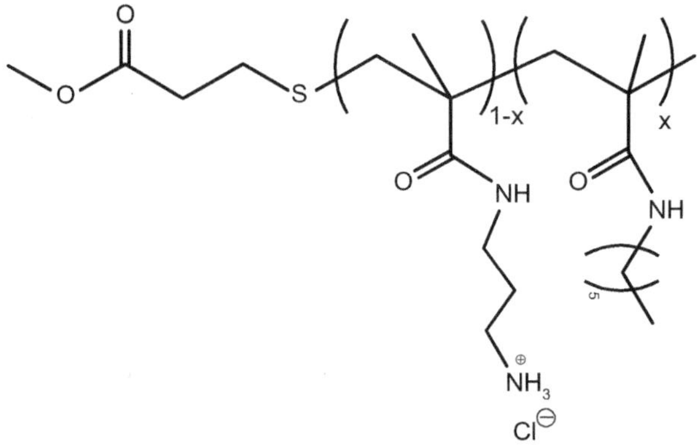		[196]
Linear random copolymers	Benzyl-2,2-bis(methylol)propionate-poly{[5-methyl-5-(3-trimethylaminopropyl chloride)oxycarbonyl 1,3-trimethylene carbonate]-*co*-(5-methyl-5-ethoxycarbonyl 1,3-trimethylene carbonate)} 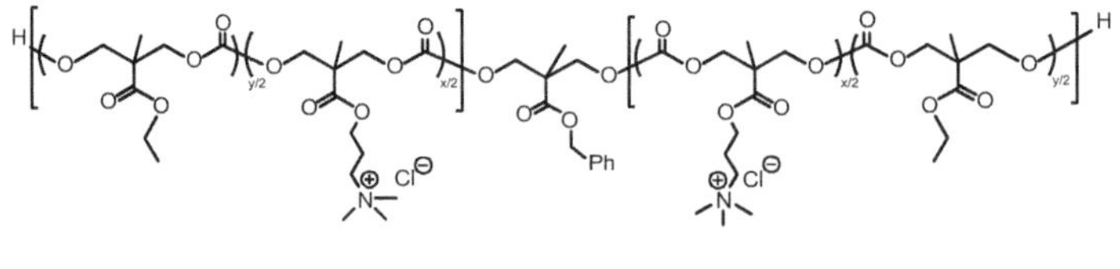 bis-MPA-p[MTC-O$(CH_2)_3N^+(CH_3)_3Cl^-$]-*co*-p[MTC-OEt]	*S. aureus*; *E. coli*; *P. aeruginosa*	[186]

(*Continued*)

Table 6.4 Main Types of Antimicrobial Polymers (Continued)

Type of Polymer by Molecular Architecture	Type of Polymer by Chemical Composition	Antimicrobial Activity	Reference(s)
Linear random copolymers	poly[(4-vinylmethylpyridinium)-co-alkyl methacrylate] 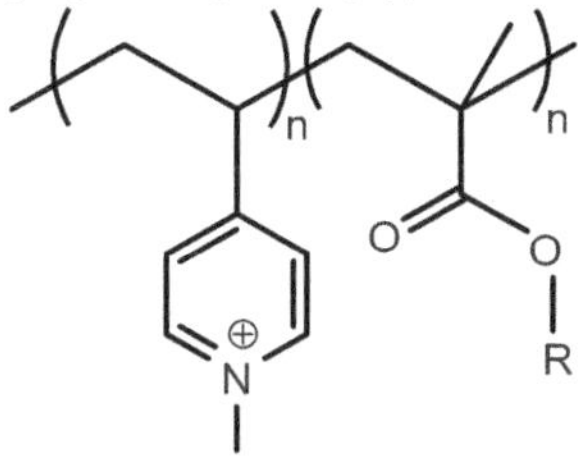 segregated centers Poly[(4-vinylalkylpyridinium)-*co*-methyl methacrylate] 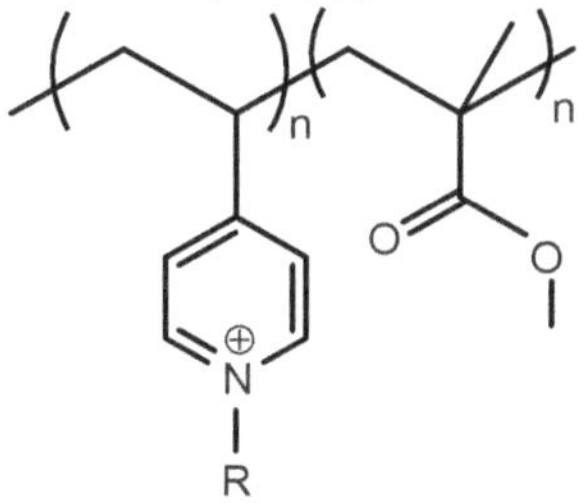 same center Poly[(4-vinylalkylpyridinium)-*co*-alkyl methacrylate] 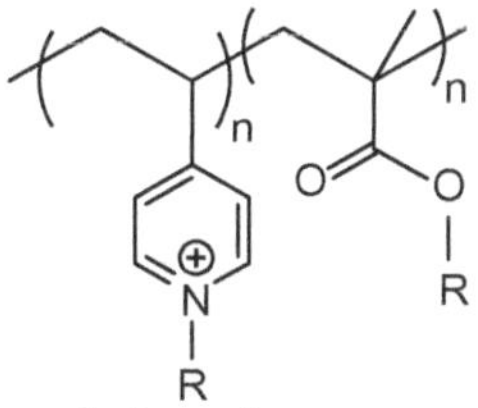 both centers	*E. coli*; *B. subtilis*	[188]
Linear homopolymers	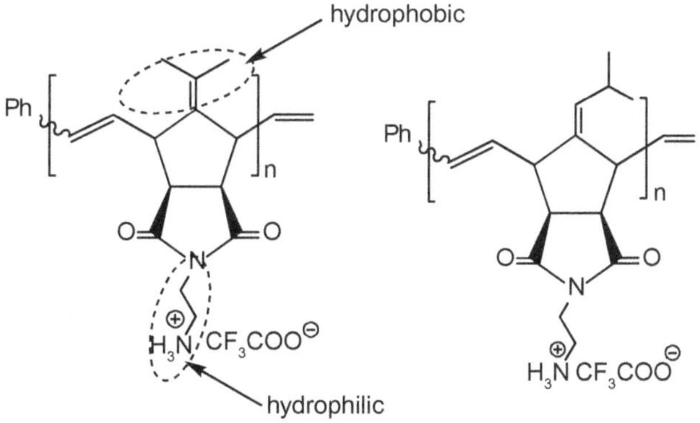 facially amphiphilic norbornene homopolymers	*S. aureus*; *E. coli*	[199,200]

(*Continued*)

Table 6.4 Main Types of Antimicrobial Polymers   (Continued)

Type of Polymer by Molecular Architecture	Type of Polymer by Chemical Composition	Antimicrobial Activity	Reference(s)
Linear random copolymers	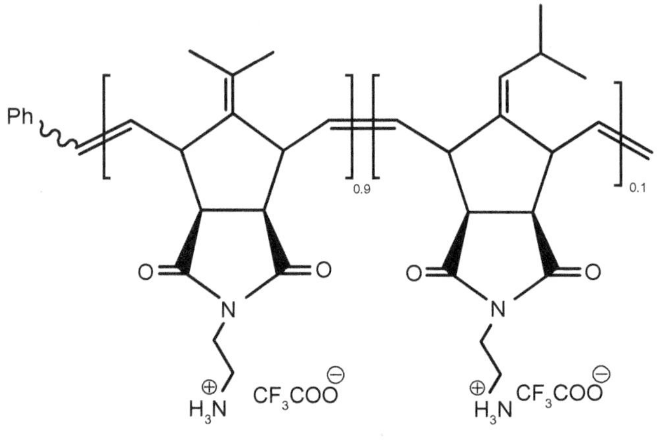 facially amphiphilic norbornene copolymers	*S. aureus; E. coli*	[199,200]
Linear homopolymers	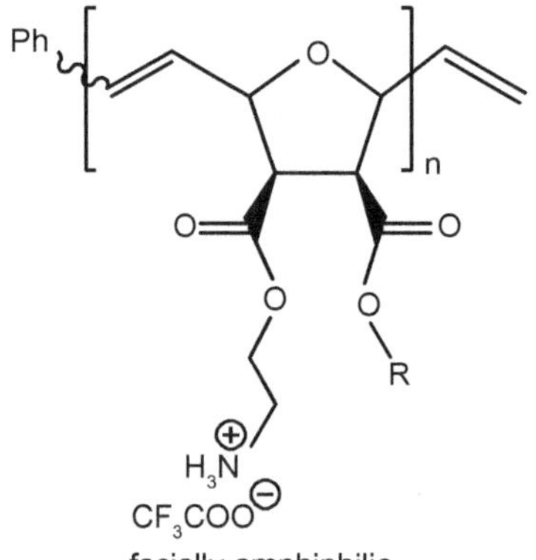 facially amphiphilic oxanorbornene homopolymers	*S. aureus; E. coli*	[187,201]

(Continued)

Table 6.4 Main Types of Antimicrobial Polymers (Continued)

Type of Polymer by Molecular Architecture	Type of Polymer by Chemical Composition	Antimicrobial Activity	Reference(s)
Linear random copolymers	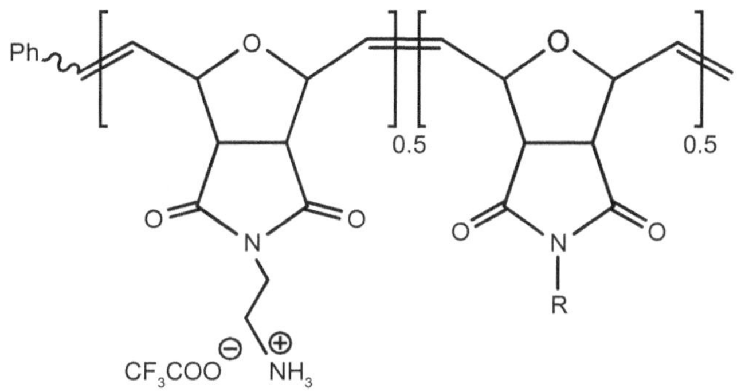 oxanorbornene copolymers with segregated charged and hydrophobic moieties	*S. aureus; E. coli*	[187,201]
Linear homopolymers	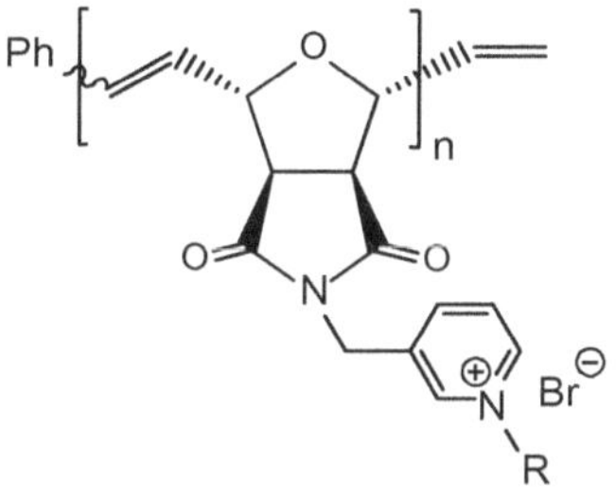 $R = C_2H_5; C_4H_9; C_6H_{13}; C_8H_{17}; C_{10}H_{21}; PhCH_2$ oxanorbornene homopolymers "same center" approach	*E. coli; B. subtilis*	[204]
Linear homopolymers	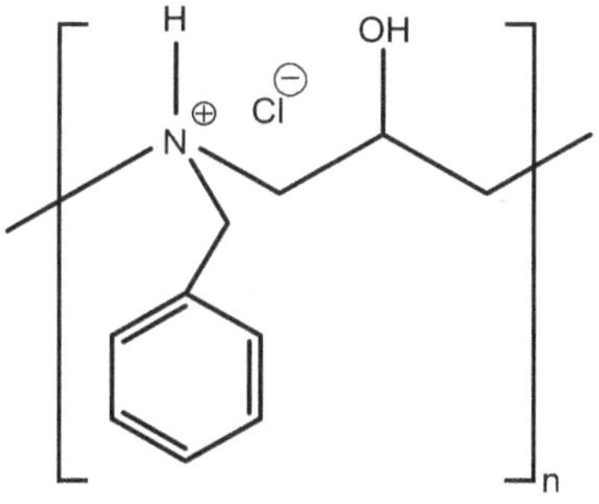 ionene polymers with quaternary ammonium groups within the main chain	*P. aeruginosa Klebsiella pneumoniae; Micrococcus luteus; B. subtilis; Enterococcus faecalis*	[205]

(Continued)

Table 6.4 Main Types of Antimicrobial Polymers (Continued)

Type of Polymer by Molecular Architecture	Type of Polymer by Chemical Composition	Antimicrobial Activity	Reference(s)
Block copolymers	Amphiphilic triblock polycarbonates of BAB type bis-MPA-p[MTC-O(CH$_2$)$_3$N$^+$(CH$_3$)$_3$Cl$^-$]-b-p[MTC-OEt]; see also the sixth entry in the table	*B. subtilis, S. aureus,* and MRSA; *Cryptococcus neoformans*	[185]
Dendrimers	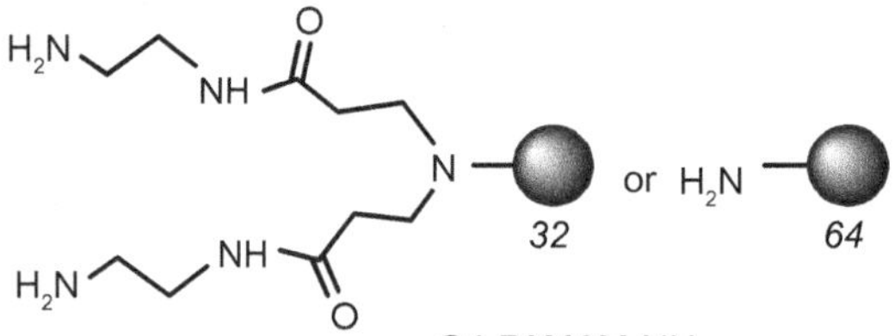 G4-PAMAM-NH$_2$ 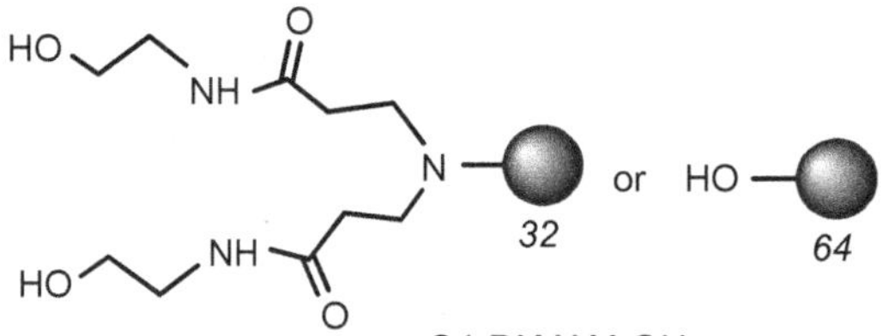 / G4-PAMAM-OH 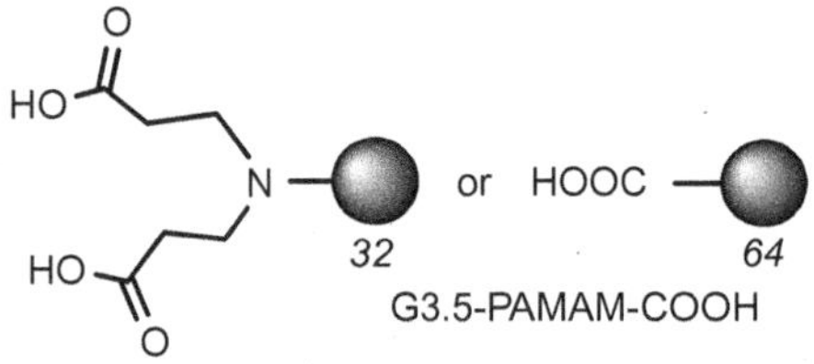 / G3.5-PAMAM-COOH	*E. coli*	[208]
	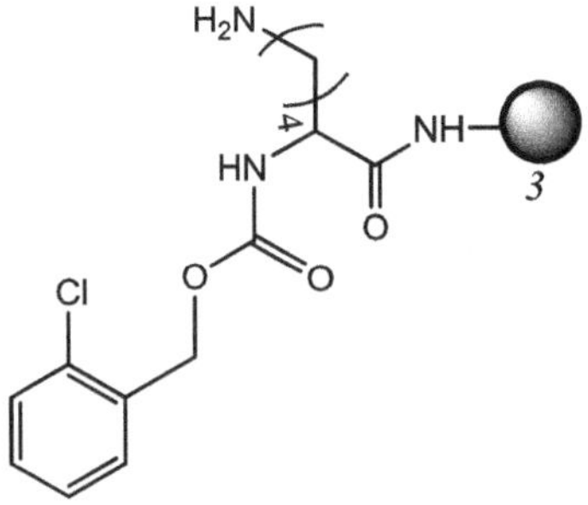 Dendritic poly(*L*-lysine)	Multiresistant strains *S. aureus* ATCC 43300 and *E. coli* ATCC BAA-198; *Candida*	[209]

(Continued)

Table 6.4 Main Types of Antimicrobial Polymers (Continued)

Type of Polymer by Molecular Architecture	Type of Polymer by Chemical Composition	Antimicrobial Activity	Reference(s)
Linear condensation polymers	Arylamide oligomers 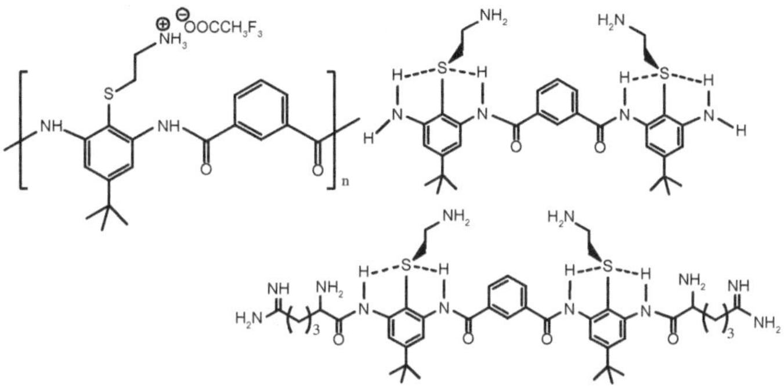	*B. subtilis; S. aureus; E. coli*	[215–218]
	Phenylene ethynylene oligomers 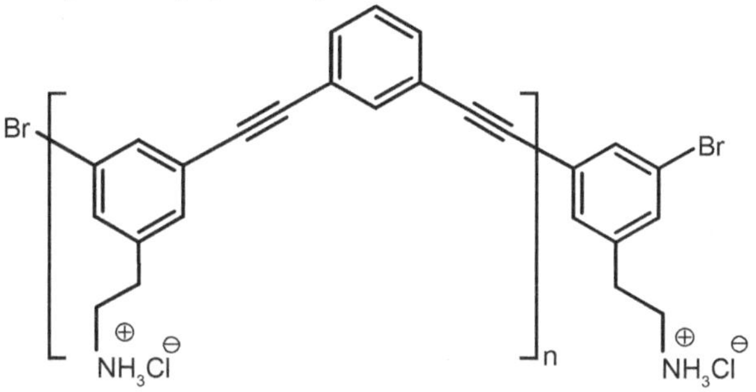	*S. aureus*	[219,220]
Linear random copolymers	Chitosan; poly[(2-acetamido-2-deoxy-*D*-glucose)-*co*-(2-amino-2-deoxy-*D*-glucose)] 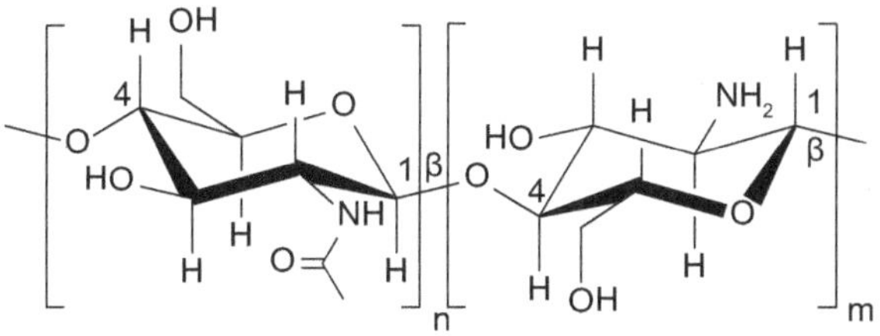	*P. aeruginosa*	[229]
	Oleoyl-chitosan 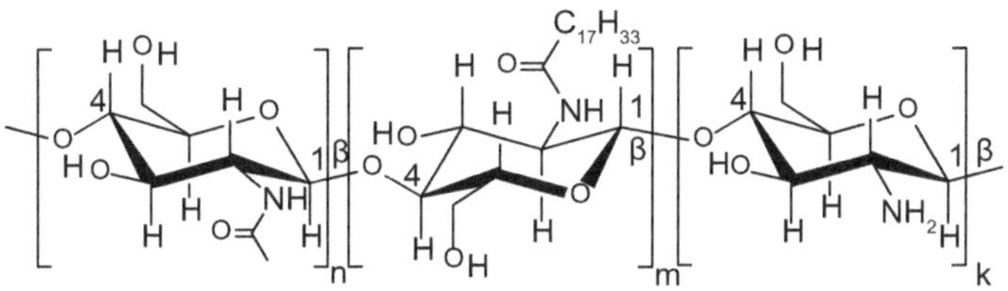	*E. coli; S. aureus*	[226–228]

application is limited because of the constrains imposed by the stability of the drug under the polymerization conditions [233].

6.6.2.1 Synthesis of Antibiotic-Polymer Conjugates From Preformed Polymers

The molecular architecture of the polymer backbone to which the antibiotic is attached can be linear, star-shaped or even dendritic. Sobczak et al. [234–236] have synthesized biodegradable linear as well as three-, four, and six-armed star polyesters by ROP of either lactide (L- and/or racemic) or ε-caprolactone initiated by poly(ethylene glycol) (PEG), glycerol, penthaerythritol (PET), and dipentaerythritol (DPET), respectively (see Scheme 6.14**a-d**). Various quinolones have been subsequently attached to the free hydroxyl groups at the ends of the polyester chains by the well documented ester bond conjugation chemistry using dicyclohexylcarbodiimide to activate the carboxyl group of quinolones as illustrated in Scheme 6.14 for the case of the six-armed DPET polyester.

Norfloxacin-conjugated linear polyurethanes have been also synthesized by Sobczak et al. [237]. To this end, they first prepared some polyesters through the ROP of caprolactone, D,L-lactide, and L-lactide. They used ethylene glycol and the nontoxic organoinitiator creatinine in order to initiate the ROP of the cyclic ester monomers that proceeded through a coordination-insertion mechanism [238]. The resulting low molecular weight polyesters possessed hydroxyl terminal groups at both ends of the polymer chain. Next, polyurethanes were prepared by reacting hexamethylene diisocyanate with the above hydroxyl terminated polyesters in a molar ratio of 2:1 as shown in Scheme 6.15. Eventually, norfloxacin was conjugated to the isocyanate-terminated polyurethane intermediates **30** to give linear antibiotic-polymer conjugates such as **31**. Unlike the previously described polyesters having the antibiotic moiety exposed at the ends of the polymer chains (see Scheme 6.14), in norfloxacin-conjugated polyurethanes **31** the drug is included as a repeating unit within the polymer backbone. Preliminary studies showed a gradual release of norfloxacin from these polyurethanes [237].

The ester end groups of the G2.5-PAMAM dendrimer **36** have been used as anchoring points for the attachment of the amino functionalized derivative of penicillin V **35** as illustrated in Scheme 6.16 [239]. The amino derivative **35** was prepared by the traditional amide bond formation chemistry using N-hydroxyscuccinimide (NHS) and dicyclohexylcarbodiimide (DCC). The activated N-hydroxysuccinimidyl ester **32** was reacted with the mono-protected diamino PEG **33** to afford the amide derivative **34**.

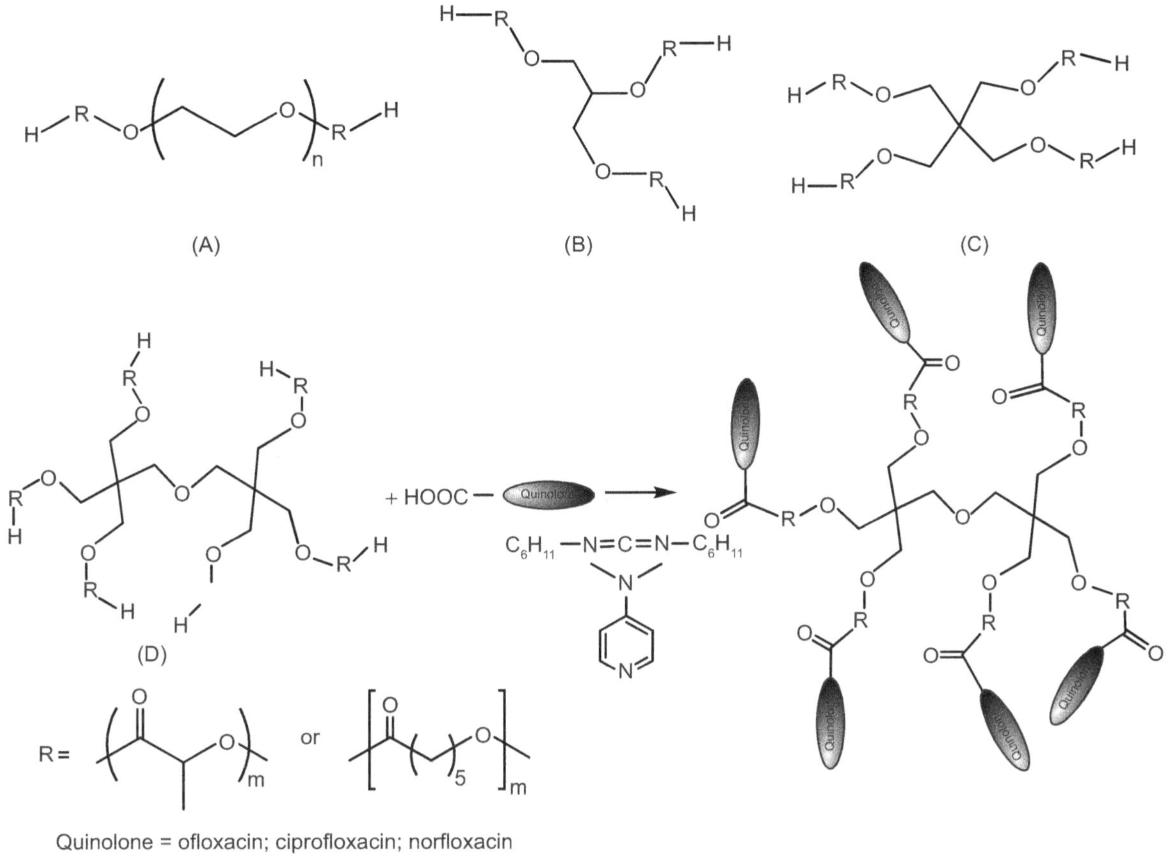

■ **SCHEME 6.14** Synthesis of linear and star-shaped polyester-quinolone conjugates.

Eventual deprotection of the second amine group on treatment with TFA led to the amino derivative **35**, which was further reacted with the terminal ester groups in 2.5G-PAMAM dendrimer to yield the penicillin V-conjugated PAMAM dendrimer **37**.

6.6.2.2 Synthesis of Antibiotic-Polymer Conjugates From Monomer Prodrugs

Turos and coworkers [240] have prepared penicillin G-bound polyacrylate nanoparticles. The syntheses of the monomer prodrugs are presented in Scheme 6.17. The polymerizable acrylic moiety has been introduced either to the amido nitrogen or to the carboxyl group in penicillin G as in the

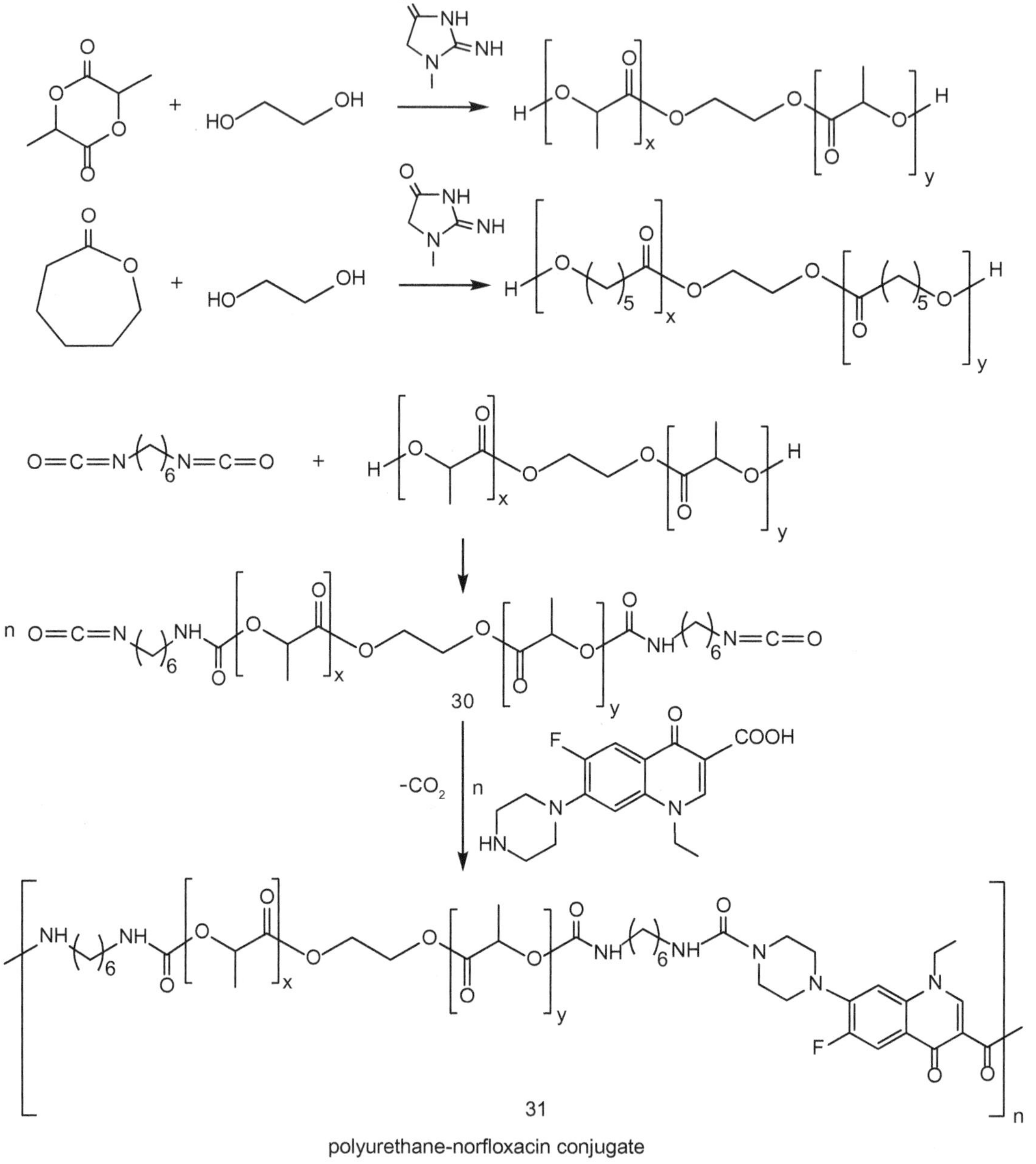

polyurethane-norfloxacin conjugate

■ **SCHEME 6.15** Synthesis of norfloxacin-polyurethane conjugates.

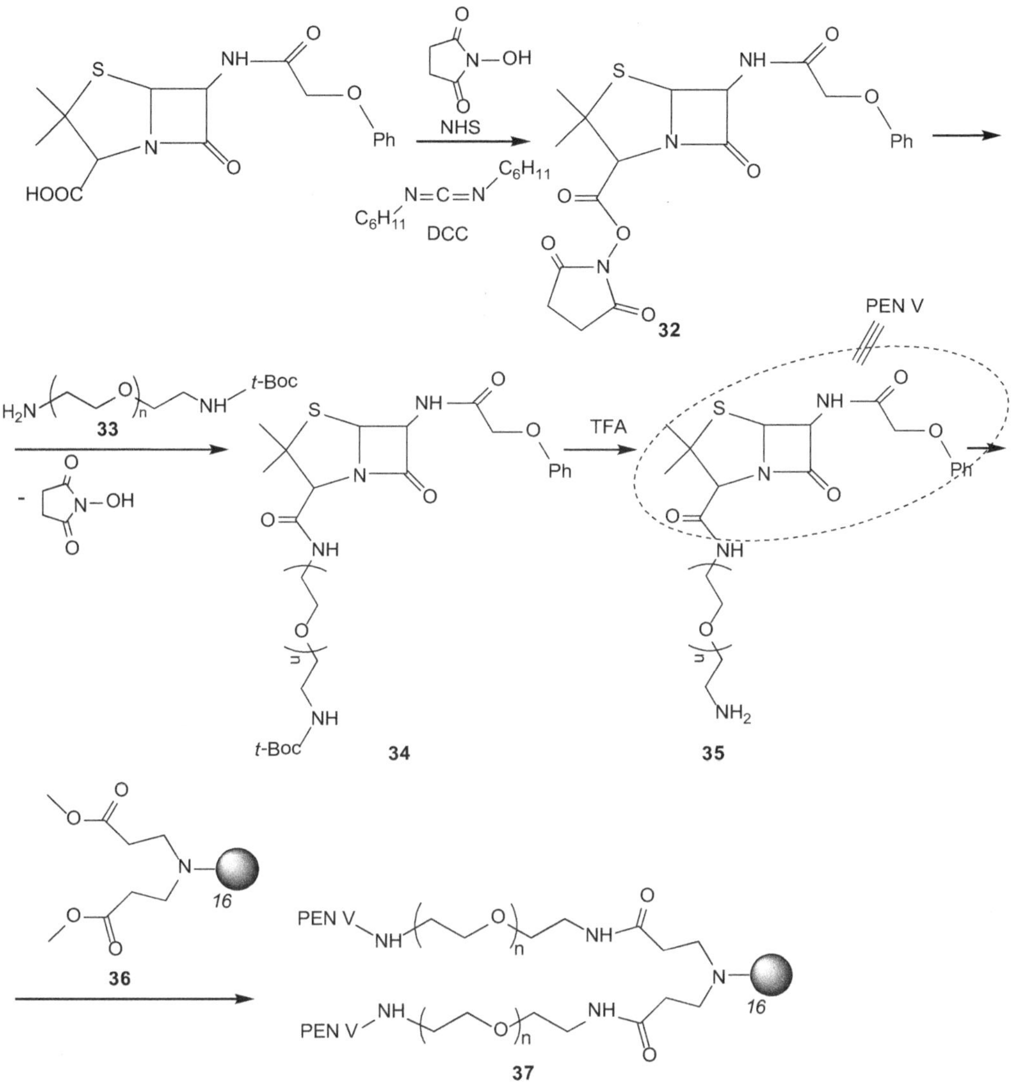

SCHEME 6.16 Synthesis of penicillin V-G2.5-PAMAM dendrimer-conjugate.

acrylic ureide derivative **38** or as in the 2-hydroxyethyl acrylate ester of penicillin G **39**, respectively. Acylation of the amide nitrogen required a previous protection of the carboxylic group through silylation. Penicillin-bound polyacrylate nanoparticles were prepared by emulsion polymerization in water. Briefly, the prodrug monomer, for instance **38**, was dissolved in a 7:3 mixture (w/w) of butyl acrylate and styrene at a concentration of 1% and the mixture was pre-emulsified in water using sodium dodecyl sulfate as a surfactant. Water-soluble potassium persulfate was added to initiate radical polymerization of the monomers and the mixture was heated at 70°C for 6 h with vigorous stirring and under nitrogen atmosphere. The so prepared penicillin G-conjugated polyacrylate nanoparticles appeared to be insensible to enzymatic degradation by β-lactamases while retaining the antibacterial activity. Broth dilution MIC assays revealed that a concentration of only 2 µg mL^{-1} of penicillin G-bound acrylate nanoparticles was enough to completely inhibit MRSA growth for 24 h. The group of Turos has also developed a novel family of N-alkylthio β-lactams with potent bacteriostatic activity toward a group of pathogenic bacteria including *S. aureus*, MRSA, *Bacillus anthracis*, and *Micrococcus luteus* [241–248]. They identified fatty acid biosynthesis as a primary cellular target of these antibacterial compounds in *S. aureus*. Encouraged by these results they also synthesized polymerizable O-acrylated-N-thiolated β-lactams like compound **41** [246–248], which is presented here as an illustrative example (see Scheme 6.18). Emulsion polymerization of a mixture of **41** and other vinyl monomers such as butyl acrylate and styrene was performed in water using sodium dodecyl sulfate as an emulsifying agent and a water soluble free radical polymerization initiator to produce antibiotic-conjugated polyacrylate nanoparticles **42**. These nanoparticles were found to have improved activity against MRSA as compared to the monomer form of the antibiotic [247].

Similarly, Moon et al. [249] have prepared an acryl monomer containing the quinolone moiety by reacting norfloxacin with glycidyl methacrylate in dimethylformamide (DMF) as depicted in Scheme 6.19. The resulted monomer **43** was suspended in dimethylformamide and AIBN was added to induce free radical polymerization. The shake flask method was used to quantify the antibacterial activity revealing excellent activity against Gram-positive bacteria as well as Gram-negative ones for both the acrylated quinolone monomer and its polymer. The polymer was compounded with PCL and the blend showed good mechanical properties and thermal stability. Taking also into account the antibacterial activity, the authors concluded that their material could find practical application in food packaging.

A multifunctional polymer-sulfamethoxazole conjugate **49** was synthesized by Chang et al. [250] through radical copolymerization of methacrylic acid (MAA) and the methacrylic monomers **43**, **44**, and **45**. The monomers were synthesized starting with methacryloyl chloride as shown in Scheme 6.20A. N-(2-hydroxypropyl)methacrylamide (HPMA) **43** was

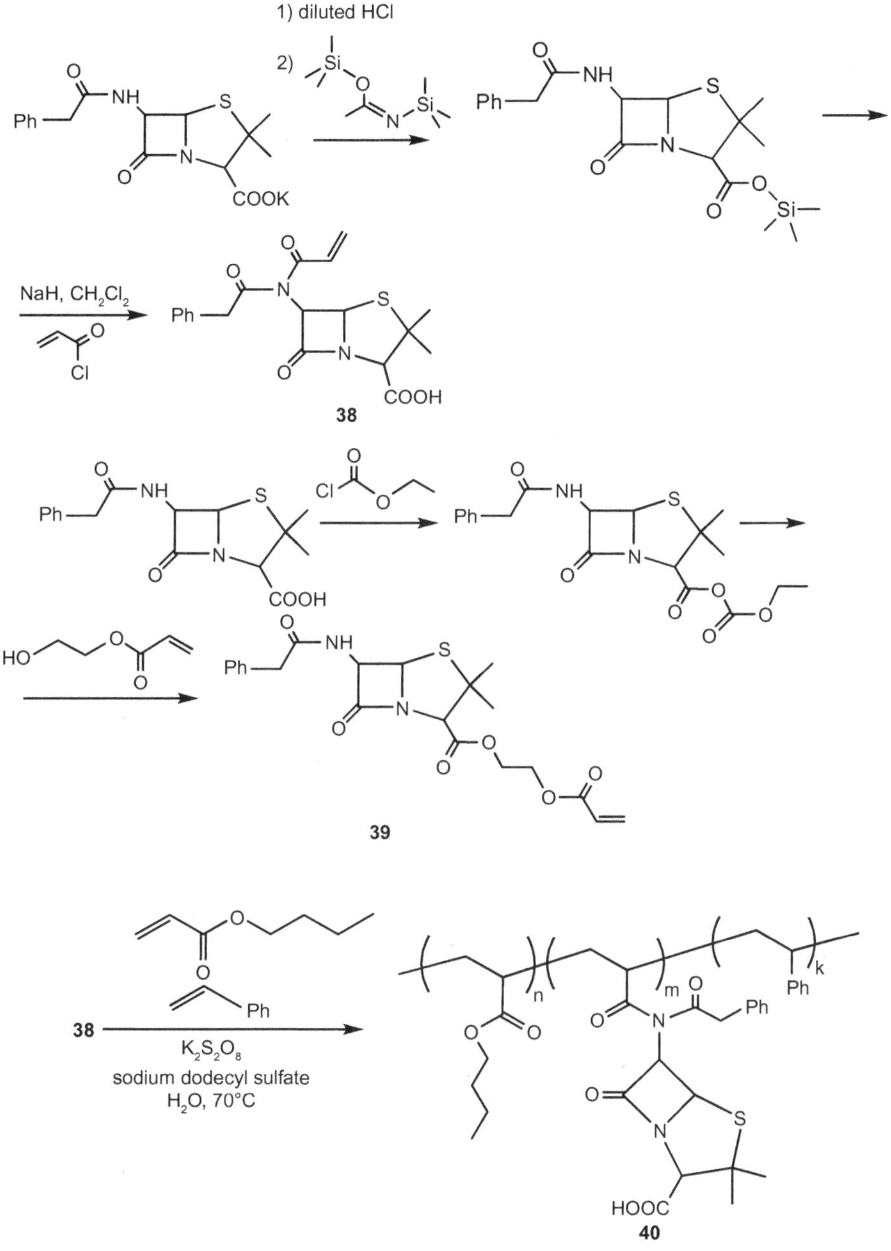

■ **SCHEME 6.17** Preparation of penicillin G–conjugated polyacrylate nanoparticles.

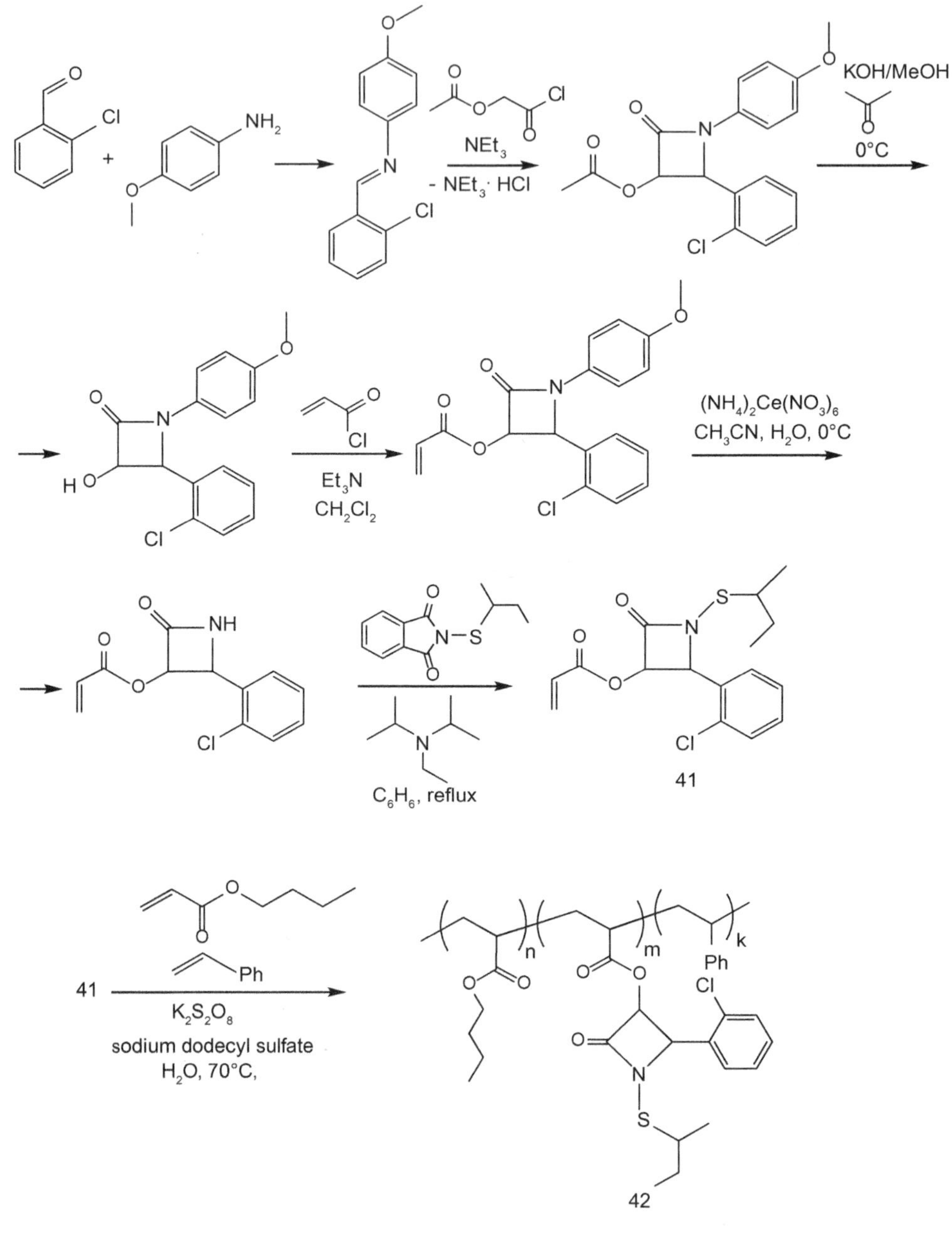

■ **SCHEME 6.18** Synthesis of *N*-thiolated β lactam antibiotic-conjugated polyacrylate nanoparticles with potent antibacterial properties against MRSA.

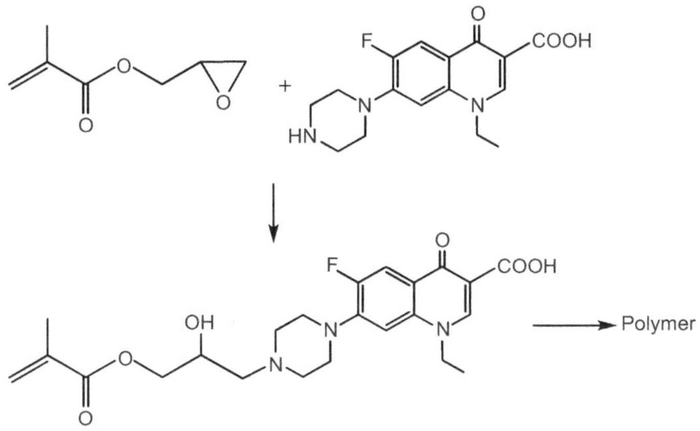

■ SCHEME 6.19 Synthesis of a quinolone-based antibacterial polymethacrylate.

synthesized by reaction of DL-1-amino-2-propanol with methacryloyl chloride in the presence of sodium bicarbonate as hydrochloric acid scavenger. *N*-Methacryloyl glycylglycyl sulfamethoxazole (MA-GG-SMX) **45** was synthesized in three steps. First, glycylglycine was acylated with methacryloyl chloride under Schotten-Baumann conditions in aqueous alkaline medium [251]. Then, the product was further esterified with *p*-nitrophenol using carbodiimide chemistry to afford methacryloyl glycylglycine 4-nitrophenyl ester (MA-GG-ONP) **44**. Eventually, a solution of MA-GG-ONP in DMF was treated with sulfamethoxazole under an inert atmosphere of nitrogen yielding MA-GG-SMX **45**. Copolymerization of methacrylic acid and methacryl amides **43**, **44**, and **45** led to random copolymer **47** (see Scheme 6.21). In a separate synthetic step, *N*-(4-aminobutyl)maleimide (AMI) **46** was prepared from 1,4-diaminobutane which was first converted to the corresponding mono-Boc-protected diamine on treatment with di-*t*-butyl dicarbonate. The latter was reacted with maleic anhydride and the intermediate maleamic acid product underwent dehydration and ring closure on heating with sodium acetate in acetic anhydride. Eventually, *N*-Boc deprotection was carried out by treatment with TFA to yield the amino-functionalized maleimide derivative **46** as shown in Scheme 6.20B. When copolymer **47** was treated with **46** the fairly good leaving group *p*-nitrophenoxide was replaced by AMI. Introduction of the maleimide linker endowed the resulted copolymer **48** with the ability to readily bind targeting nanobodies such as an antibody fragment. Copolymer **48** also showed significant growth inhibitory effect against *E. coli*. Bearing pendant carboxylic acid moieties, copolymer **48** exhibited pH-sensitive behavior.

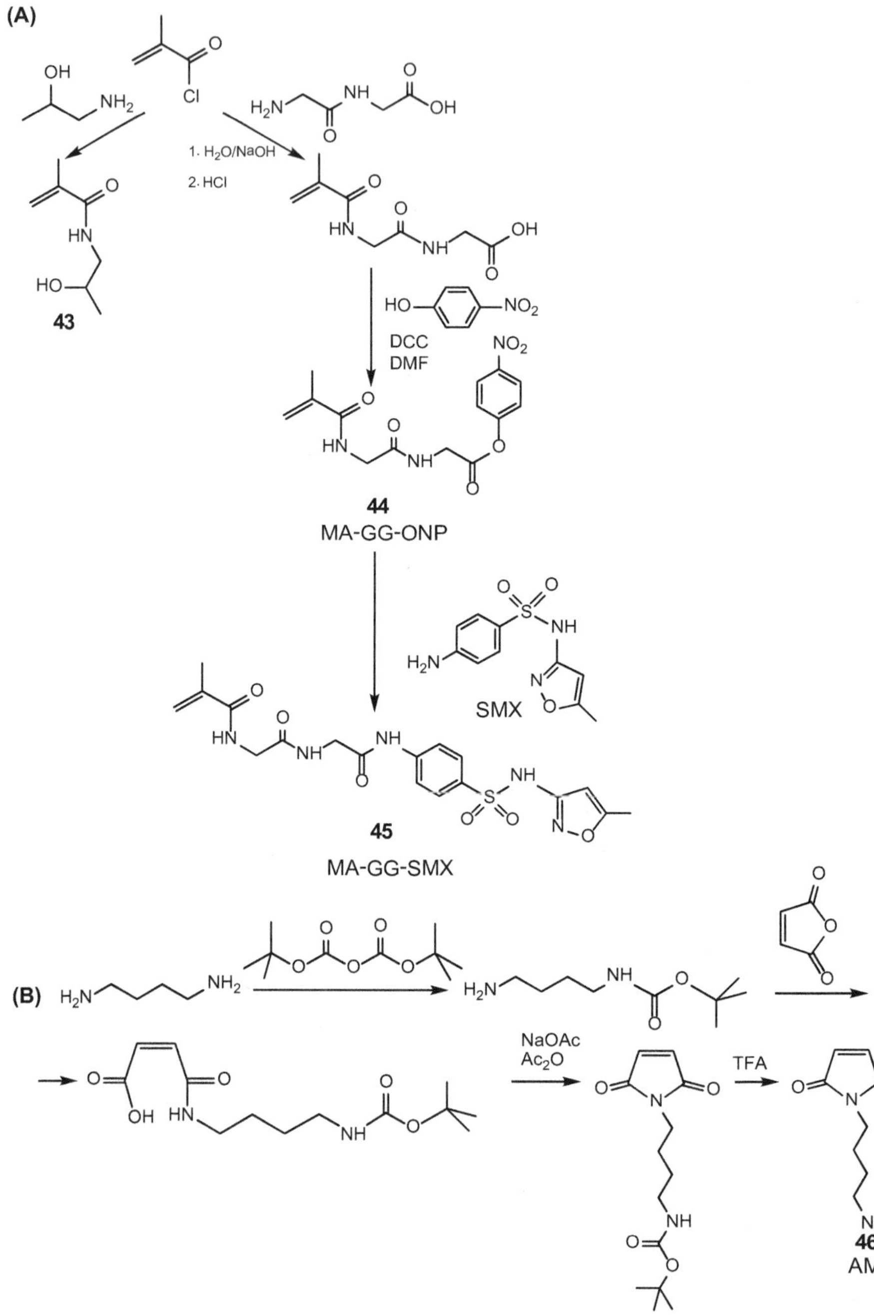

■ **SCHEME 6.20** Syntheses of methacryl amide monomers **43**, **44**, and **45** (A) and of *N*-(4-aminobutyl) maleimide (AMI) **46** (B).

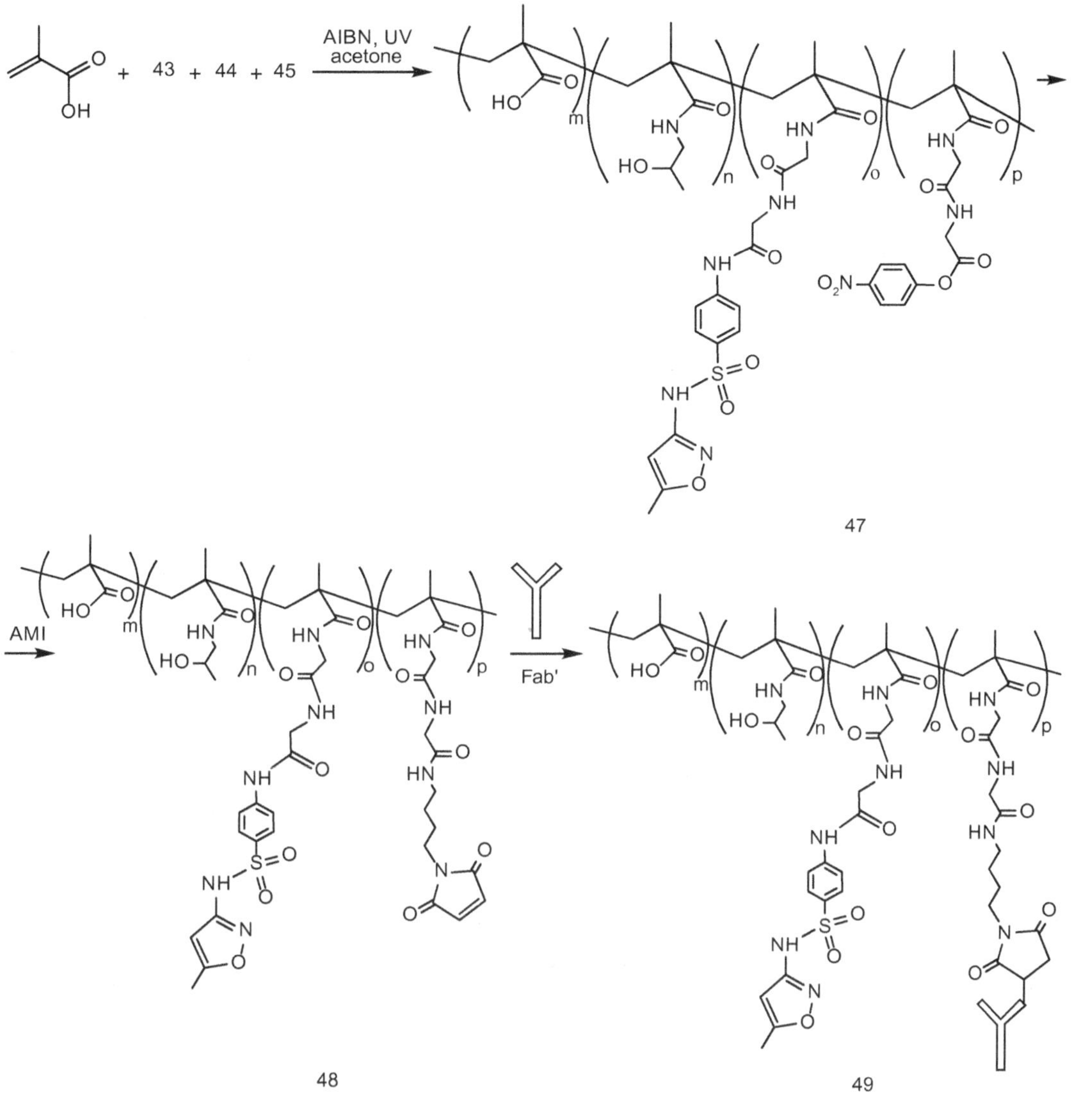

■ SCHEME 6.21 Synthesis of water-soluble pH-responsive tetrapolymer-sulfomethoxazole conjugate **48** and its functionalization by coupling to a targeting antibody fragment.

When the pH was increased, the antibacterial activity decreased due to ionization of the carboxylic groups resulting in weakened interaction with the negatively charged *E. coli* outer membrane. TEM imaging showed that at the normal physiological pH, the tetrapolymer **48** formed spherical micelles in water with a diameter ranging from 50 to 150 nm with the carboxylic acid groups being solvated and exposed on the outer surface of the nanospheres while the hydrophobic sulfonamide units are hidden being buried inside the nanoparticles. Decreasing the pH value, the carboxyl groups became protonated and hydrophobic intrapolymer chain-chain interactions overcame the intramolecular hydrogen bonding. As a result micelles collapsed and partially precipitated undergoing a reversible pH-dependent phase transition.

The data within this section are summarized in Table 6.5.

6.6.3 Smart Supramolecular Nanovehicles for Antibiotic Delivery Based on Organic Polymers and Enhancement of Antimicrobial Activity Through Nanoencapsulation

The stimuli-responsiveness behavior of smart polymers relies on their capacity to undergo reversible self-assembly in a particular given environment (water being the most relevant medium for biological applications) through intra- and/or intermolecular noncovalent bonds such as hydrogen bonding, electrostatic interactions (ion-ion; ion-dipole, and dipole-dipole), van der Waals forces, π-π stacking, hydrophobic interactions, and

Table 6.5 Polymer-Antibiotic Conjugates

Synthetic Strategy	Molecular Architecture of the Polymer-Antibiotic Conjugate	Antibiotic/Reference(s)
Conjugation of the antibiotic to the end groups of a preformed polymer	Linear BAB type triblock copolymer PLA-*b*-PEG-*b*-PLA-quinolone conjugates	Ofloxacin/[234] Ciprofloxacin/[235] Norfloxacin/[236]

(Continued)

Table 6.5 Polymer-Antibiotic Conjugates (Continued)

Synthetic Strategy	Molecular Architecture of the Polymer-Antibiotic Conjugate	Antibiotic/Reference(s)
	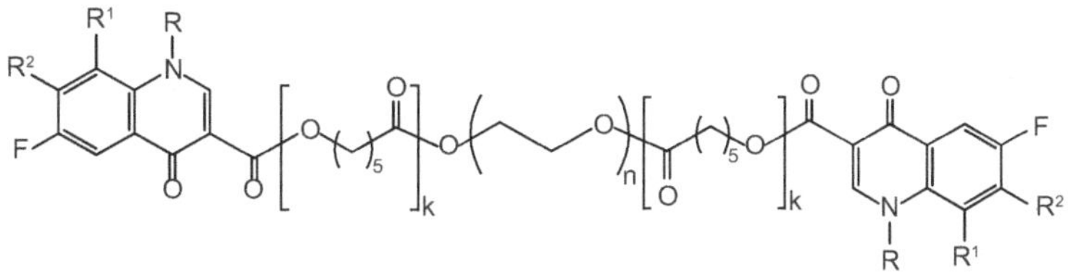 Linear BAB type triblock copolymer PCL-*b*-PEG-*b*-PCL-quinolone conjugates	Ofloxacin/[234] Ciprofloxacin/[235] Norfloxacin/[236]
	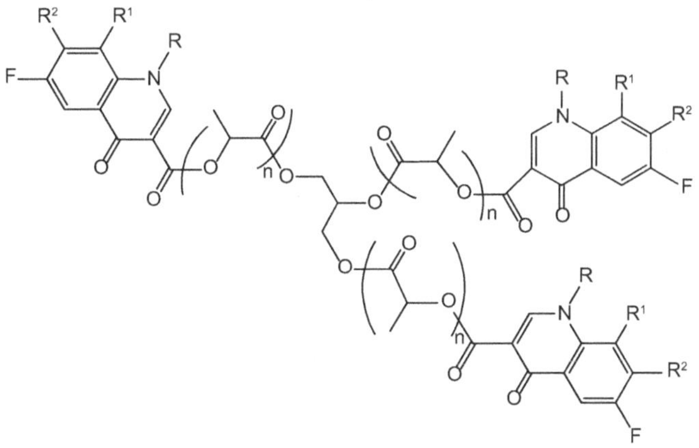 3-arm star-shaped PLA-quinolone conjugates	Ofloxacin/[234] Ciprofloxacin/[235] Norfloxacin/[236]
	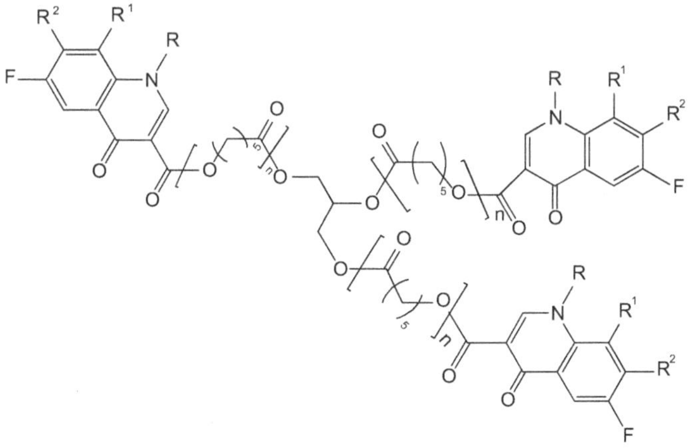 3-arm star-shaped PCL-quinolone conjugates	Ofloxacin/[234] Ciprofloxacin/[235] Norfloxacin/[236]

(Continued)

Table 6.5 Polymer-Antibiotic Conjugates (Continued)

Synthetic Strategy	Molecular Architecture of the Polymer-Antibiotic Conjugate	Antibiotic/Reference(s)
	4-arm star-shaped PLA-quinolone conjugates	Ofloxacin/ [234] Ciprofloxacin/[235] Norfloxacin/[236]
	4-arm star-shaped PCL-quinolone conjugates	Ofloxacin/[234] Ciprofloxacin/[235] Norfloxacin/[236]

(Continued)

Table 6.5 Polymer-Antibiotic Conjugates (Continued)

Synthetic Strategy	Molecular Architecture of the Polymer-Antibiotic Conjugate	Antibiotic/Reference(s)
	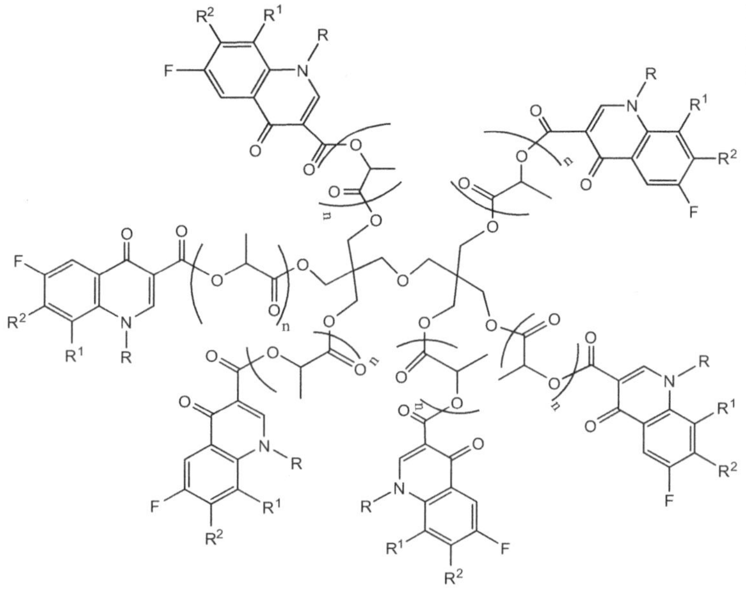 6-arm star-shaped PLA-quinolone conjugates	Ofloxacin/[234] Ciprofloxacin/[235] Norfloxacin/[236]
	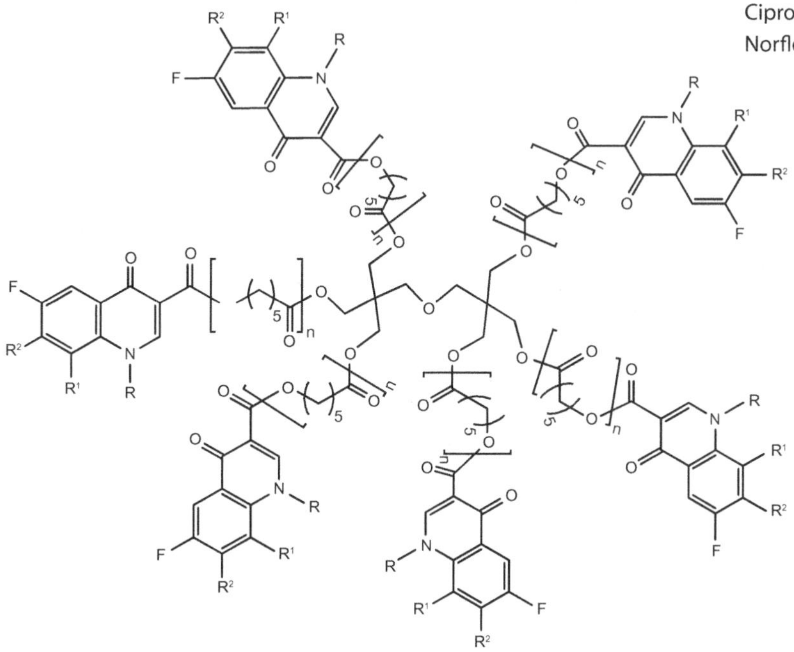 6-arm star-shaped PCL-quinolone conjugates	Ofloxacin/[234] Ciprofloxacin/[235] Norfloxacin/[236]

(Continued)

Table 6.5 Polymer-Antibiotic Conjugates (Continued)

Synthetic Strategy	Molecular Architecture of the Polymer-Antibiotic Conjugate	Antibiotic/Reference(s)
	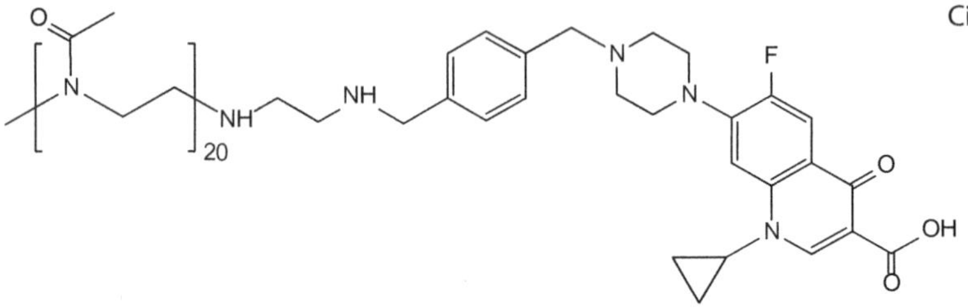 Linear poly(2-methyloxazoline)-ciprofloxacin conjugate	Ciprofloxacin/[252]
	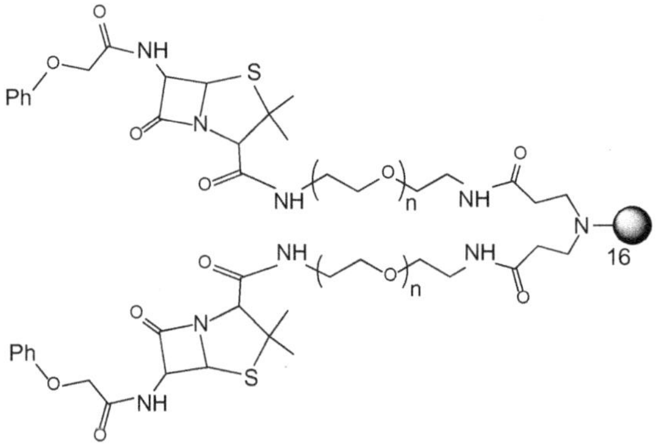 Penicillin V-G2.5-PAMAM dendrimer-conjugate	Penicillin V/[239]
Inclusion of the drug as a repeating unit within the polymer backbone	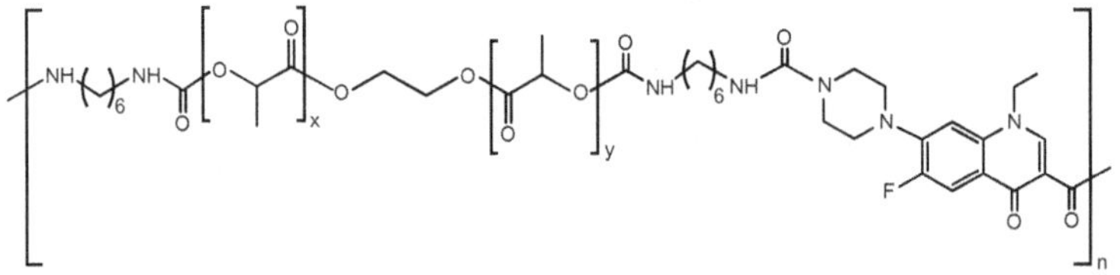	Norfloxacin/[237]
Polymerization of a monomer prodrug	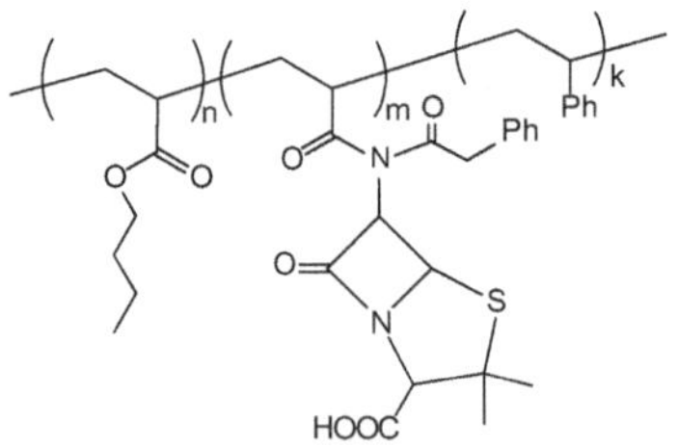 Linear random copolymer poly(butyl acrylate)-*co*-poly(6-aminopenicillanic acid acryl amide)-*co*-polystyrene	Penicillin G/[240]

(Continued)

Table 6.5 Polymer-Antibiotic Conjugates (Continued)

Synthetic Strategy	Molecular Architecture of the Polymer-Antibiotic Conjugate	Antibiotic/Reference(s)
	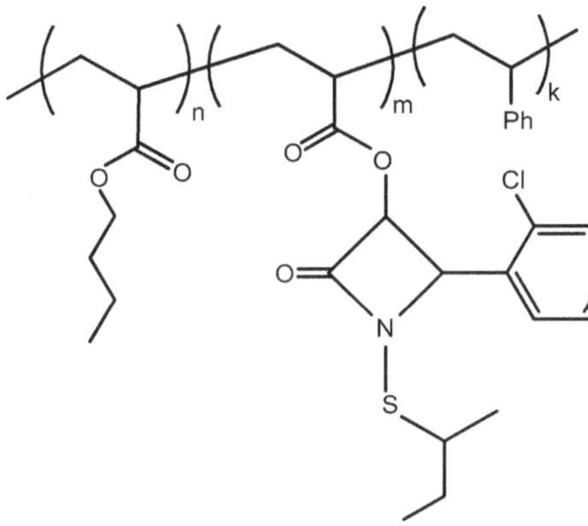 Linear random copolymer poly(butyl acrylate)-*co*-poly[3-acryloyloxy-*N*-*sec*-butylthio-4-(2-chlorophenyl)-2-azetidinone]-*co*-polystyrene	*N*-alkylthio β lactams/[246–248]
	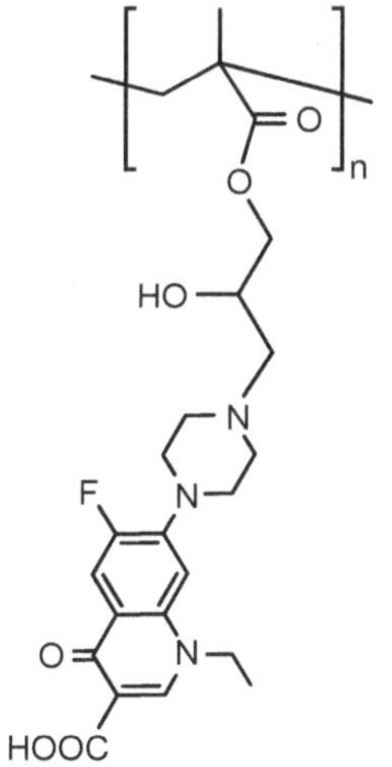 Linear homopolymer of 1-ethyl-6-fluoro-7-{4-[2-hydroxy-3-(2-methylacryloyloxy) propyl] piperazin-1-yl}-4-oxo-1,4-dihyroquinolin-3-carboxylic acid	Norfloxacin/[249]

(*Continued*)

Table 6.5 Polymer-Antibiotic Conjugates (Continued)

Synthetic Strategy	Molecular Architecture of the Polymer-Antibiotic Conjugate	Antibiotic/Reference(s)
	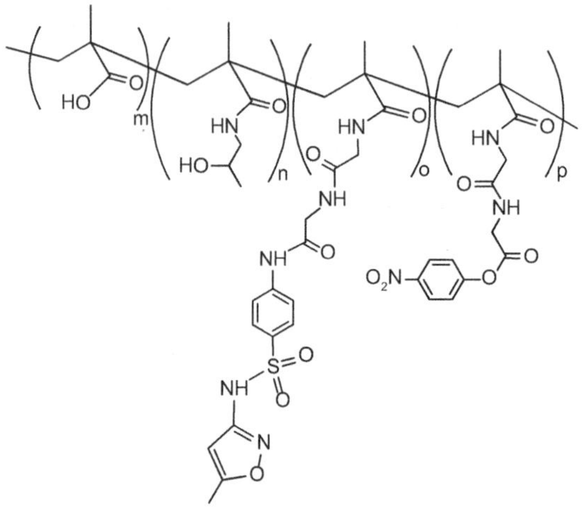 Linear tetrapolymer poly(MAA-*co*-HPMA-*co*-MA-GG-SMX-*co*-MA-GG-ONP)	Sulfamethoxazole/ [250]

metal coordination. Any changes in polymer environment altering the pattern of these weak noncovalent interactions will also modify the properties derived from a particular internal packing structure of the polymer backbone and the behavior of the polymer in that particular environment thereby eliciting a response from the polymer. For instance, in response to specific changes in its environment, a polymer can modify its conformation, its hydration state, its solubility, and its interaction patterns with other molecular structures and chemical entities present in the environment. Environmental pH or ionic strength changes will alter the charge-based interactions and will elicit a response from polyelectrolytes. Modification of the hydrophobic/hydrophilic balance along the backbone of an amphiphilic polymer will modify polymer solubility causing either dissolution of the polymer or by contrary aggregation and phase separation. Similarly, enzyme-induced alteration of the hydrophobic/hydrophilic balance may produce destabilization of supramolecular drug nanovehicles such as polymeric micelles, which eventually collapse releasing their drug payload. Temperature responsiveness relies on the temperature-dependent changes in the solvation state and conformation of a thermosensitive polymer block in aqueous media. Below LCST, the polymer forms many hydrogen bonds with water, being in a hydrated and swollen state and adopting a random coil conformation. Increasing temperature above LCST, the hydrogen bonds between the hydrophilic moieties on the polymer backbone and water are broken, the hydrophobic chain-chain interactions taking over and provoking polymer phase transition to an insoluble collapsed globular state.

The aim of this section is to highlight some of the newest and most relevant achievements in the field of smart delivery of antibiotics. For more comprehensive treatment of smart multifunctional drug delivery the reader can consult bibliographical references [52,66,67,118,253–263]. A special paragraph presents some results of our own research studies regarding the capacity of nontoxic, biocompatible and biodegradable polymeric nanostructured materials to improve the activity of currently used antibiotics against both planktonic and biofilm-growing bacteria.

6.6.3.1 *pH-Responsive Antibiotic Delivery*

Radovic-Moreno et al. [264] have synthesized vancomycin encapsulated amphiphilic triblock copolymer nanoparticles with surface charge switching properties for selective targeting of bacterial cell walls. Briefly, the authors started their synthetic strategy (see Scheme 6.22) with the oligohistidine Lys-His$_{20}$-Cys **50** having N-terminal lysine and C-terminal cysteine residues, respectively, for further conjugation. The C-terminal cysteine residue was used for covalent attachment of the hydrophylic methoxy poly(ethylene glycol) (MPEG) segment of the triblock copolymer on treatment with the 3-(2-pyridyldithio) propionic acid functionalized MPEG polymer (MPEG-OPSS), **51**. The hydrophobic poly(lactide-*co*-glycolide) (PLGA) block was added to the intermediate diblock copolymer poly(L-histidine)-*b*-MPEG (MPEG-SS-PLH-NH$_2$) **52** as follows. First, the carboxyl-terminated PLGA copolymer was activated as the *N*-hydroxysuccinimidyl ester PLGA-NHS **53** on treatment with *N*-hydroxysuccinimide (NHS) in the presence of 1-ethyl-3-(3-dimethylaminopropyl)carbodiimide (EDC). Eventual coupling between the activated ester **53** and the amino group of the N-terminal lysine residue of MPEG-SS-PLH-NH$_2$ resulted in the formation of the amphiphilic triblock copolymer poly(D,L-lactide-*co*-glycolide)-*b*-poly(L-histidine)-*b*-poly(ethylene glycol) (PLGA-PLH-MPEG), **54**. As one can see, each segment in **54** has a distinct role: (1) the hydrophilic MPEG block confers water solubility and imparts "stealth" properties to the nanoparticles acting as a protective shield against opsonins in the bloodstream thereby assuring long circulation times upon administration; (2) the hydrophobic PLGA block forms the core of the nanoparticles and the drug depot; (3) the PLH block provides pH-sensitivity. Amphiphilic triblock copolymer **54**

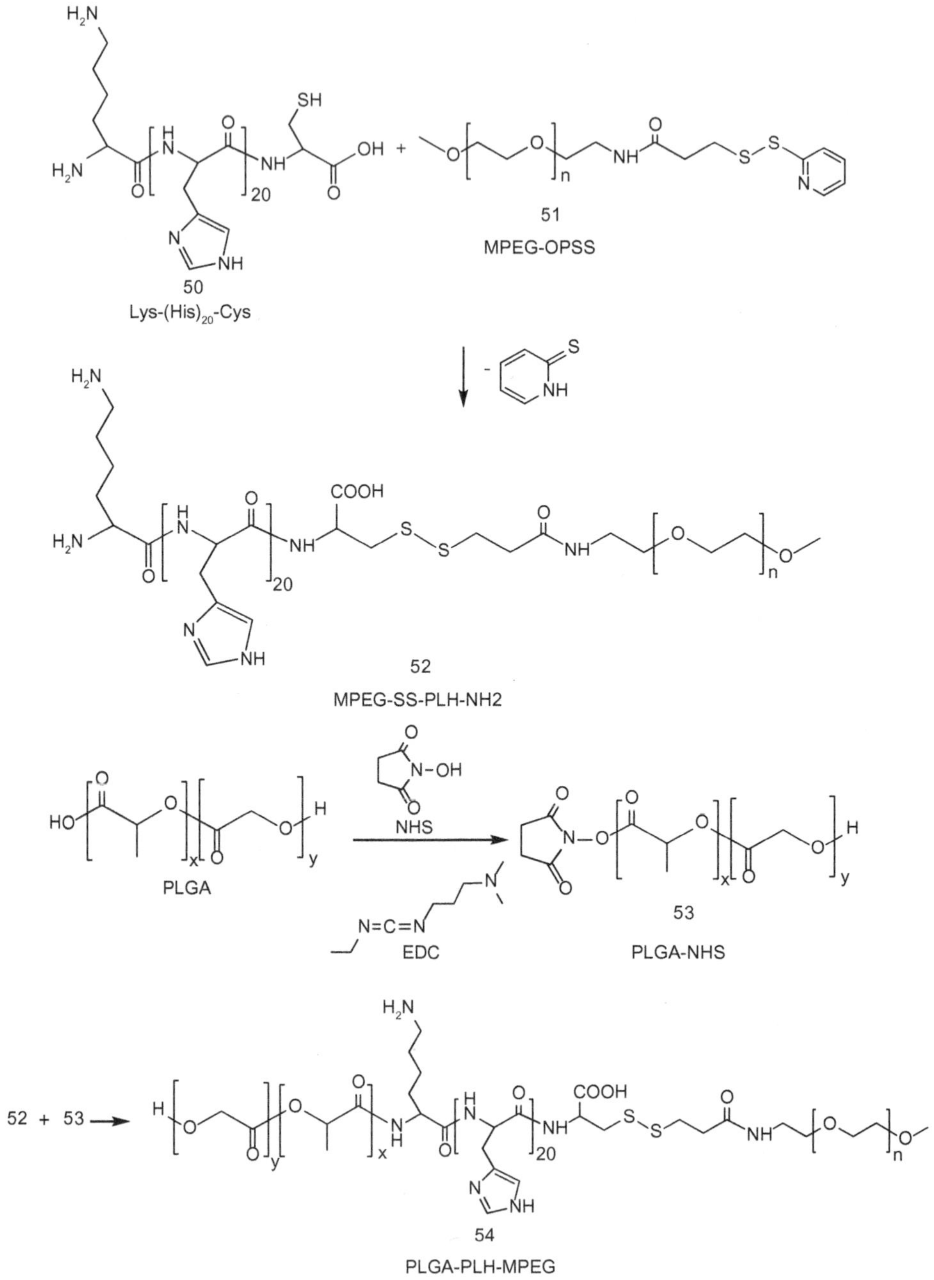

■ **SCHEME 6.22** Synthesis of PLGA-PLH-MPEG nanoparticles with pH-triggered surface charge-switching properties for targeted delivery of antibiotics.

was formulated as vancomycin encapsulating nanoparticles using a modified (W/O)/W double emulsion-solvent evaporation technique. The zeta potential of the resulted nanoparticles was monitored as a function of pH and a switch in the surface charge from negative values at pH 7.4 to positive ones with reduction in pH was revealed, the transition occurring as early as pH 7.0. This switch was due to the progressive protonation of the imidazole moieties in the polyhistidine block. The negative zeta potential at pH 7.4 was attributed to residual negative charge from acid groups in PLGA precursor and to partial hydrolysis of PLGA chains. As the bacterial cell wall is negatively charged, switching the nanoparticles surface charge from negative to positive would increase nanoparticles binding to the bacterial cell walls and this assumption was experimentally proved in binding studies performed on fluorescently labeled nanoparticles. Since it is well known that infected sites are characterized by a lower pH as compared to healthy tissues, the pH-triggered switch of the surface charge of PLGA-PLH-MPEG nanoparticles would allow the latter to selectively target bacteria without binding non-target healthy cells. The authors have tested this hypothesis by choosing the antibiotic vancomycin as a model drug payload because it is known that acidic media inhibit the antibacterial activity of vancomycin. Although vancomycin encapsulated PLGA-PLH-MPEG nanoparticles exhibited increased MIC values as compared to the free drug in *S. aureus*, encapsulation also resulted in significant reduction of the acid-induced loss of antibiotic activity, recommending these nanoparticles as promising candidates for smart antibiotic delivery.

A novel nanoDDS able to effectively combat *Helicobacter pylori* by local antibiotic delivery to infection sites on the gastric epithelium has been developed by Lin et al. [265]. *Helicobacter pylori* colonizes about 50% of the world's population and it is recognized today that infection with this Gram-negative pathogen is a key factor in the etiology of gastritis and peptic ulcers and the strongest known risk factor of developing gastric cancer [266,267]. The efficacy of in vivo antibiotic therapy is hampered by the poor stability of the antibiotic in the highly acidic gastric juice, by the poor permeation of the antibiotic across the gastric mucus layer, and by the short time the antibiotic remains in the stomach. The nanoDDS developed by Lin and coworkers is based on formation via electrostatic attractions of stable polyelectrolyte complexes between chitosan and heparin in the pH range 1.2–2.5. Heparin is a linear polysaccharide constituted by alternating disaccharide sequences of a uronic acid i.e., 2-O-sulfated-L-iduronic acid (IdoASO$_3^-$) and D-glucuronic acid (GlcA) and an aminosugar which is 6-O-sulfated-2-N-sulfated-2-amino-2-deoxy-D-glucose (GlcNSO$_3^-$). In heparin the IdoA uronic acid moieties and the aminosugar GlcN moieties

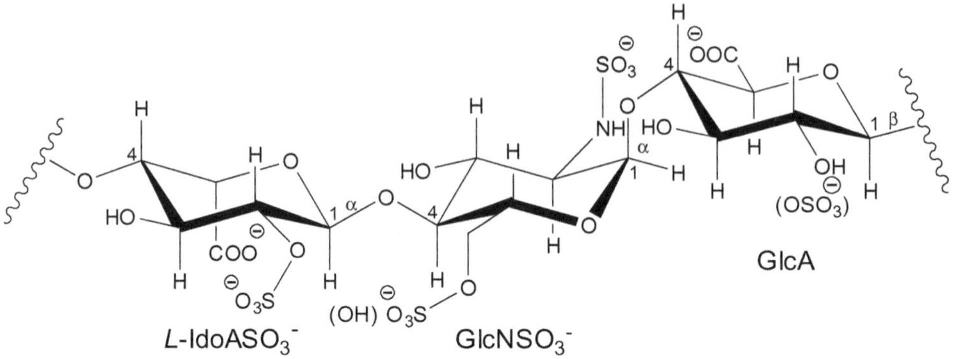

■ FIGURE 6.8 Monosaccharide components of heparin and their mode of linkage.

are linked α 1→4 to the next residue in the chain, while the GlcA moieties are linked β 1→4 as shown in Fig. 6.8. Preparation of chitosan/heparin nanoparticles is quite easy involving ionic gelation of chitosan with heparin. Heparin contains three types of acidic groups; two of them, the sulfate ester and the sulfamido groups, have very low pK_a values (0.5–1.5), while the third, namely the carboxylate groups, are less acidic with pK_a values ranging from 2 to 4. Chitosan on the other hand bears basic amino groups. In the pH range of 1.2–6.5, the positively charged chitosan and the negatively charged heparin associate with each other to form polyelectrolyte matrices of spherical shape with pH-tunable dimensions and stability. At pH 1.2–2.5 (simulating the pH of gastric acid), the sulfate ester and the sulfamido groups in heparin and the amino groups in chitosan are ionized, while the carboxylate groups in heparin are protonated and the nanoparticles are moderately swollen. Increasing the pH to 4.5–6.5 (simulating the environment of the gastric mucosa), some of the carboxylic groups in heparin become gradually deprotonated and as a consequence the attractive electrostatic force to the positively charged amino groups in chitosan increases and the nanoparticles shrink in size. Further pH increase results in deprotonation of the amino groups in chitosan accompanied by collapse of nanoparticles at pH 7.0. These results demonstrate that chitosan/heparin nanoparticles are able to effectively protect encapsulated antibiotics from being inactivated or degraded by the gastric acid (pH 1.2–2.5). The authors have used fluorescently labeled nanoparticles in order to track cellular internalization. Their results prove the capacity of pH-sensitive chitosan/heparin nanoparticles to adhere to the gastric mucosa and infiltrate the mucus layer, thereby prolonging gastric residence time. Furthermore, the positive zeta potential of these nanoparticles favors interaction with the negatively charged sites on the cell surfaces and tight junctions thereby facilitating direct contact with the regions of *H. pylori* infection on the gastric mucosa. Since the gastric mucosal pH is above 7.0, chitosan/heparin nanoparticles are destabilized and concomitantly release their antibiotic payload. Moreover, another important advantage of this novel nanoDDS is its construction from safe, nontoxic building blocks.

Recently, Convertine and coworkers reported the RAFT synthesis of polymer brushes which reversibly self-assemble into pH-responsive polymersomes [268]. The third generation cephalosporin antibiotic ceftazidime was successfully encapsulated within the above polymersomes and a pH-dependent drug release behavior was proved. Endosomal pH-triggered polymersomes disassembly followed by antibiotic release provides an effective way of targeting intracellular bacteria. The synthesis of pH-responsive polymersomes is depicted in Scheme 6.23. First, a polymer scaffold with pendant hydroxyl groups for further grafting of the chain transfer agents (CTA) was prepared by copolymerization of 2-hdroxyethyl methacrylate (HEMA) and poly(ethylene glycol) methyl ether methacrylate (PEGMA). For the polymer scaffold synthesis, 4-cyano-4-phenylcarbonothioylthio) pentanoic acid (CTP) and 4,4′-azobis(4-cyanovaleric acid) (ABCVA) were used as the RAFT agent and initiator, respectively. Kinetic studies showed that the addition of HEMA to the growing chain is favored the result being therefore the formation of a polymer scaffold with a gradient structure. Next, by using carbodiimide (DCC) chemistry, the CTA 4-cyano-4-(ethylsulfanylthiocarbonyl)sulfanylpentanoic acid (ECT) was coupled to the pendant hydroxyl groups of the polymer scaffold to obtain the macro-RAFT agent **55**. Eventual sequential RAFT polymerization of dimethylaminoethyl methacrylate (DMAEMA) and butyl methacrylate (BM) results in formation of polymer brushes with diblock copolymer arms **56**. These polymer brushes self-assemble in water to form pH-responsive polymersomes. The PEGMA segments in the polymer scaffold stabilize the polymersomes in aqueous solution, while the short DMAEMA and BM segments are responsible for the destabilization of the nanostructure in the endosomal acidic milieu. In experiments carried out on blood macrophages (RAW 264.7 cell line) infected with *Burkholderia thailandensis*, very high level of bacteria inhibition were revealed following treatment with ceftazidime-loaded pH-responsive polymersomes. Moreover, a cell proliferation colorimetric MTS (3-(4,5-dimethylthiazol-2-yl)-5-(3-carboxymethoxyphenyl)-2-(4-sulfophenyl)-2H-tetrazolium) assay proved the low intrinsic cytotoxicity of the polymer brushes **56**.

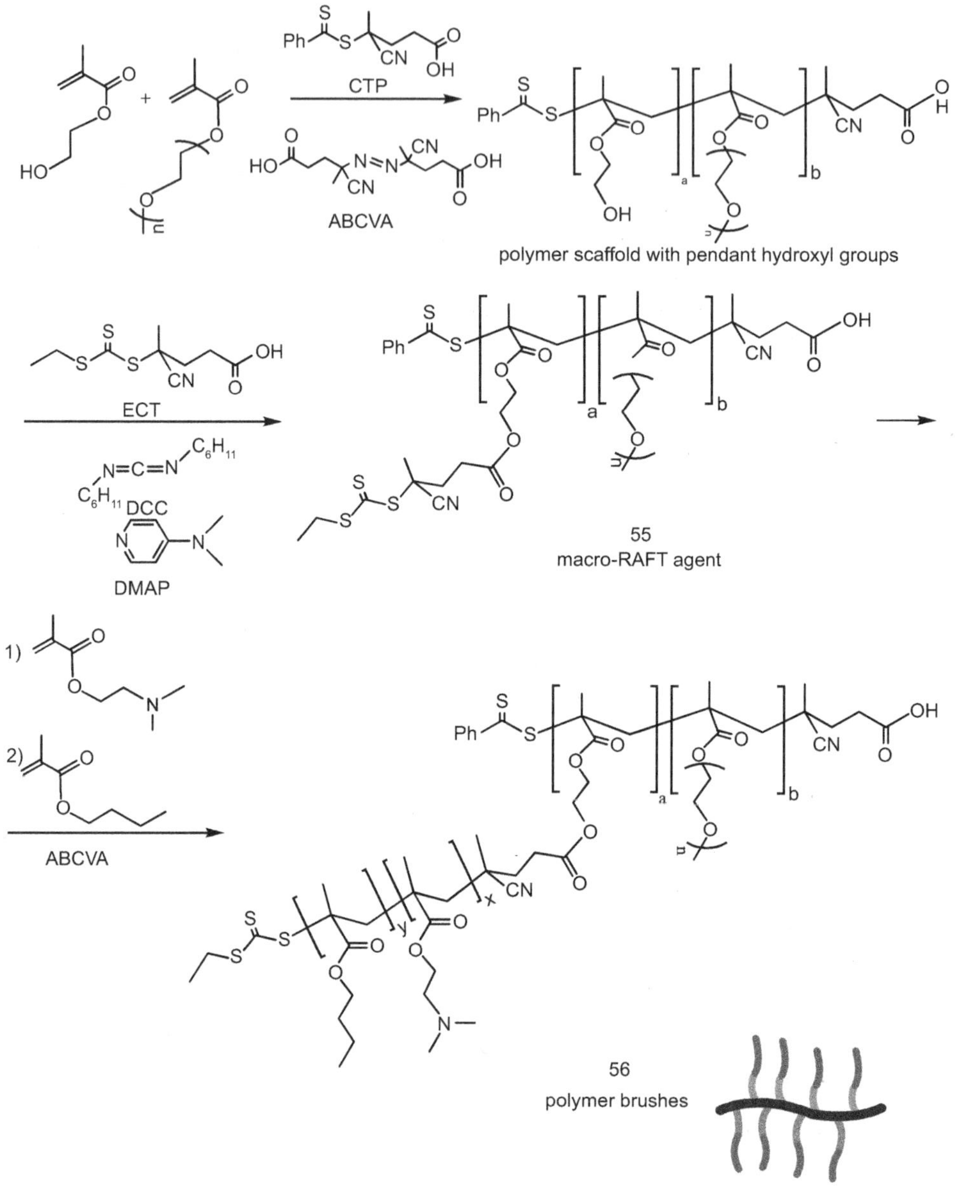

SCHEME 6.23 Synthesis of polymer brushes **56** able to reversible form self-assembled pH-responsive polymersomes.

6.6.3.2 Enzyme-Sensitive Antibiotic Delivery

Smart targeted vancomycin delivery at infection sites has been achieved by Wang and coworkers who used a lipase-sensitive triple-layered nanogel as the drug vehicle [269,270]. They have prepared a nanogel star polymer with hyperbranched polyphosphoester cross-linked core and amphiphilic diblock copolymer arms. The authors used the "arm first" approach in the synthesis of the star copolymers. The terminal hydroxyl groups of the amphiphilic diblock copolymer MPEG-*b*-PCL initiated ROP of the bifunctional cross-linker 3-oxapentane-1,5-diyl bis(ethylene phosphate) **57** as depicted in Scheme 6.24. The latter was obtained in a separate synthetic step from 2-chloro-2-oxo-1,3,2-dioxaphospholane (COP) and diethylene glycol. When placed in water, the hydrophobic PCL segments of the diblock copolymer arms of the star copolymer collapsed forming a shell that surrounded the polyphosphoester core and acted as a protective fence avoiding premature degradation of the encapsulated antibiotic. The hydrophilic MPEG segments of the diblock copolymer arms formed an outer hydrophilic corona imparting both water solubility and "stealth" properties to the nanovehicle. When reaching the targeted infection sites, enzymatic hydrolysis by bacterial lypases and phosphatases causing breakdown of the PCL layer and the polyphosphoester core resulted in "on demand" release of the encapsulated antibiotic.

6.6.3.3 Improving the Antimicrobial Activity of Currently Used Antibiotics Through Nanoencapsulation

Recently, our group has started a series of research studies with the aim to enhance the therapeutic efficiency of antibiotics through nanoencapsulation into nontoxic, biocompatible and biodegradable polymers. Balaure and coworkers reported the fabrication of a novel drug delivery system based on a chitosan/alginate polyelectrolyte matrix and silica network [271]. Various antibiotics were loaded onto the synthesized biocomposites and tested against *P. aeruginosa* and *E. coli* as reference strains. The obtained results showed increased antibacterial activity of cefazolin (KZ), cefaclor (CEC), cefuroxime (CXM), ceftriaxone (CRO), cefoxitin (FOX), trimethoprim/sulfamethoxazole (SXT) against *E. coli* ATCC 25922 strain and of piperacillin/tazobactam (TZP) against *P. aeruginosa* ATCC 27853 strain when encapsulated in the aforesaid biocomposite carrier.

In a different approach, we have synthesized water-dispersible nanocomposite magnetic materials based on ferrite (Fe_3O_4) core and cross-linked chitosan shell for magnetically targeted delivery of aminoglycoside antibiotics [272]. The MIC values for the encapsulated neomycin (N) and kanamycin (K) were lowered for both Gram-positive (*S. aureus*) and Gram-negative (*P. aeruginosa*) reference strains. In Fig. 6.9 an illustrative schematic representation of the architecture of the hybrid inorganic/organic aminoglycoside nanocarrier is presented, while Fig. 6.10 shows SEM micrographs of bare Fe_3O_4/cross-liked chitosan nanoparticles

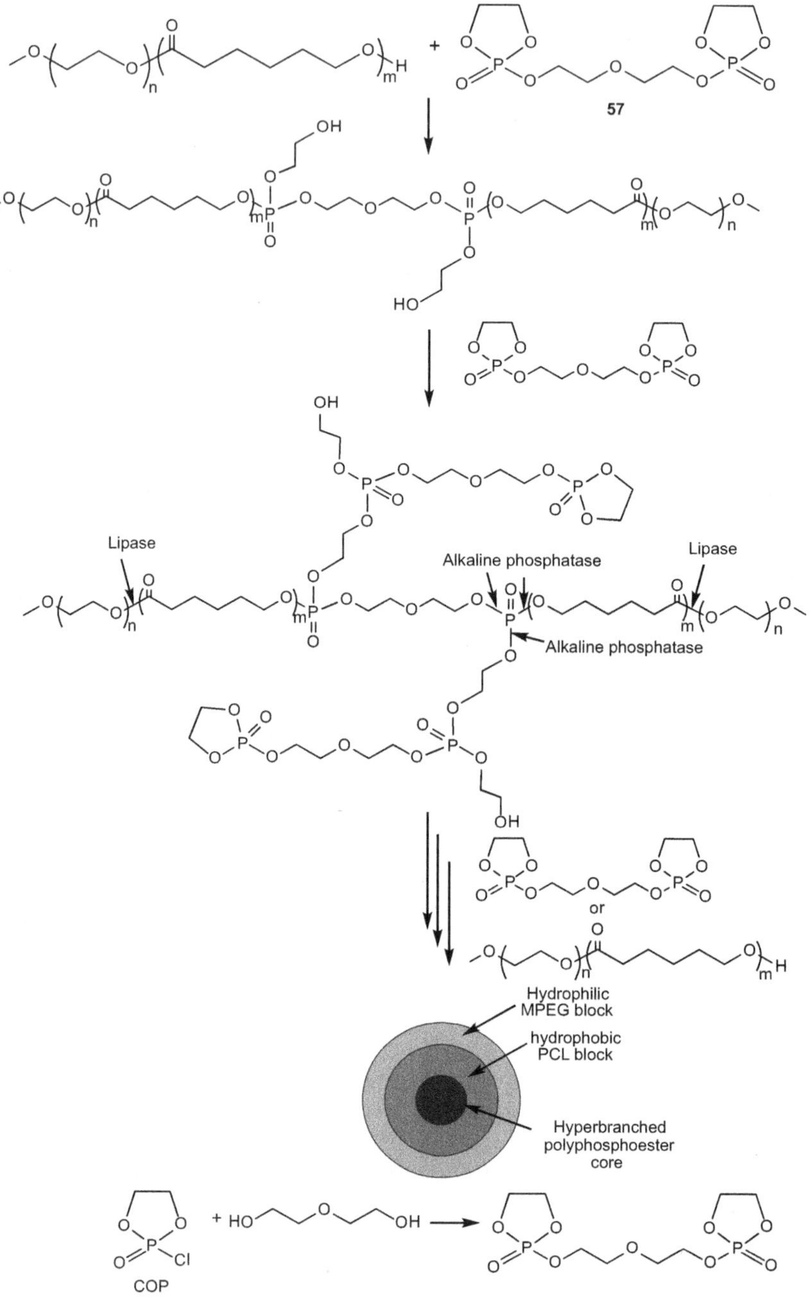

■ **SCHEME 6.24** Synthesis of enzyme-sensitive triple-layered nanogel for targeted vancomycin delivery.

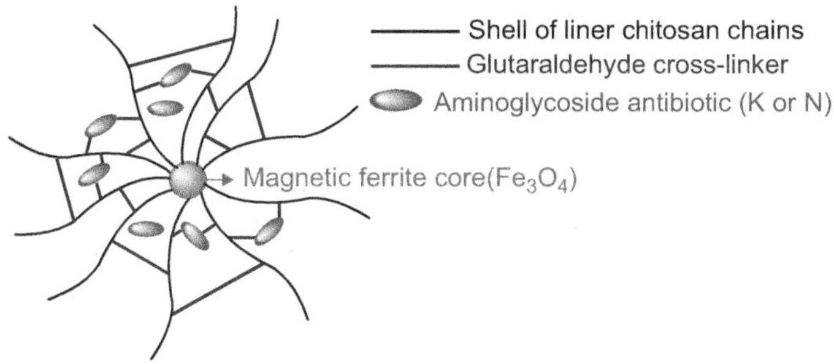

■ **FIGURE 6.9**   Schematic illustration of aminoglycoside-encapsulated Fe_3O_4/cross-linked chitosan core/shell magnetic nanocarrier.

(Fe_3O_4@CS) and antibiotic encapsulated magnetic nanoparticles (Fe_3O_4@ CS-K, and Fe_3O_4@CS-N, respectively).

We have also prepared cephalosporins-loaded magnetic chitosan microspheres revealing that embedded antibiotics exhibited significantly lowered MIC values as compared to free antibiotics when tested on *E. coli* ATCC 259922 and *S. aureus* ATCC 25923 bacterial strains. The enhancement of the antimicrobial activity was ascribed to the intimate interaction of the polycationic vehicle with the negatively charged bacterial surface [273,274]. Similarly, replacing chitosan with diethylaminoethyl-cellulose we obtained another type of magnetic core/polysaccharide shell nanocarrier. Performing release profile studies we were able to reveal the controlled release of various cephalosporins from the above magnetic nanovehicles [275]. Polysaccharides we used in these studies are highly biocompatible and bear a lot of free amino and hydroxyl groups that can function as anchoring points for further conjugation of some biologically relevant moieties such as targeting vitamins or aptamers. These structural features can be nicely exploited not only in drug and gene delivery, but also in magnetic resonance imaging, tissue engineering and enzyme immobilization. Chitosan beads and diethylaminoethyl-cellulose fibers covered with magnetic Fe_3O_4 nanoparticles have been also prepared and they provided tailored rifampicin release kinetics since the release rate can be modified by changing the polymer type [276]. In another study we proved the potentiating effect of magnetic dextran microspheres on the antimicrobial activity of large spectrum antibiotics belonging to penicillins, aminoglycosides, rifampicines and quinolones classes [277].

Using the matrix assisted pulsed laser evaporation (MAPLE) technique we were able to deposit thin coatings of usnic acid-loaded polylactide-polyvinyl alcohol microspheres on the surface of a Ti support [278]. The coatings

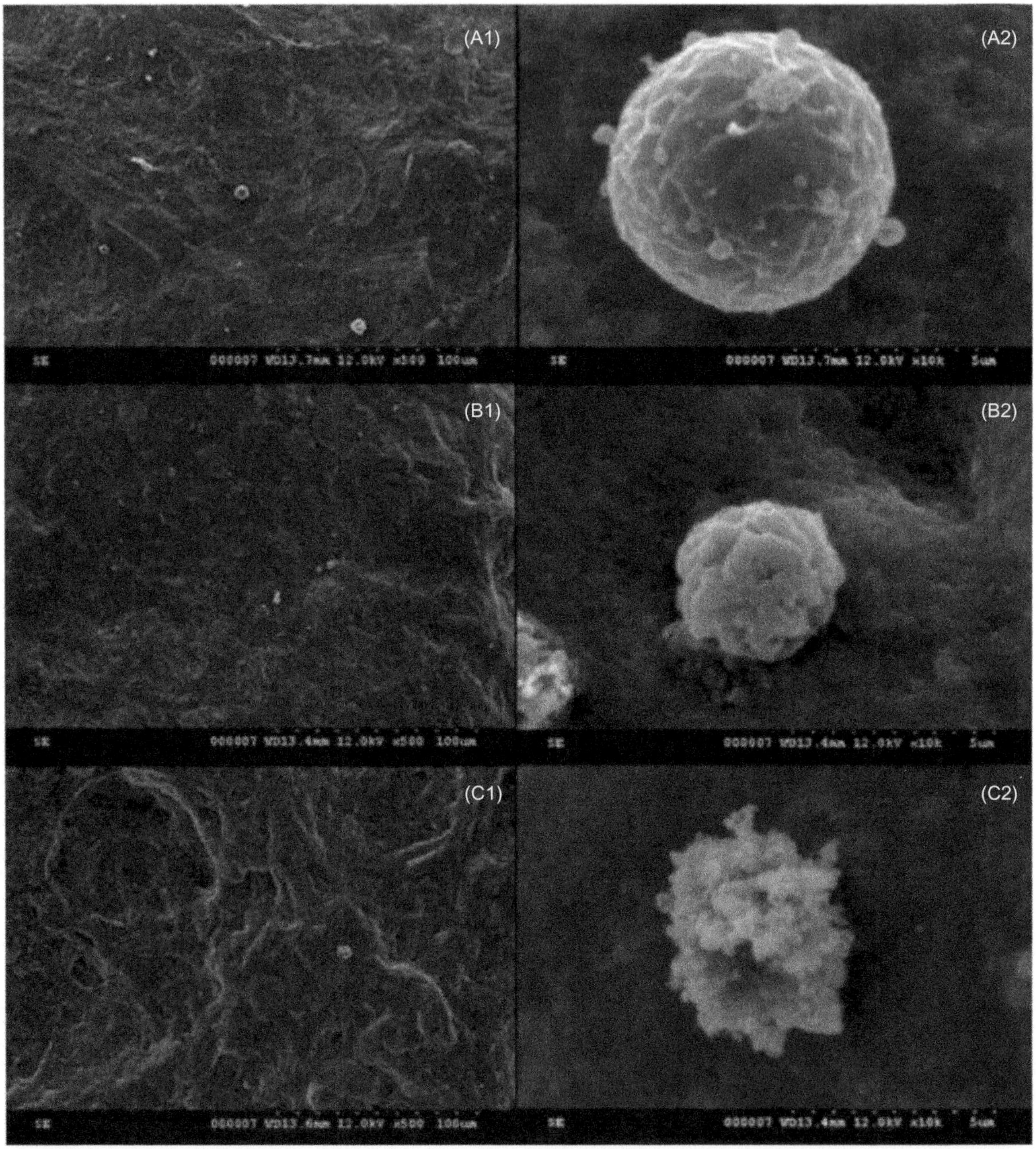

■ **FIGURE 6.10** SEM micrographs (A) Fe$_3$O$_4$@CS, (B) Fe$_3$O$_4$@CS-K and (C) Fe$_3$O$_4$@CS-N. *Reproduced with permission from [272] Grumezescu AM, Andronescu E, Holban AM, Ficai A, Ficai D, Voicu G, et al. Water dispersible cross-linked magnetic chitosan beads for increasing the antimicrobial efficiency of aminoglycoside antibiotics. Int J Pharm 2013;454:233–40.*

Table 6.6 Representative Examples of Dendrimer Nanocarriers for Smart Delivery of Antibiotics

Dendrimer	Type of Interaction	Antibiotic	Reference
Polyamidoamine (PAMAM)	Covalent binding	Penicillin	[140]
		Azithromycin	[280]
	Noncovalent binding	Nadifloxacin	[281]
Polypropylenimine (PPI)	Noncovalent binding	Amphotericin B	[282]
		Rifampicin	[283]

showed excellent antimicrobial properties inhibiting *S. aureus* adherence and preventing biofilm development. The antibiofilm activity corroborated with the good biocompatibility of these coatings recommends them as a promising solution in prevention of Ti prosthesis-related infections.

6.6.4 Dendrimers

Antibiotics can be attached not only to the outer surface of dendrimers through covalent conjugation to the end groups as previously presented in literature [239], but they can also be physically entrapped within the hydrophobic cavities inside the dendritic architecture [279]. The latter mode of binding has the advantage of a better protection of the drug but, on the other hand, can reduce the interaction between the antibiotic and its target.

Table 6.6 presents some applications of dendrimers as nanocarriers for antibiotic delivery.

6.7 CONCLUDING REMARKS AND FUTURE PERSPECTIVES

Intelligent nanoDDS have a series of remarkable advantages over their classic counterparts. They are able to:

1. Increase water solubility of hydrophobic antibiotics
2. Selectively direct the transport of the drug to the targeted infected sites
3. Prolong the circulation times upon administration and prevent antibiotic degradation prior to reaching its target
4. Provide the possibility to administer hydrophilic and hydrophobic drugs simultaneously to exploit interesting synergetic effects

5. Interfere with the mechanism through which bacteria develop resistance to antibiotics, disrupting or side-stepping bacterial drug-resistance mechanisms
6. Realize "on demand" release, the release process being triggered in sharp response to a localized internal or external stimulus
7. Increase the antimicrobial activity of antibiotics
8. Diminish harmful side effects of the medication
9. Reduce administered dose or the administration frequency
10. Increase patient compliance

Modern macromolecular engineering techniques such as controlled polymerization methods, post-polymerization functionalization methods and routes to control the macromolecular self-assembly gained strong impulse within the past decades and this development in turn allowed fabrication of smart nanoDDS based on self-assembled supramolecular polymeric materials.

However, there are much fewer reports on nanoencapsulation of anti-infectious agents as compared to other conditions such as cancer or cardiovascular diseases. Moreover, incorporation of more than one antibiotic or incorporation of an antibiotic along with an efflux pump inhibitor in order to exploit the synergistic co-delivery capability of nanoDDS would greatly increase the therapeutic efficacy [284].

In the field of smart nanoDDS we think that the future will belong to the systems able to orthogonally respond to different stimuli, such as pH and temperature, pH and redox potential, or any other combination of interest [52].

On the other hand, a viable and promising alternative to the therapy with antibiotics are the antimicrobial peptides and the synthetic mimics of antimicrobial peptides (SMAMPs), since resistance development toward these type of anti-infectious agents is much less likely than development of resistance toward antibiotics. However, in the case of SMAMPs more efforts remain to be made with the aim of increasing selectivity by maintaining high antibacterial activity while reducing their intrinsic hemolytic activity.

We can conclude that despite the significant advances made in the fight against superbugs, the field is still in its infancy and a lot still remains to be done since only a few nanoantibiotics have made it to the pharmaceutical market.

We hope that the present book chapter provides a general comprehensive view of this very important research field that greatly impacts on public health and will stimulate researchers especially the young ones to get involved in the complex interdisciplinary domain of the fight against antibiotic-resistant infections.

REFERENCES

[1] Ehrlich P, Hata S. Die experimentelle chemotherapie der spirilosen. Berlin: Julius Springer; 1910.

[2] Fleming A. On antibacterial action of culture of *Penicillium*, with special reference to their use in isolation of *B. influenzae*. Br J Exp Pathol 1929;10:226–36.

[3] Lambert PA. Bacterial resistance to antibiotics: modified target sites. Adv Drug Deliver Rev 2005;57:1471–85.

[4] Connell SR, Tracz DM, Nierhaus KH, Taylor DE. Ribosomal protection proteins and their mechanism of tetracycline resistance. Antimicrob Agents Chemother 2003;47(12):3675–81.

[5] Nikaido H. Multidrug resistance in bacteria. Annu Rev Biochem 2009;78:119–46.

[6] Wright GD. Bacterial resistance to antibiotics: enzymatic degradation and modification. Adv Drug Deliver Rev 2005;57:1451–70.

[7] Kumar A, Schweizer HP. Bacterial resistance to antibiotics: active efflux and reduced uptake. Adv Drug Deliver Rev 2005;57:1486–513.

[8] Sun J, Deng Z, Yan A. Bacterial multidrug efflux pumps: mechanisms, physiology and pharmacological exploitations. Biochem Biophys Res Commun 2014;453:254–67.

[9] Fernández L, Hancock REW. Adaptive and mutational resistance: role of porins and efflux pumps in drug resistance. Clin Microbiol Rev 2012;25(4):661–81.

[10] Tenover FC. Mechanisms of antimicrobial resistance in bacteria. Am J Med 2006;116(6):S3–S10.

[11] Cohen ML. Changing patterns of infectious diseases. Nature 2000;406:762–7.

[12] Gold HS, Moellering RC. Drug therapy: antimicrobial-drug resistance. New Engl J Med 1996;335:1445–53.

[13] Perichon B, Courvalin P. VanA-type vancomycin-resistant *Staphylococcus aureus*. Antimicrob Agents Chemother 2009;53(11):4580–7.

[14] a. Page MGP, Walsh CT. Antimicrobials. Curr Opin Microbiol 2005;58:495–7.
 b. Liu Y-Y, Wang Y, Walsh TR, Yi L-X, Zhang R, Spencer J, et al. Emergence of plasmid-mediated colistin resistance mechanism MCR-1 in animals and human beings in China: a microbiological and molecular biological study. Lancet Infect Dis 2015 published on line November 18, 2015.

[15] Seil JT, Webster TJ. Antimicrobial applications of nanotechnology: methods and literature. Int J Nanomed 2012;7:2767–81.

[16] Antimicrobial resistance. Accessed from: <http://www.who.int/mediacentre/factsheets/fs194/en/ >.

[17] Kalhapure RS, Suleman N, Mocktar C, Seedat N, Govender T. Nanoengineered drug delivery systems for enhancing antibiotic therapy. J Pharm Sci 2014;104(3):872–905.

[18] Antibiotic/Antimicrobial resistance. Accessed from: <http://www.cdc.gov/drugresistance/ >.

[19] Chen M, Yu Q, Sun H. Novel strategies for the prevention and treatment of biofilm related infections. Int J Mol Sci 2013;14:18488–501.

[20] Wu H, Moser C, Wang H-Z, Høiby N, Song Z-J. Strategies for combating bacterial biofilm infections. Int J Oral Sci 2015;7:1–7.

[21] Hetrick EM, Schoenfisch MH. Reducing implant-related infections: active release strategies. Chem Soc Rev 2006;35:780–9.

[22] Annous BA, Fratamico PM, Smith JL. Quorum sensing in biofilms: why bacteria behave the way they do they do? J Food Sci 2009;74(1):R24–37.

[23] Pesci EC, Milbank JBJ, Pearson JP, Mcknight S, Kende AS, Greenbergi EP, et al. Quinolone signaling in the cell-to-cell communication system of *Pseudomonas aeruginosa*. Proc Natl Acad Sci USA 1999;96:11229–34.

[24] Parsek MR, Greenberg EP. Sociomicrobiology: the connections between quorum sensing and biofilms. Trends Microbiol. 2005;13:27–33.

[25] Lewis K. Riddle of biofilm resistance. Antimicrob Agents Chemother 2001;45(4):999–1007.

[26] Wu P, Grainger DW. Drug/device combinations for local drug therapies and infection prophylaxis. Biomaterials 2006;27:2450–67.

[27] Khardori N, Yassien M. Biofilms in device-related infections. J Ind Microbiol 1995;15:141–7.

[28] Smith AW. Biofilms and antibiotic therapy: is there a role for combating bacterial resistance by the use of novel drug delivery systems? Adv Drug Deliver Rev 2005;57:1539–50.

[29] Donlan RM. Biofilms and device-associated infections. Emerg Infect Dis 2001;7:277–81.

[30] Tran PL, Lowry N, Campbell T, Reid TW, Webster DR, Tobin E, et al. An organoselenium compound inhibits Staphylococcus aureus biofilms on hemodialysis catheters in vivo. Antimicrob Agents Chemother 2012;56(2):972–8.

[31] Tollefson DF, Bandyk DF, Kaebnick HW, Seabrook GR, Towne JB. Surface biofilm disruption enhanced recovery of microorganisms from vascular prostheses. Arch Surg 1987;122(1):38–43.

[32] Fux CA, Quigley M, Worel AM, Post C, Zimmerli S, Ehrlich G, et al. Biofilm-related infections of cerebrospinal fluid shunts. Clin Microbiol Infect 2006;12(4):331–7.

[33] Song Z, Borgwardt L, Høiby N, Wu H, Sandberg Sørensen T, Borgwardt A. Prosthesis infections after orthopedic joint replacement: the possible role of bacterial biofilms. Orthop Rev (Pavia) 2013;5(2):65–71.

[34] Santos AP, Watanabe E, Andrade D. Biofilm on artificial pacemaker: fiction or reality? Arq Bras Cardiol 2011;97(5):e113–20.

[35] Dasgupta MK. Biofilms and infection in dialysis patients. Semin Dial 2002;15(5):338–46.

[36] Auler ME, Morreira D, Rodrigues FFO, Abrão MS, Margarido PFR, Matsumoto FE, et al. Biofilm formation on intrauterine devices in patients with recurrent vulvovaginal candidiasis. Med Mycol 2010;48(1):211–6.

[37] Donelli G, Vuotto C, Cardines R, Mastrantonio P. Biofilm-growing intestinal anaerobic bacteria. FEMS Immunol Med Microbiol 2012;65(2):318–25.

[38] Murakami M, Nishi Y, Seto K, Kamashita Y, Nagaoka E. Dry mouth and denture plaque microflora in complete denture and palatal obturator prosthesis wearers. Gerodentology 2015;32:188–94.

[39] Rieger UM, Mesina J, Kalbermatten DF, Haug M, Frey HP, Pico R, et al. Bacterial biofilms and capsular contracture in patients with breast implants. Br J Surg 2013;100(6):768–74.

[40] Abidi SH, Sherwani SK, Siddiqui TR, Bashir A, Kazmi SU. Drug resistance profile and biofilm forming potential of *Pseudomonas aeruginosa* isolated from contact lenses in Karachi-Pakistan. BMC Ophthalmol 2013;13:57.

[41] Agarwal A, Prasad Singh K, Jain A. Medical significance and management of staphylococcal biofilm. FEMS Immunol Med Microbiol 2010;58:147–60.

[42] Maki DG, Kluger DM, Crnich CJ. The risk of bloodstream infection in adults with different intravascular devices: a systematic review of 200 published prospective studies. Mayo Clin Proc 2006;81(9):1159–71.

[43] Bryers JD. Medical biofilms. Biotechnol Bioeng 2008;100(1):1–18.

[44] Wenzel RP. Health care-associated infections: major issues in the early years of the 21st century. Clin Infect Dis 2007;45(Suppl 1):S85–8.

[45] Healthcare-associated infections. Accessed from: <http://ecdc.europa.eu/en/healthtopics/Healthcare-associated_infections/Pages/index.aspx > .

[46] Moellering RC. Discovering new antimicrobial agents. Int J Antimicrob Agents 2011;37:2–9.

[47] Huh AJ, Kwon YJ. "Nanoantibiotics": a new paradigm for treating infectious diseases using nanomaterials in the antibiotics resistant era. J Control Release 2011;156:128–45.

[48] Davies J, Davies D. Origins and evolution of antibiotic resistance. Microbiol Mol Biol Rev 2010;74(3):417–33.

[49] So AD, Gupta N, Cars O. Tackling antibiotic resistance. BMJ-Brit Med J 2010;340:1091–2.

[50] a. Boucher HW, Talbot GH, Bradley JS, Edwards JE, Gilbert D, Rice LB, et al. Bad bugs, no drugs: no eskape! An update from the Infectious Diseases Society of America. Clin Infect Dis 2009;48(1):1–12.

b. U.S. Department of Health and Human Services, Centers for Disease Control and Prevention Antibiotic resistance threats in the United States, 2013.

c. CDDEP. The Center for Disease Dynamics, Economic & Policy. Antibiotic resistance. Accessed at <http://resistancemap.cddep.org/resmap/resistance>.

d. The bacterial challenge: time to react. A call to narrow the gap between multidrug-resistant bacteria in the EU and the development of new antibacterial agents. EMEA doc. ref. EMEA/576176/2009, ISBN 978-92-9193-193-4, Stockholm, September 2009, <http://dx.doi.org/10.2900/2518>.

e. European Centre for Disease Prevention and Control (ECDC). Antimicrobial resistance surveillance in Europe. Annual report of the European Antimicrobial Resistance Surveillance Network (EARS-Net) 2013. ISNB 978-92-9193-603-8, Stockholm: ECDC, 2014, <http://dx.doi.org/10.2900/3977>.

[51] Cameron DR, Howden BP, Peleg AY. The Interface between antibiotic resistance and virulence in *Staphylococcus aureus* and its impact upon clinical outcomes. Clin Infect Dis 2011;53(6):576–82.

[52] Balaure PC, Grumezescu AM. Smart synthetic polymer nanocarriers for controlled and site-specific drug delivery. Curr Top Med Chem 2015;15:1424–90.

[53] Pelgrift RY, Friedman AJ. Nanotechnology as a therapeutic tool to combat microbial resistance. Adv Drug Deliver Rev 2013;65:1803–15.

[54] Blecher K, Nasir A, Friedman A. The growing role of nanotechnology in combating infectious disease. Virulence 2011;2(5):395–401.

[55] Zhang L, Pornpattananangkul D, Hu C-MJ, Huang C-M. Development of nanoparticles for antimicrobial drug delivery. Curr Med Chem 2010;17(6):585–94.

[56] Friedman AJ, Phan J, Schairer DO, Champer J, Qin M, Pirouz A, et al. Antimicrobial and anti-inflammatory activity of chitosan–alginate nanoparticles: a targeted therapy for cutaneous pathogens. J Investig Dermatol 2013;133(5):1231–9.

[57] Maeda H, Wu J, Sawa T, Matsumura Y, Hori K. Tumor vascular permeability and the EPR effect in macromolecular therapeutics: a review. J Control Release 2000;65:271–84.

[58] Maeda H. Macromolecular therapeutics in cancer treatment: the EPR effect and beyond. J Control Release 2012;164:138–44.

[59] Ruggeri V, Francolini I, Donelli G, Piozzi A. Synthesis, characterization, and *in vitro* activity of antibiotic releasing polyurethanes to prevent bacterial resistance. J Biomed Mater Res A 2006;81:287–98.

[60] Dave RN, Joshi HM, Venugopalan VP. Novel Biocatalytic polymer-based antimicrobial coatings as potential ureteral biomaterial: preparation and *in vitro* performance evaluation. Antimicrob Agents Chemother 2011;55(2):845–53.

[61] Colombo MI. Pathogens and autophagy: subverting to survive. Cell Death Differ 2005;12:1481–3.

[62] Huang CM, Chen CH, Pornpattananangkul D, Zhang L, Chan M, Hsieh MF, et al. Eradication of drug resistant *Staphylococcus aureus* by liposomal oleic acids. Biomaterials 2011;32(1):214–21.

[63] Toti US, Guru BR, Hali M, McPharlin CM, Wykes SM, Panyam J, et al. Targeted delivery of antibiotics to intracellular chlamydial infections using PLGA nanoparticles. Biomaterials 2011;32:6606–13.

[64] Kale AA, Torchilin VP. Environment-responsive multifunctional liposomes. Methods Mol Biol 2010;605:213–42.

[65] Mura S, Nicolas J, Couvreur P. Stimuli-responsive nanocarriers for drug delivery. Nat Mater 2013;12:991–1003.

[66] Jabr-Milane L, van Vlerken L, Devalapally H, Shenoy D, Komareddy S, Bhavsar M, et al. Multi-functional nanocarriers for targeted delivery of drugs and genes. J Control Release 2008;130:121–8.

[67] Kim S, Kim J-H, Jeon O, Kwon IC, Park K. Engineered polymers for advanced drug delivery. Eur J Pharm Biopharm 2009;71:420–30.

[68] Li Q, Mahendra S, Lyon DY, Brunet L, Liga MV, Li D, et al. Antimicrobial nanomaterials for water disinfection and microbial control: potential applications and implications. Water Res 2008;42:4591–602.

[69] Cohen Stuart MA, Huck WTS, Genzer J, Müller M, Ober C, Stamm M, et al. Emerging applications of stimuli-responsive polymer materials. Nat Mater 2010;9(2):101–13.

[70] Bawa P, Pillay V, Choonara YE, du Toit LC. Stimuli-responsive polymers and their applications in drug delivery. Biomed Mater 2009;4:022001.

[71] de las Heras Alarcón C, Pennadam S, Alexander C. Stimuli responsive polymers for biomedical applications. Chem Soc Rev 2005;34:276–85.

[72] Tian H, Tang Z, Zhuang X, Chen X, Jing X. Biodegradable synthetic polymers: Preparation, functionalization and biomedical application. Prog Polym Sci 2012;37:237–80.

[73] You J-O, Almeda D, Ye GJC, Auguste DT. Bioresponsive matrices in drug delivery. J Biol Eng 2010;4:15.

[74] Hrubý M, Filippov SK, Štěpánek P. Smart polymers in drug delivery systems on crossroads: Which way deserves following? Eur Polym J 2015;65:82–97.

[75] Honey PJ, Rijo J, Anju A, Anoop KR. Smart polymers for the controlled delivery of drugs – a concise overview. Acta Pharm Sin B 2014;4(2):120–7.

[76] Ward MA, Georgiou TK. Thermoresponsive polymers for biomedical applications. Polymers 2011;3:1215–42.

[77] Ward MA, Georgiou TK. Thermoresponsive terpolymers based on methacrylate monomers: effect of architecture and composition. J Polym Sci Pol Chem 2010;48:775–83.

[78] Qiu LY, Bae YH. Polymer architecture and drug delivery. Pharm Res 2006;23(1):1–30.

[79] Wang J-S, Matyjaszewski K. Controlled/ "living" radical polymerization. Halogen atom transfer radical polymerization promoted by a Cu(I)/Cu(II) redox process. Macromolecules 1995;28:7901–10.

[80] Moad G, Rizzardo E, Solomon DH. Selectivity of the reaction of free radicals with styrene. Macromolecules 1982;15:909–14.

[81] Georges MK, Veregin RPN, Kazmaier PM, Hamer GK. Narrow molecular weight resins by a free-radical polymerization process. Macromolecules 1993;26:2987–8.

[82] Chiefari J, Chong YKB, Ercole F, Krstina J, Jeffery J, Le TPT, et al. Living free-radical polymerization by reversible addition-fragmentation chain transfer: the RAFT process. Macromolecules 1998;31:5559–62.

[83] Endo T. General mechanisms in ring-opening polymerization. ISBN 978-3-527-31953-4. Dubois P, Coulembier O, Raquez J-M, editors. Handbook of ring opening polymerization. Weinheim: WILEY-VCH Verlag GmbH & Co. KGaA; 2009. p. 53–6.

[84] Endo T. General mechanisms in ring-opening polymerization. ISBN 978-3-527-31953-4. Dubois P, Coulembier O, Raquez J-M, editors. Handbook of ring opening polymerization. Weinheim: WILEY-VCH Verlag GmbH & Co. KGaA; 2009. p. 56–60.

[85] Endo T. General mechanisms in ring-opening polymerization. ISBN 978-3-527-31953-4. Dubois P, Coulembier O, Raquez J-M, editors. Handbook of ring opening polymerization. Weinheim: WILEY-VCH Verlag GmbH & Co. KGaA; 2009. p. 60–1.

[86] Dechy-Cabaret O, Martin-Vaca B, Bourissou D. Polyesters from dilactones. ISBN 978-3-527-31953-4. Dubois P, Coulembier O, Raquez J-M, editors. Handbook of ring opening polymerization. Weinheim: WILEY-VCH Verlag GmbH & Co. KGaA; 2009. p. 256–8.

[87] Buchmeiser MR. Ring-opening metathesis polymerization. ISBN 978-3-527-31953-4. Dubois P, Coulembier O, Raquez J-M, editors. Handbook of ring opening polymerization. Weinheim: WILEY-VCH Verlag GmbH & Co. KGaA; 2009. p. 197–226.

[88] Newkome GR, Shreiner CD. Poly(amidoamine), polypropylenimine, and related dendrimers and dendrons possessing different 1→2 branching motifs: an overview of the divergent procedures. Polymer 2008;49:1–173.

[89] Grayson SM, Fréchet JMJ. Convergent dendrons and dendrimers: from synthesis to applications. Chem Rev 2001;101:3819–67.

[90] Matyjaszewski K, Xia J. Atom transfer radical polymerization. Chem Rev 2001;101:2921–90.

[91] Nicolas J, Guillaneuf Y, Lefay C, Bertin D, Gigmes D, Charleux B. Nitroxide-mediated polymerization. Prog Polym Sci 2013;38:63–235.

[92] Gregory A, Stenzel MH. Complex polymer architectures via RAFT polymerization: from fundamental process to extending the scope using click chemistry and nature's building blocks. Prog Polym Sci 2012;37:38–105.

[93] Balaure PC, Grumezescu AM. Methods for synthesizing the macromolecular constituents of smart nanosized carriers for controlled drug delivery. Curr Med Chem 2014;21:3333–74.

[94] Ghaemy M, Ziaei S, Alizadeh R. Synthesis of pH-sensitive amphiphilic penta-block copolymers via combination of ring-opening and atom transfer radical polymerization for drug delivery. Eur Polym J 2014;58:103–14.

[95] Nelson-Mendez A, Aleksanian S, Oh M, Lim H-S, Oh JK. Reductively degradable polyester-based block copolymers prepared by facile polycondensation and ATRP: synthesis, degradation, and aqueous micellization. Soft Matter 2011;7:7441–52.

[96] Zhang W, He J, Liu Z, Ni P, Zhu X. Biocompatible and pH-responsive triblock copolymer MPEG-*b*-PCL-*b*-PDMAEMA: synthesis, self-assembly, and application. J Polym Sci A: Polym Chem 2010;48:1079–91.

[97] Xu Y, Bolisetty S, Drechsler M, Fang B, Yuan J, Ballauff M, et al. pH and salt responsive poly(*N,N*-dimethylaminoethyl methacrylate) cylindrical brushes and their quaternized derivatives. Polymer 2008;49:3957–64.

[98] Lee H, Pietrasik J, Sheiko SS, Matyjaszewski K. Stimuli-responsive molecular brushes. Prog Polym Sci 2010;35:24–44.

[99] Handké N, Trimaille T, Luciani E, Rollet M, Delair T, Verrier B, et al. Elaboration of densely functionalized polylactide nanoparticles from n-acryloxysuccinimide-based block copolymers. J Polym Sci A: Polym Chem 2011;49:1341–50.

[100] Delplace V, Tardy A, Harrisson S, Mura S, Gigmes D, Guillaneuf Y, et al. Degradable and comb-like PEG-based copolymers by nitroxide-mediated radical ring-opening polymerization. Biomacromolecules 2013;14(10):3769–79.

[101] Delplace V, Harrisson S, Tardy A, Gigmes D, Guillaneuf Y, Nicolas J. Nitroxide-mediated radical ring-opening copolymerization: chain-end investigation and block copolymer synthesis. Macromol Rapid Commun 2014;35:484–91.

[102] Tunca U, Ozyurek Z, Erdogan T, Hizal G. Novel miktofunctional initiator for the preparation of an ABC-type miktoarm star polymer via a combination of controlled polymerization techniques. J Polym Sci A: Polym Chem 2004;42:4228–36.

[103] Bernaerts KV, Du Prez FE. Dual/heterofunctional initiators for the combination of mechanistically distinct polymerization techniques. Prog Polym Sci 2006;31:671–722.

[104] Song Z, Xu Y, Yang W, Cui L, Zhang J, Liu J. Graphene/tri-block copolymer composites prepared via RAFT polymerizations for dual controlled drug delivery via pH stimulation and biodegradation. Eur Polym J 2015;69:559–72.

[105] Li Y, Liu T, Zhang G, Ge Z, Liu S. Tumor-targeted redox-responsive nonviral gene delivery nanocarriers based on neutral-cationic brush block copolymers. Macromol Rapid Commun 2014;35:466–73.

[106] Du Y, Chen W, Zheng M, Meng F, Zhong Z. pH-sensitive degradable chimaeric polymersomes for the intracellular release of doxorubicin hydrochloride. Biomaterials 2012;33:7291–9.

[107] Sun J-T, Yu Z-Q, Hong C-Y, Pan C-Y. Biocompatible zwitterionic sulfobetaine copolymer-coated mesoporous silica nanoparticles for temperature-responsive drug release. Macromol Rapid Commun 2012;33:811–8.

[108] Parwe SP, Chaudhari PN, Mohite KK, Selukar BS, Nande SS, Garnaik B. Synthesis of ciprofloxacin-conjugated poly(L-lactic acid) polymer for nanofiber fabrication and antibacterial evaluation. Int J Nanomed 2014;9:1463–77.

[109] Förster N, Pöppler A-C, Stalke D, Vana P. Photocrosslinkable star polymers via RAFT-copolymerizations with N-ethylacrylate-3,4-dimethylmaleimide. Polymers 2013;5:706–29.

[110] Förster N, Schmidt S, Vana P. Tailoring confinement: nano-carrier synthesis via Z-RAFT star polymerization. Polymers 2015;7:695–716.

[111] Lienkamp K, Madkour AE, Kumar K-N, Nűsslein K, Tew GN. Antimicrobial polymers prepared by ring-opening metathesis polymerization: manipulating antimicrobial properties by organic counterion and charge density variation. Chem Eur J 2009;15:11715–22.

[112] Yi X, Zhang Q, Dong H, Zhao D, Xu J-q, Zhuo R, et al. One-pot synthesis of crosslinked amphiphilic polycarbonates as stable but reduction-sensitive carriers for doxorubicin delivery. Nanotechnology 2015;26:395602.

[113] Khoee S, Kavand A. Preparation, co-assembling and interfacial crosslinking of photocurable and folate-conjugated amphiphilic block copolymers for controlled and targeted drug delivery: smart armored nanocarriers. Eur J Med Chem 2014;73:18–29.

[114] Khoee S, Hemati K. Synthesis of magnetite/polyamino-ester dendrimer based on PCL/PEG amphiphilic copolymers via convergent approach for targeted diagnosis and therapy. Polymer 2013;54:5574–85.

[115] Solaro R, Chiellini F, Battisti A. Targeted delivery of protein drugs by nanocarriers. Materials 2010;3:1928–80.

[116] Ringsdorf H. Structure and properties of pharmacologically active polymers. J Polym Sci Symp 1975;51:135Y153.

[117] Xiao Y, Zhu J, Zheng L. Application of chitosan-based gels in pharmaceuticals Yao K, Li J, Yao F, Yin Y, editors. Chitosan-based hydrogels. Functions and applications. USA: CTC Press Taylor & Francis Group; 2012. p. 324.

[118] Yoon H-J, Jang W-D. Polymeric supramolecular systems for drug delivery. J Mater Chem 2010;20:211–22.

[119] Vauthier C, Bouchemal K. Methods for the preparation and manufacture of polymeric nanoparticles. Pharm Res 2009;26(5):1025–58.

[120] Mora-Huertasa CE, Fessi H, Elaissari A. Polymer-based nanocapsules for drug delivery. Int J Pharm 2010;385:113–42.

[121] Devissaguet J-P, Fessi H, Puisieux F Process for the preparaton of dispersible colloidal systems of a substance in the form of nanocapsules. US Patent 5049322, September 17 1991.

[122] Gao P, Nie X, Zou M, Shi Y, Cheng G. Recent advances in materials for extended-release antibiotic delivery system. J Antibiot 2011;64:625–34.

[123] Letchford K, Burt H. A review of the formation and classification of amphiphilic block copolymer nanoparticle structures: micelles, nanospheres, nanocapsules and polymersomes. Eur J Pharm Biopharm 2007;65:259–69.

[124] O'Reilly RK, Hawker CJ, Wooley KL. Cross-linked block copolymer micelles: functional nanostructures of great potential and versatility. Chem Soc Rev 2006;35:1068–83.

[125] Riess G. Micellization of block copolymers. Prog Polym Sci 2003;28:1107–70.

[126] Balsara NP, Tirrell M, Lodge TP. Micelle formation of BAB triblock copolymers in solvents that preferentially dissolve the A block. Macromolecules 1991;24(8):1975–86.

[127] Xiao RZ, Zeng ZW, Zhou GL, Wang JJ, Li FZ, Wang AM. Recent advances in PEG-PLGA block copolymer nanoparticles. Int J Nanomedicine 2010;5:1057–65.

[128] Riley T, Stolnik S, Heald CR, Xiong CD, Garnett MC, Illum L, et al. Physicochemical evaluation of nanoparticles assembled from poly(lactic acid)- poly(ethylene glycol) (PLA-PEG) block copolymers as drug delivery vehicles. Langmuir 2001;17:3168–74.

[129] Heald CR, Stolnik S, Kujawinski KS, De Matteis C, Garnett MC, Illum L, et al. Poly(lactic acid)-poly(ethylene oxide) (PLA-PEG) nanoparticles: NMR studies of the central solid-like PLA core and the liquid PEG corona. Langmuir 2002;18:3669–75.

[130] Owens DE, Peppas NA. Opsonization, biodistribution, and pharmacokinetics of polymeric nanoparticles. Int J Pharm 2006;307:93–102.

[131] Ruysschaert T, Germain M, Pereira da Silva Gomes JF, Fournier D, Sukhorukov GB, Meier W, et al. Liposome-based nanocapsules. IEEE Trans NanoBioScience 2004;3:49–55.

[132] Schillen K, Bryskhe K, Mel'nikova YS. Vesicles formed from a poly(ethylene oxide)-poly(propylene oxide)-poly(ethylene oxide) triblock copolymer in dilute aqueous solution. Macromolecules 1999;32:6885–8.

[133] Liu G, Ma S, Li S, Cheng R, Meng F, Liu H, et al. The highly efficient delivery of exogenous proteins into cells mediated by biodegradable chimaeric polymersomes. Biomaterials 2010;31:7575–85.

[134] Valente I, del Valle LJ, Casas MT, Franco L, Rodriguez-Galan A, Puiggali J, et al. Nanospheres and nanocapsules of amphiphilic copolymers constituted by methoxypolyethylene glycol cyanoacrylate and hexadecyl cyanoacrylate units. Express Polym Lett 2013;7:2–20.

[135] Chern CS. Emulsion polymerization mechanisms and kinetics. Prog Polym Sci 2006;31:443–86.

[136] Fattal E, Youssef M, Couvreur P, Andremont A. Treatment of experimental salmonellosis in mice with ampicillin-bound nanoparticles. Antimicrob Agents Chemother 1989;33(9):1540–3.

[137] Vauthier C, Dubernet C, Fattal E, Pinto-Alphandary H, Couvreur P. Poly(alkylcyanoacrylates) as biodegradable materials for biomedical applications. Adv Drug Deliver Rev 2003;55:519–48.

[138] Vauthier C, Labarre D, Ponchel G. Design aspects of poly(alkylcyanoacrylate) nanoparticles for drug delivery. J Drug Target 2007;15(10):641–63.

[139] Yordanov G. Poly(alkyl cyanoacrylate) nanoparticles as drug carriers: 33 years later. Bulg J Chem 2012;1:61–73.

[140] Kisich KO, Gelperina S, Higgins MP, Wilson S, Shipulo E, Oganesyan E, et al. Encapsulation of moxifloxacin within poly(butyl cyanoacrylate) nanoparticles enhances efficacy against intracellular *Mycobacterium tuberculosis*. Int J Pharm 2007;345:154–62.

[141] Yordanov G, Abrashev N, Dushkin C. Poly(*n*-butylcyanoacrylate) submicron particles loaded with ciprofloxacin for potential treatment of bacterial infections. Progr Colloid Polym Sci 2010;137:53–9.

[142] Peracchia MT, Vauthier C, Popa MI, Puisieux F, Couvreur P. Investigation of the formation of sterically stabilized poly(ethylene glycol/isobutylcyanoacrylate) nanoparticles by chemical grafting of polyethylene glycol during the polymerization of isobutyl cyanoacrylate. STP Pharma Sci 1997;7:513–20.

[143] Schork FJ, Luo Y, Smulders W, Russum JP, Butté A, Fontenot K. Miniemulsion polymerization. Adv Polym Sci 2005;175:129–255.

[144] Weiss CK, Ziener U, Landfester K. A route to nonfunctionalized and functionalized poly(n-butylcyanoacrylate) nanoparticles: preparation in miniemulsion. Macromolecules 2007;40:928–38.

[145] Nagavarma BVN, Yadav HKS, Ayaz A, Vasudha LS, Shivakumar HG. Different techniques for preparation of polymeric nanoparticles- A review. Asian J Pharm Clin Res 2012;5(Suppl. 3):16–23.

[146] Reis CP, Neufeld RJ, Ribeiro AJ, Veiga F. Nanoencapsulation I. methods for preparation of drug-loaded polymeric nanoparticles. Nanomed Nanotechnol Biol Med 2006;2:8–21.

[147] Lambert G, Fattal E, Pinto-Alphandary H, Gulik A, Couvreur P. Polyisobutylacrylate nanocapsules containing an aqueous core as a novell colloidal carrier for the delivery of oligonucleotides. Pharm Res 2000;17(6):707–14.

[148] Montasser I, Briançon S, Fessi H. The effect of monomers on the formulation of polymeric nanocapsules based on polyureas and polyamides. Int J Pharm 2007;335:176–9.

[149] Ngwuluka N. Application of in situ polymerization for design and development of oral drug delivery systems. AAPS PharmSciTech 2010;11(4):1603–11.

[150] Lee Y-D, Huang C-Y, Chen C-M Drug-loaded poly(alkyl-cyanoacrylate) nanoparticles and process for the preparation thereof. US Patent 8679539 B2, March 25, 2014.

[151] Wang S, Kim G, Lee Y-EK, Hah HJ, Ethirajan M, Pandey RK, et al. Multifunctional biodegradable polyacrylamide nanocarriers for cancer theranostics - a "see and treat" strategy. ACS Nano 2012;6(8):6843–51.

[152] Tong R, Cheng J. Drug-initiated, controlled ring-opening polymerization for the synthesis of polymer-drug conjugates. Macromolecules 2012;45(5):2225–32.

[153] Mohammadi G, Valizadeh H, Barzegar-Jalali M, Lotfipour F, Adibkia K, Milani M, et al. Development of azithromycin–PLGA nanoparticles: Physicochemical characterization and antibacterial effect against *Salmonella typhi*. Colloid Surf B: Biointerfaces 2010;80:34–9.

[154] Azhdarzadeh M, Lotfipour F, Zakeri-Milani P, Mohammadi G, Valizadeh H. Antibacterial performance of azithromycin nanoparticles as colloidal drug delivery system against different gram-negative and gram-positive bacteria. Adv Pharm Bull 2012;2(1):17–24.

[155] Cheow WS, Hadinoto K. Enhancing encapsulation efficiency of highly water-soluble antibiotic in poly(lactic-co-glycolic acid) nanoparticles: modifications of standard nanoparticle preparation methods. Colloid Surf A: Physicochem Eng 2010;370:79–86.

[156] Ghari T, Mortazavi SA, Khoshayand MR, Kobarfard F, Gilani K. Preparation, optimization, and in vitro evaluation of azithromycin encapsulated nanoparticles by using response surface methodology. J Drug Deliv Sci Technol 2014;24(4):352–60.

[157] Moinard-Chécot D., Y. Chevalier, Briançon S., Beney L., Fessi H. Mechanism of nanocapsules formation by the emulsion–diffusion process. J Colloid Interface Sci. 2008, 317, 458-468.

[158] Swarnakar NK, Jain AK, Singh RP, Godugu C, Das M, Jain S. Oral bioavailability, therapeutic efficacy and reactive oxygen species scavenging properties of coenzyme Q10-loaded polymeric nanoparticles. Biomaterials 2011;32:6860–74.

[159] Murakami H, Kobayashi M, Takeuchi H, Kawashima Y. Preparation of poly(DL-lactide-co-glycolide) nanoparticles by modified spontaneous emulsification solvent diffusion method. Int J Pharm 1999;187:143–52.

[160] Esmaeili F, Hosseini-Nasr M, Rad-Malekshahi M, Samadi N, Atyabi F, Dinarvand R. Preparation and antibacterial activity evaluation of rifampicin-loaded poly lactide-co-glycolide nanoparticles. Nanomed Nanotechnol Biol Med 2007;3:161–7.

[161] Koopaei MN, Maghazei MS, Mostafavi SH, Jamalifar H, Samadi N, Amini M, et al. Enhanced antibacterial activity of roxithromycin loaded pegylated poly lactide-co-glycolide nanoparticles. DARU 2012;20:92.

[162] Grigoriev DO, Miller R. Mono- and multilayer covered drops as carriers. Curr Opin Colloid Interf Sci 2009;14:48–59.

[163] Jeong Y-I, Na H-S, Seo D-H, Kim D-G, Lee H-C, Jang M-K, et al. Ciprofloxacin-encapsulated poly(dl-lactide-co-glycolide) nanoparticles and its antibacterial activity. Int J Pharm 2008;352:317–23.

[164] Dillen K, Vandervoort J, Van den Mooter G, Ludwig A. Evaluation of ciprofloxacin-loaded Eudragit® RS100 or RL100/PLGA nanoparticles. Int J Pharm 2006;314:72–82.

[165] Zakeri-Milani P, Loveymi BD, Jelvehgari M, Valizadeh H. The characteristics and improved intestinal permeability of vancomycin PLGA-nanoparticles as colloidal drug delivery system. Colloid Surf B: Biointerfaces 2013;103:174–81.

[166] du Toit LC, Pillay V, Choonara YE, Iyuke SE. Formulation and evaluation of a salted-out isoniazid-loaded nanosystem. AAPS PharmSciTech 2008;9(1): 174–81.

[167] Allouche J. Synthesis of organic and bioorganic nanoparticles: An overview of the preparation methods ISBN 978-1-4471-4212-6 Brayner R, Fiévet F, Coradin T, editors. Nanomaterials: A danger or a promise? A chemical and biological perspective. London: Springer-Verlag; 2013. p. 27–74.

[168] Jeong Y-I, Cho C-S, Kim S-H, Ko K-S, Kim S-I, Shim Y-H, et al. Preparation of poly(dl-lactide-co-glycolide) nanoparticles without surfactant. J Appl Polym Sci 2001;80:2228–36.

[169] Jain SK, Gupta Y, Ramalingam L, Jain A, Jain A, Khare P, et al. Lactose-conjugated PLGA nanoparticles for enhanced delivery of rifampicin to the lung for effective treatment of pulmonary tuberculosis. PDA J Pharm Sci Technol 2010;64(3):278–87.

[170] Hu X, Liu S, Zhou G, Huang Y, Xie Z, Jing X. Electrospinning of polymeric nanofibers for drug delivery applications. J Control Release 2014;185:12–21.

[171] Taylor G. Electrically driven jets. Proc R Soc Lond A 1969;313:453–75.

[172] Hong Y, Fujimoto K, Hashizume R, Guan J, Stankus JJ, Tobita K, et al. Generating elastic, biodegradable polyurethane/poly(lactide-co-glycolide) fibrous sheets with controlled antibiotic release via two-stream electrospinning. Biomacromolecules 2008;9(4):1200–7.

[173] Zahedi P, Karami Z, Rezaeian I, Jafari S-H, Mahdaviani P, Abdolghaffari Preparation and performance evaluation of tetracycline hydrochloride loaded wound dressing mats based on electrospun nanofibrous poly(lactic acid)/poly(ε-caprolactone) blends. J Appl Polym Sci 2012;124:4174–83.

[174] Gentile P, Chiono V, Tonda-Turo C, Ferreira AM, Ciardelli G. Polymeric membranes for guided bone regeneration. Biotechnol J 2011;6:1187–97.

[175] Xue J, Niu Y, Gong M, Shi R, Chen D, Zhang L, et al. Electrospun microfiber membranes embedded with drug-loaded clay nanotubes for sustained antimicrobial protection. ACS Nano 2015;9(2):1600–12.

[176] Prabhakaran MP, Zamani M, Felice B, Ramakrishna S. Electrospraying technique for the fabrication of metronidazole contained PLGA particles and their release profile. Mater Sci Eng C: Mater Biol Appl 2015;56:66–73.

[177] Ungaro F, d'Angelo I, Coletta C, d'Emmanuele di Villa Bianca R, Sorrentino R, Perfetto B, et al. Dry powders based on PLGA nanoparticles for pulmonary delivery of antibiotics: modulation of encapsulation efficiency, release rate and lung deposition pattern by hydrophilic polymers. J Control Release 2012;157:149–59.

[178] AL-Quadeib BT, Radwan MA, Siller L, Horrocks B, Wright MC. Stealth Amphotericin B nanoparticles for oral drug delivery: in vitro optimization. Saudi Pharm J 2015;23:290–302.

[179] Awasthi R, Kulkarni GT. Development and characterization of amoxicillin loaded floating microballoons for the treatment of *Helicobacter pylori* induced gastric ulcer. Asian J Pharm Sci 2013;8:174–80.

[180] Dorati R, DeTrizio A, Genta I, Grisoli P, Merelli A, Tomasi C, et al. An experimental design approach to the preparation of pegylated polylactide-co-glicolide gentamicin loaded microparticles for local antibiotic delivery. Mater Sci Eng C: Mater Biol Appl 2016;58:909–17.

[181] Nandakumar V, Geetha V, Chittaranjan S, Doble M. High glycolic poly (DL lactic co glycolic acid) nanoparticles for controlled release of meropenem. Biomed Pharmacother 2013;67:431–6.

[182] Jaeger W, Bohrisch J, Laschewsky A. Synthetic polymers with quaternary nitrogen atoms-synthesis and structure of the most used type of cationic polyelectrolytes. Prog Polym Sci 2010;35:511–77.

[183] Muñoz-Bonilla A, Fernández-García M. Polymeric materials with antimicrobial activity. Prog Polym Sci 2012;37:281–339.

[184] Engler AC, Wiradharma N, Ong ZY, Coady DJ, Hedrick JL, Yang Y-Y. Emerging trends in macromolecular antimicrobials to fight multi-drug-resistant infections. Nano Today 2012;7:201–22.

[185] Nederberg F, Zhang Y, Tan JPK, Xu K, Wang H, Yang C, et al. Biodegradable nanostructures with selective lysis of microbial membranes. Nat Chem 2011;3:409–14.

[186] Qiao Y, Yang C, Coady DJ, Ong ZY, Hedrick JL, Yang Y-Y. Highly dynamic biodegradable micelles capable of lysing Gram-positive and Gram-negative bacterial membrane. Biomaterials 2012;33:1146–53.

[187] Gabriel GJ, Maegerlein JA, Nelson CF, Dabkowski JM, Eren T, Nüsslein K, et al. Comparison of facially amphiphilic versus segregated monomers in the design of antibacterial copolymers. Chem Eur J 2009;15:433–9.

[188] Sambhy V, Peterson BR, Sen A. Antibacterial and hemolytic activities of pyridinium polymers as a function of the spatial relationship between the positive charge and the pendant alkyl tail. Angew Chem Int Ed 2008;47:1250–4.

[189] Timofeeva LM, Kleshcheva NA, Moroz AF, Didenko LV. Secondary and tertiary polydiallylammonium salts: Novel polymers with high antimicrobial activity. Biomacromolecules 2009;10(11):2976–86.

[190] Timofeeva LM, Vasilieva YA, Kleshcheva NA, Gromova GL, Topchiev DA. Radical polymerization of diallylamine compounds: from quantum chemical modeling to controllable synthesis of high-molecular-weight polymers. Int J Quantum Chem 2002;88:531–41.

[191] Kuroda K, DeGrado WF. Amphiphilic polymethacrylate derivatives as antimicrobial agents. J Am Chem Soc 2005;127(12):4128–9.

[192] Kuroda K, Caputo GA, DeGrado WF. The role of hydrophobicity in the antimicrobial and hemolytic activities of polymethacrylate derivatives. Chem Eur J 2009;15:1123–33.

[193] Sovadinova I, Palermo EF, Huang R, Thoma LM, Kuroda K. Mechanism of polymer-induced hemolysis: nanosized pore formation and osmotic lysis. Biomacromolecules 2011;12:260–8.

[194] Punia A, Lee K, He E, Mukherjee S, Mancuso A, Banerjee P, et al. Effect of relative arrangement of cationic and lipophilic moieties on hemolytic and antibacterial activities of pegylated polyacrylates. Int J Mol Sci 2015;16:23867–80.

[195] Punia A, Mancuso A, Banerjee P, Yang N-L. Nonhemolytic and antibacterial acrylic copolymers with hexamethyleneamine and poly(ethylene glycol) side chains. ACS Macro Lett 2015;4(4):426–30.

[196] Palermo EF, Sovadinova I, Kuroda K. Structural determinants of antimicrobial activity and biocompatibility in membrane-disrupting methacrylamide random copolymers. Biomacromolecules 2009;10(11):3098–107.

[197] Nederberg F, Lohmeijer BGG, Leibfarth F, Pratt RC, Choi J, Dove AP, et al. Organocatalytic ring opening polymerization of trimethylene carbonate. Biomacromolecules 2007;8(1):153–60.

[198] Helou M, Miserque O, Brusson J-M, Carpentier J-F, Guillaume SM. Organocatalysts for the controlled "immortal" ring-opening polymerization of six-membered-ring cyclic carbonates: a metal-free, green process. Chem Eur J 2010;16:13805–13.

[199] Ilker MF, Schule H, Coughlin EB. Modular norbornene derivatives for the preparation of well-defined amphiphilic polymers: study of the lipid membrane disruption activities. Macromolecules 2004;37:694–700.

[200] Ilker MF, Nűsslein K, Tew GN, Coughlin EB. Tuning the hemolytic and antibacterial activities of amphiphilic polynorbornene derivatives. J Am Chem Soc 2004;126:15870–5.

[201] Lienkamp K, Madkour AE, Musante A, Nelson CF, Nüsslein K, Tew GN. Antimicrobial polymers prepared by ROMP with unprecedented selectivity: a molecular construction kit approach. J Am Chem Soc 2008;130(30):9836–43.

[202] AL-Badri ZM, Tew GN. Well-defined acetylene-functionalized oxanorbornene polymers and block copolymers. Macromolecules 2008;41:4173–9.

[203] Love JA, Morgan JP, Trnka TM, Grubbs RH. A practical and highly active ruthenium-based catalyst that effects the cross metathesis of acrylonitrile. Angew Chem Int Ed 2002;41(21):4035–7.

[204] Eren T, Som A, Rennie JR, Nelson CF, Urgina Y, Nűsslein K, et al. Antibacterial and hemolytic activities of quaternary pyridinium functionalized polynorbornenes. Macromol Chem Phys 2008;209:516–24.

[205] Cakmak I, Ulukanli Z, Tuzcu M, Karabuga S, Genctav K. Synthesis and characterization of novel antimicrobial cationic polyelectrolytes. Eur Polym J 2004;40:2373–9.

[206] Tomalia DA, Baker H, Dewald J, Hall M, Kallos G, Martin S, et al. A new class of polymers: starburst-dendritic macromolecules. Polym J 1985;17(1):117–32.

[207] Hawker CJ, Fréchet JMJ. Unusual macromolecular architectures: the convergent growth approach to dendritic polyesters and novel block copolymers. J Am Chem Soc 1992;114(22):8405–13.

[208] Wang B, Navath RS, Menjoge AR, Balakrishnan B, Bellair R, Dai H, et al. Inhibition of bacterial growth and intramniotic infection in a guinea pig model of chorioamnionitis using PAMAM dendrimers. Int J Pharm 2010;395:298–308.

[209] Polcyn P, Jurczak M, Rajnisz A, Solecka J, Urbanczyk-Lipkowska Z. Design of antimicrobially active small amphiphilic peptide dendrimers. Molecules 2009;14:3881–905.

[210] Zasloff M. Magainins, a class of antimicrobial peptides from *Xenopus* skin: Isolation, characterization of two active forms, and partial cDNA sequence of a precursor. Proc Natl Acad Sci USA 1987;84:5449–553.

[211] de Leeuw E, Lu W. Human defensins: turning defense into offense? Infect Disord Drug Targets 2007;7:67–70.

[212] Dai C, Ma Y, Zhao Z, Zhao R, Wang Q, Wu Y, et al. Mucroporin, the first cationic host defense peptide from the venom of *Lychas mucronatus*. Antimicrob Agents Chemother 2008;52(11):3967–72.

[213] Lo JCY, Lange D. Current and potential applications of host-defense peptides and proteins in urology. Biomed Res Int 2015;2015:189016.

[214] Bahar AA, Ren D. Antimicrobial Peptides. Pharmaceuticals 2013;6:1543–75.

[215] Liu D, Choi S, Chen B, Doerksen RJ, Clements DJ, Winkler JD, et al. Nontoxic membrane-active antimicrobial arylamide oligomers. Angew Chem Int Ed 2004;43:1158–62.

[216] Tew GN, Liu D, Chen B, Doerksen RJ, Kaplan J, Carroll PJ, et al. *De novo* design of biomimetic antimicrobial polymers. Proc Natl Acad Sci USA 2002;99(8):5110–4.

[217] Tew GN, Scott RW, Klein ML, DeGrado WF. De novo design of antimicrobial polymers, foldamers, and small molecules: from discovery to practical applications. Acc Chem Res 2010;43(1):30–9.

[218] a. Lienkamp K, Madkour AE, Tew GN. Antibacterial peptidomimetics: polymeric synthetic mimics of antimicrobial peptides.. Adv Polym Sci 2013;251.
b. In: Abe A, Kausch H-H, Möller M, Pasch H., editors. Polymer composites—polyolefin fractionation—polymeric peptidomimetics—collagens. ISBN 978-3-642-34329-2 141-72, p. 141–72.

[219] Arnt L, Nűsslein K, Tew GN. Nonhemolytic abiogenic polymers as antimicrobial peptide mimics. J Polym Sci Pol Chem 2004;42:3860–4.

[220] Tew GN, Clements D, Tang H, Arnt L, Scott RW. Antimicrobial activity of an abiotic host defense peptide mimic. Biochim Biophys Acta: Biomembr 2006;1758:1387–92.

[221] Arnt L, Tew GN. New poly(phenyleneethynylene)s with cationic, facially amphiphilic structures. J Am Chem Soc 2002;124(26):7664–5.

[222] Breitenkamp RB, Arnt L, Tew GN. Facially amphiphilic phenylene ethynylenes. Polym Adv Technol 2005;16:189–94.

[223] Fernandes JC, Tavaria FK, Soares JC, Ramos OS, Monteiro MJ, Pintado ME, et al. Antimicrobial effects of chitosans and chitooligosaccharides, upon *Staphylococcus aureus* and *Escherichia coli*, in food model systems. Food Microbiol 2008;25:922–8.

[224] Chen J, Wang F, Liu Q, Du J. Antibacterial polymeric nanostructures for biomedical applications. Chem Commun 2014;50:14482–93.

[225] Gilbert P, Moore LE. Cationic antiseptics: diversity of action under a common epithet. J Appl Microbiol 2005;99:703–15.

[226] Li Y-Y, Chen X-G, Yu L-M, Wang S-X, Sun G-Z, Zhou H-Y. Aggregation of hydrophobically modified chitosan in solution and at the air-water interface. J Appl Polym Sci 2006;102:1968–73.

[227] Xing K, Chen XG, Kong M, Liu CS, Cha DS, Park HJ. Effect of oleoyl-chitosan nanoparticles as a novel antibacterial dispersion system on viability, membrane permeability and cell morphology of *Escherichia coli* and *Staphylococcus aureus*. Carbohydr Polym 2009;76:17–22.

[228] Xing K, Chen XG, Liu CS, Cha DS, Park HJ. Oleoyl-chitosan nanoparticles inhibits *Escherichia coli* and *Staphylococcus aureus* by damaging the cell membrane and putative binding to extracellular or intracellular targets. Int J Food Microbiol 2009;132:127–33.

[229] Tin S, Sakharkar KR, Lim CS, Sakharkar MK. Activity of chitosans in combination with antibiotics in *Pseudomonas aeruginosa*. Int J Biol Sci 2009;5: 153–60.

[230] Straccia MC, Gomez d'Ayala G, Romano I, Oliva A, Laurienzo P. Alginate hydrogels coated with chitosan for wound dressing. Mar Drugs 2015;13:2890–908.

[231] Shen XL, Wu JM, Chen Y, Zhao G. Antimicrobial and physical properties of sweet potato starch films incorporated with potassium sorbate or chitosan. Food Hydrocolloids 2010;24:285–90.

[232] Liu J, Liu W, Weitzhandler I, Bhattacharyya J, Li X, Wang J, et al. Ring-opening polymerization of prodrugs: a versatile approach to prepare well-defined drug-loaded nanoparticles. Angew Chem 2015;127:1016–20.

[233] Stebbins ND, Ouimet MA, Uhrich KE. Antibiotic-containing polymers for localized, sustained drug delivery. Adv Drug Deliver Rev 2014;78:77–87.

[234] Sobczak M, Nałęcz-Jawecki G, Kołodziejski WL, Goś P, Żółtowska K. Synthesis and study of controlled release of ofloxacin from polyester conjugates. Int J Pharm 2010;402:37–43.

[235] Sobczak M. Synthesis and characterization of polyester conjugates of ciprofloxacin. Eur J Med Chem 2010;45:3844–9.

[236] Sobczak M, Witkowska E, Olędzka E, Kolodziejski W. Synthesis and structural analysis of polyester prodrugs of norfloxacin. Molecules 2008;13:96–106.

[237] Sobczak M, Nurzyńska K, Kolodziejski W. Seeking polymeric prodrugs of norfloxacin. Part 2. Synthesis and structural analysis of polyurethane conjugates. Molecules 2010;15:842–56.

[238] Wang C, Li H, Zhao X. Ring opening polymerization of l-lactide initiated by creatinine. Biomaterials 2004;25:5797–801.

[239] Yang H, Lopina ST. Penicillin V-conjugated PEG-PAMAM star polymers. J Biomater Sci Polymer Edn 2003;14(10):1043–56.

[240] Turos E, Reddy GSK, Greenhalgh K, Ramaraju P, Abeylath SC, Jang S, et al. Penicillin-bound polyacrylate nanoparticles: restoring the activity of β-lactam antibiotics against MRSA. Bioorg Med Chem Lett 2007;17:3468–72.

[241] Turos E, Long TE, Heldreth B, Leslie JM, Reddy GSK, Wang Y, et al. N-Thiolated β-lactams: a new family of anti-*Bacillus* agents. Bioorg Med Chem Lett 2006;16:2084–90.

[242] Revell KD, Heldreth B, Long TE, Jang S, Turos E. N-thiolated β-lactams: Studies on the mode of action and identification of a primary cellular target in *Staphylococcus aureus*. Bioorg Med Chem 2007;15:2453–67.

[243] Turos E, Konaklieva MI, Ren RX-F, Shi H, Gonzalez J, Dickey S, et al. N-Thiolated bicyclic and monocyclic β-lactams. Tetrahedron 2000;56:5571–8.

[244] Prosen KR, Carroll RK, Burda WN, Krute CN, Bhattacharya B, Dao ML, et al. The impact of fatty acids on the antibacterial properties of N-thiolated β-lactams. Bioorg Med Chem Lett 2011;21:5293–5.

[245] Coates C, Long TE, Turos E, Dickey S, Lim DV. N-Thiolated β-lactam antibacterials: defining the role of unsaturation in the C_4 side chain. Bioorg Med Chem 2003;11:193–6.

[246] Heldreth B, Long TE, Jang S, Reddy GSK, Turos E, Dickey S, et al. N-Thiolated β-lactam antibacterials: effects of the N-organothio substituent on anti-MRSA activity. Bioorg Med Chem 2006;14:3775–84.

[247] Turos E, Shim J-Y, Wang Y, Greenhalgh K, Reddy GSK, Dickey S, et al. Antibiotic-conjugated polyacrylate nanoparticles: new opportunities for development of anti-MRSA agents. Bioorg Med Chem Lett 2007;17:53–6.

[248] Bhattachary B, Turos E. Synthesis and biology of N-thiolated β-lactams. Tetrahedron 2012;68:10665–85.

[249] Moon W-S, Kim JC, Chung K-H, Park E-S, Kim M-N, Yoon J-S. Antimicrobial activity of a monomer and its polymer based on quinolone. J Appl Polym Sci 2003;90:1797–801.

[250] Chang H-P, Chen J-Y, Zhong P-S, Chang Y-H, Liang M. Synthesis and characterization of a new polymer-drug conjugate with pH-induced activity. Polymer 2012;53:3498–507.

[251] Kopecek J. Reactive copolymers of *N*-(2-hydroxypropyl)methacrylamide with *N*-methacryloylated derivatives of L-leucine and L-phenylalanine. 1. Preparation, characterization, and reactions with diamines. Makromol Chem Macromol Chem Phys 1977;178(8):2169–83.

[252] Schmidt M, Harmuth S, Barth ER, Wurm E, Fobbe R, Sickmann A, et al. Conjugation of ciprofloxacin with poly(2-oxazoline)s and polyethylene glycol via end groups. Bioconjugate Chem. 2015;26(9):1950–62.

[253] Haag R. Supramolecular drug-delivery systems based on polymeric core–shell architectures. Angew Chem Int Ed 2004;43:278–82.

[254] Schmaljohann D. Thermo- and pH-responsive polymers in drug delivery. Adv Drug Deliver Rev 2006;58:1655–70.

[255] Chan A, Orme RP, Fricker RA, Roach P. Remote and local control of stimuli responsive materials for therapeutic applications. Adv Drug Deliver Rev 2013;65:497–514.

[256] Gao W, Thamphiwatana S, Angsantikul P, Zhang L. Nanoparticle approaches against bacterial infections. Wiley Interdiscip Rev: Nanomed Nanobiotechnol 2014;6(6):532–47.

[257] Rijcken CJ, Snel CJ, Schiffelers RM, von Nostrum CF, Hennink WE. Hydrolysable core-crosslinked thermosensitive polymeric micelles: synthesis, characterization and in vivo studies. Biomaterials 2007;28:5581–93.

[258] Neradovic D, van Nostrum CF, Hennink WE. Thermoresponsive polymeric micelles with controlled instability based on hydrolytically sensitive *N*-isopropylacrylamide copolymers. Macromolecules 2001;34:7589–91.

[259] Jhaveri AM, Torchilin VP. Multifunctional polymeric micelles for delivery of drugs and siRNA. Front Pharmacol 2014;5:77.

[260] Torchilin V. Multifunctional and stimuli-sensitive pharmaceutical nanocarriers. Eur J Pharm Biopharm 2009;71:431–44.

[261] Torchilin VP. Multifunctional nanocarriers. Adv Drug Deliver Rev 2012;64:302–15.

[262] Sawant RR, Torchilin VP. Multifunctional nanocarriers and intracellular drug delivery. Curr Opin Solid State Mat Sci 2012;16:269–75.

[263] Cheng R, Meng F, Deng C, Klok H-A, Zhong Z. Dual and multi-stimuli responsive polymeric nanoparticles for programmed site-specific drug delivery. Biomaterials 2013;34:3647–57.

[264] Radovic-Moreno AF, Lu TK, Puscasu VA, Yoon CJ, Langer R, Farokhzad OC. Surface charge-switching polymeric nanoparticles for bacterial cell wall-targeted delivery of antibiotics. ACS Nano 2012;6(5):4279–87.

[265] Lin Y-H, Chang C-H, Wu Y-S, Hsu Y-M, Chiou S-F, Chen Y-J. Development of pH-responsive chitosan/heparin nanoparticles for stomach-specific anti-*Helicobacter pylori* therapy. Biomaterials 2009;30:3332–42.

[266] Kusters JG, van Vliet AHM, Kuipers EJ. Pathogenesis of *Helicobacter pylori* infection. Clin Microbiol Rev 2006;19(3):449–90.

[267] Wroblewski LE, Peek Jr. RM, Wilson KT. *Helicobacter pylori* and gastric cancer: factors that modulate disease risk. Clin Microbiol Rev 2010;23(4):713–39.

[268] Lane DD, Su FY, Chiu DY, Srinivasan S, Wilson JT, Ratner DM, et al. Dynamic intracellular delivery of antibiotics via pH-responsive polymersomes. Polym Chem 2015;6(8):1255–66.

[269] Xiong M-H, Bao Y, Yang X-Z, Wang Y-C, Sun B, Wang J. Lipase-sensitive polymeric triple-layered nanogel for "on-demand" drug delivery. J Am Chem Soc 2012;134:4355–62.

[270] Xiong M-H, Li Y-J, Bao Y, Yang X-Z, Hu B, Wang J. Bacteria-responsive multifunctional nanogel for targeted antibiotic delivery. Adv Mater 2012;24:6175–80.

[271] Balaure PC, Andronescu E, Grumezescu AM, Ficai A, Huang K-S, Yang C-H, et al. Fabrication, characterization and in vitro profile based interaction with eukaryotic and prokaryotic cells of alginate-chitosan-silica biocomposite. Int J Pharm 2013;441:555–61.

[272] Grumezescu AM, Andronescu E, Holban AM, Ficai A, Ficai D, Voicu G, et al. Water dispersible cross-linked magnetic chitosan beads for increasing the antimicrobial efficiency of aminoglycoside antibiotics. Int J Pharm 2013;454:233–40.

[273] Chifiriuc CM, Grumezescu AM, Saviuc C, Croitoru C, Mihaiescu DE, Lazar V. Improved antibacterial activity of cephalosporins loaded in magnetic chitosan microspheres. Int J Pharm 2012;436(1–2):201–5.

[274] Grumezescu AM, Saviuc C, Holban A, Hristu R, Croitoru C, Stanciu G, et al. Magnetic chitosan for drug targeting and in vitro drug delivery response. Biointerface Res Appl Chem 2011;1(5):160–5.

[275] Mihaiescu DE, Grumezescu AM, Balaure PC, Mogosanu DE, Traistaru V. Magnetic scaffold for drug targeting: evaluation of cefalosporins controlled release profile. Biointerface Res Appl Chem 2011;1(5):191–5.

[276] Grumezescu AM, Mihaiescu DE, Tamaş D. Hybrid materialsfor drug delivery of rifampicin: evaluation of release profile. Biointerface Res Appl Chem 2011;1(6):229–35.

[277] Andronescu E, Grumezescu AM, Ficai A, Gheorghe I, Chifiriuc M, Mihaiescu DE, et al. *In vitro* efficacy of antibiotic magnetic dextran microspheres complexes against *Staphylococcus aureus* and *Pseudomonas aeruginosa* strains. Biointerface Res Appl Chem 2012;2(3):332–8.

[278] Grumezescu V, Socol G, Grumezescu AM, Holban AM, Ficai A, Truşcă R, et al. Functionalized antibiofilm thin coatings based on PLA-PVA microspheres loaded with usnic acid natural compounds fabricated by MAPLE. Appl Surf Sci 2014;302:262–7.

[279] Kesharwani P, Jain K, Jain NK. Dendrimer as nanocarrier for drug delivery. Prog Polym Sci 2014;39:268–307.

[280] Mishra MK, Kotta K, Hali M, Wykes S, Gerard HC, Hudson AP, et al. PAMAM dendrimer-azithromycin conjugate nanodevices for the treatment of *Chlamydia trachomatis* infections. Nanomed Nanotechnol Biol Med 2011;7:935–44.

[281] Cheng Y, Qu H, Ma M, Xu Z, Xu P, Fang Y, et al. Polyamidoamine (PAMAM) dendrimers as biocompatible carriers of quinolone antimicrobials: an in vitro study. Eur J Med Chem 2007;42:1032–8.

[282] Gupta U, Bharat H, Narendra A, Jain K. Polypropylene imine dendrimer mediated solubility enhancement: effect of ph and functional groups of hydrophobes. J Pharm Pharmaceut Sci 2007;10(3):358–67.

[283] Kumar PV, Agashe H, Dutta T, Jain NK. PEGylated dendritic architecture for development of a prolonged drug delivery system for an antitubercular drug. Curr Drug Deliv 2007;4:11–19.

[284] Abed N, Couvreur P. Nanocarriers for antibiotics: a promising solution to treat intracellular bacterial infections. Int J Antimicrob Agents 2014;43:485–96.

Chapter 7

Bacteriocins and Nanotechnology

L.M.T. Dicks, A.D.P. van Staden and B. Klumperman
Stellenbosch University, Stellenbosch, South Africa

CHAPTER OUTLINE

7.1 INTRODUCTION

Bacteriocins are defined as ribosomally synthesized peptides with antibacterial activity. Those produced by lactic acid bacteria are grouped into two classes, mainly based on the presence of lanthionine ring structures [1–3]. Each class is further divided into subclasses, based on the presence of specific amino acids, cross-links between amino acids, receptors required for cell adhesion and mode of action [3–6].

Most bacteriocins interact with phospholipids in the cell membrane and destabilize the proton motive force (PMF), or bind to lipid II and interfere with peptidoglycan synthesis [6–8]. Nisin, a class I bacteriocin (lantibiotic) uses the pyrophosphate linkage in lipid II as a docking molecule and then span across the double phospholipid membrane to form a pore [7]. The transmembrane orientation of nisin is facilitated by the binding of eight nisin molecules to four lipid II molecules [2,7]. The epidermin family of lantibiotics has the same A and B ring structures as nisin and also bind to lipid II. However, due to their shorter structures, they do not have the required length to span the bacterial cell membrane and do not form pores [8]. Their mode of activity lies in the removal of lipid II from the cell division site (septum), resulting in the blocking of peptidoglycan synthesis [9]. The lantibiotics duramycin and cinnamycin bind to phosphatidylethanolamine in cell membranes and inhibit the activity of phospholipase A2 [10]. Most of the class II bacteriocins form pores, leading to dissipation of the PMF. Some bacteriocins in this class depend on a mannose permease of the phosphotransferase system (PTS) as target [6,11]. A few bacteriocins with antiviral activity have been reported [12–22]. Subtilosin, a cyclical peptide isolated from *Bacillus amyloliquefaciens*, acted against herpes simplex virus type 2 (HSV-2) by inhibiting late stages of the viral replicative cycle, specifically the intracellular transport of viral glycoprotein gD [22]. Similar unexpected findings were that nisin and subtilosin have spermicidal properties [23–25].

Unlike antibiotics, bacteria seldom develop resistance to bacteriocins, however resistance can still develop with frequent use. Resistance mechanisms include changes in the cell-wall and -membrane, which can include changes in the charge of the cell wall or changes in the phospholipid composition of the cell membrane [26–29]. Increasing the D-alanyl composition of the cell wall reduces the net negative charge, which may prevent cationic peptides like bacteriocins to sufficiently bind and interact with their target sites [26]. The dlt operon, found in Gram positive bacteria, encodes proteins responsible for the D-alanylation of lipoteichoic acids and has been shown to be associated with nisin-resistance [30,31]. Increased levels of neutrally phosphatidylethanolamine (PE) in the cell membranes of *L. monocytogenes* is associated with resistance [29]. The increase in neutrally charged PE and the decrease in anionic phospholipids, such as phosphatidylglycerol and cardiolipin, may also alter the capability of peptides to interact with the cell membrane. Some strains of *S. aureus* have a mutation in the *nsaS* (nisin susceptibility-associated sensor) gene that encodes a putative histidine kinase [32]. Both the putative cognate response regulator (NsaR) and the histidine kinase (NsaS), encoded by *nsaR* and *nsaS* respectively, are located on the same operon, upstream of genes encoding a permease and an ATP-binding domain of a putative ABC transporter [32]. A mutation in the operon prevented nisin from entering susceptible cells [32,33]. Changes in cell wall thickness may also prevent the interaction between nisin and lipid II [34]. *Listeria monocytogenes* obtained resistance to mesentericin Y105 by inactivating the *rpoN* gene that encodes the σ^{54} subunit of bacterial RNA polymerase [35]. *Enterococcus faecalis*, strain JH2-2, developed resistance to divercin V41, a pediocin-like bacteriocin, by inactivating genes encoding a glycerophosphoryl diester phosphodiesterase (GlpQ) and a protein with a putative phosphodiesterase function [36]. In the case of *Lactococcus lactis*, a mutation in or close to the uppP region (encoding an undecaprenyl pyrophosphate phosphatase) resulted in resistance to lactococcin G and enterocin 1071, both class IIb bacteriocins [37]. Nonproteolytic nisin-inactivating enzymes with putative dehydroalanine reductase activity have been described for several *Bacillus* spp.[38,39].

The lantibiotics duramycin, duramycin B and C, and cinnamycin act as anti-inflammatory agents by isolating phosphatidylethanolamine, which indirectly inhibits the activity of phospholipase A2 [10,40,41]. Ancovenin, a cinnamycin-like lantibiotic, inhibits the activity of the angiotensin-converting enzyme [42–44] and may be used in the treatment of high blood pressure [45]. Some antimicrobial peptides have also shown antitumor activity [46,47].

Although most bacteriocins, at least those produced by lactic acid bacteria, are considered safe, a few papers reported low levels of toxicity. Mutacin 1140, rapidly injected into the bloodstream, caused a hypersensitive reaction however this reaction could be blocked by administration with diphenhydramine [51]. Symptoms from the rapid infusion subsided within 20 minutes and was not observed in subsequent administrations. Pediocin PA-1, nisin and colicin E6 displayed some cytotoxicity against Vero monkey kidney cells and simian virus 40-transfected human colon cells (SVHC), with the human SVHC demonstrating greater sensitivity to the bacteriocins [52]. The proliferation of human fibroblasts (HNF, ATCC CCL-28) decreased when treated with high concentrations (5000- and 1000-AU/mL) of a bacteriocin produced by *Lactococcus* sp. HY 449 [54]. Enterocin S37, a bacteriocin produced by *E. faecalis* S37, showed dose-dependent toxicity when tested against undifferentiated Caco-2/TC7 cells [55]. However, no significant effects were seen on differentiated Caco-2/TC7 cells. Bacteriocin Bcn2–5 was cytotoxic in a dose-dependent manner against peritoneal mouse macrophages [56]. Treatment of HeLa cells with microcin E492, produced by *Klebsiella pneumoniae,* resulted in biochemical and morphological changes at low concentrations (5–10 µg/mL) indicative of apoptosis and at higher concentrations (>20 µg/mL) resulted in a necrotic phenotype [57]. Subtle changes in amino acid composition of a peptide can also influence its hemolytic activity. Subtilosin A variants, with an isoleucine at position 6 instead of a threonine, showed higher levels of hemolytic activity [58].

The intention of this chapter is not to review the practical applications of bacteriocins, but to discuss methods that may be used to protect these peptides from destruction when administered orally or intravenously. The possibility of using nanotechnology to develop target-specific bacteriocin delivery systems will also be addressed.

7.2 STABILITY OF BACTERIOCINS

Several bacteriocins have been tested against bacteria associated with infection, including several *in vivo* studies (listed in Table 7.1).

Table 7.1 Bacteriocins With Activity Against Bacterial Pathogens

Bacteriocin	Producer	Application	References
Nisin A	*Lactococcus lactis* subsp. *lactis*	Treatment of dental plaque and gingivitis	[48]
		Prevention of intramammary streptococcal and staphylococcal infections	[59–62]
		Treatment of diarrhea caused by *Clostridium botulinum, Clostridium tyrobutyricum* and *Clostridium difficile*	[63–65]
		Treatment of stomach ulcers, caused by *Helicobacter pylori*, and oral mucositis	[63,66]
		Inactivation of sperm	[25]
		Coating of biomedical implants	[67,68]
Nisin A in combination with polymyxin E and clarithromycin		Treatment of *Pseudomonas aeruginosa* infections	[69]
Nisin A in combination with lysostaphin		Prevention of intramammary infections caused by *Staphylococcus aureus, Streptococcus agalactiae*, and *Streptococcus uberis*	[70]
Nisin F	*Lactococcus lactis* subsp. *lactis*	Treatment of intranasal infections caused by *S. aureus*	[71]
		Controls subcutaneous infections of *S. aureus*	[72]
		Suppresses the growth of *S. aureus* in the intraperitoneal cavity	[73]
Lacticin 3147	*Lactococcus lactis* subsp. *lactis*	Prevention of dental decay caused by *Streptococcus mutans*	[74]
		Treatment of *S. aureus* and MRSA infections	[74]
		Treatment of bovine mastitis caused by mastitic staphylococci and streptococci	[76]
Salivaricin A2 and B	*Streptococcus salivarius*	Treatment of bad breath	[77]
Macedocin ST91KM	*Streptococcus gallolyticus* subsp. *macedonicus*	Treatment of mastitis in dairy cows, caused by *S. agalactiae, Streptococcus dysgalactiae* subsp. *dysgalactiae, S. uberis, S. aureus*, and *Staphylococcus epidermidis*	[78]
Mutacin 1140	*Streptococcus mutans*	Prevention of tooth decay caused by *S. mutans*	[79]
Mutacin B-Ny266	*Streptococcus mutans*	Controls *S. aureus* infection in the peritoneal cavity	[80]
Abp118	*Lactobacillus salivarius*	Treatment of *Listeria* infections	[81]

(Continued)

Table 7.1 Bacteriocins With Activity Against Bacterial Pathogens (Continued)

Bacteriocin	Producer	Application	References
Peptide ST4SA	*Enterococcus mundtii*	Prevention of middle ear bacterial infections	[82]
E50–52	*Enterococcus faecium*	Inhibits the intracellular growth of *Mycobacterium tuberculosis*	[56]
Pediocin PA-1	*Pediococcus acidilactici*	Treatment of *Listeria monocytogenes* infections	[84]
Piscicolin 126	*Carnobacterium piscicola*	Treatment of *Listeria* infections	[53]
Divercin V41	*Carnobacterium divergens*	Treatment of *Listeria* infections	[85]
Subtilosin[a]	*Bacillus subtilis* and *Bacillus amyloliquefaciens*	Treatment of bacterial vaginosis. Inactivation of sperm	[24]
Mersacidin[a]	*Bacillus* spp.	Treatment of MRSA infections in the nasal cavity	[50]
Cinnamycin[a]	*Streptomyces cinnamoneus*	Regulation of blood pressure and fluid balance	[10]
Ancovenin[a]	*Streptomyces* spp.	Regulation of blood pressure and fluid balance	[42,45]
Bacteriocin Pep5[a] and epidermin	*Staphylococcus epidermidis*	Prevents growth of staphylococci and/or enterococci in and on catheter tubing	[90]
Planosporicin[a]	*Planomonospora* sp.	Controls *S. pyogenes*-induced septicemia	[49]
Philipimycin[a]	*Actinoplanes philippinensis*	Controls *S. aureus* infection	[75]
Thiazomycin[a]	*Amycolatopsis fastidiosa*	Controls *S. aureus* infection	[110]
Nosiheptide[a]	*Streptomyces actuosus*	Controls methicillin-resistant *S. aureus* infection	[91]
Microcin J25[a]	*Escherichia coli*	Decrease *Salmonella* numbers in the liver and spleen	[92]

[a]Not produced by lactic acid bacteria.

Bacteriocins are labile to degradation in the GIT, tissue, serum, and organs such as the liver and kidneys [93,94]. However, a few reports showed that some bacteriocins may survive these *in vivo* conditions and still inhibit bacterial growth. Brand et al. [73] showed that nisin F remained active against *S. aureus* for 15 min in the peritoneal cavity and that nisin may have a stabilizing effect on the bacterial population in the GIT when injected [95]. De Kwaadsteniet et al. [71] used nisin F in the treatment of respiratory infections caused by *S. aureus* and later in the treatment of subcutaneous skin infections caused by *S. aureus* [96]. Granger et al. [97]

showed that *Enterococcus mundtii* cells exposed to conditions that simulated the GIT, were still able to express active bacteriocin. Using a more direct approach with real time-PCR, Ramiah et al. [98] showed that the structural gene encoding plantaricin 423 was still expressed when cells were exposed to conditions simulating the GIT.

Nisin A inhibited the growth of *Streptococcus pneumoniae* in mice when administered intraperitoneally [62]. Similarly, mutacin B-Ny266, was also able to protect mice from *S. aureus* Smith infection when injected intraperitoneally [80]. The lantibiotic microbisporicin is also able to prevent *S. aureus* infection when administered intravenously and subcutaneously to mice, with low acute toxicity at high doses (>200 mg/kg) [99]. Lacticin 3147 prevented the systemic spread of *S. aureus* Xen 29 in mice [100].

Pediocin PA-1, piscolin 126, and recombinant divercin RV41 reduced cell numbers of *L. monocytogenes* in the liver and spleen of mice [53,84,85]. Bacteriocin Abp118, produced *in vivo* by *Lactobacillus salivarius* UCC118, showed good activity against *L. monocytogenes* in infected mice, while a mutant lacking the ability to produce this bacteriocin was ineffective in protecting mice [81].

The lantibiotics, nisin, clausin and amyloliquecidin have been shown to be effective in the treatment of *S. aureus* induced skin infections in mice. Furthermore, the lantibiotics did not have a determental effect on wound closure and healing [101]. Additionally, lantibiotics such as gallidermin, epidermin, mersacidin, and lacticin 3147 (all lantibiotics) are active against bacteria associated with skin infections [74,112,113].

Bacteriocins can be modified to be more stable under physiological conditions, including increased solubility and protection from protease degradation. The modification machinery used for the incorporation of lanthionine ring structures in nisin (i.e. NisB and NisC) can also be used to modify unrelated peptides, thereby increasing their stability [102–105]. Further stabilization can be achieved by incorporation of D-amino acids and cyclization [106–108]. Rink et al. [109] combined the two approaches to create peptides with enhanced *ex vivo* stability.

7.3 IMMUNE RESPONSE TO BACTERIOCINS

Continuous *in vivo* injection of nisin F into the peritoneal cavity of mice did not elicit an immune response [114]. Active and inactive nisin F stimulated the activity of cytokines interleukin-6, interleukin-10, and tumor necrosis factor, but the overall immune response was too small to trigger an abnormally high antigenic immune reaction [114]. Begde et al. [115], on the other hand, reported toxic side effects to human lymphocytes and neutrophils when exposed to a commercial sample containing a combination of nisin A and Z. After administration of Nisaplin (a commercial form of nisin A) to mice for 30, 75 and 100 days, an increase in CD4 and CD8 T-lymphocytes and a decrease in B-lymphocytes were observed after short-term administration [116]. However, after long-term administration T-cell counts returned to normal levels, with B-lymphocyte levels only normalizing in mice that received Nisaplin at a lower dose. The authors also found enhanced phagocytic activity of peritoneal cells after long-term administration of Nisaplin. In another study, nisin showed immunostimulatory effects on head kidney macrophages in fish [117]. When administered to rats as a vaginal microbicide, RP-HPLC purified nisin proved to be non-toxic to host cells [118]. Nisin, pediocin, and peptide AS-48 had immunogenic properties in antibody studies [119–121].

7.4 PRECLINICAL TRIALS CONDUCTED ON BACTERIOCINS

Most of the clinical studies focused on peptide antibiotics, of which the 18-amino acid protegrin-like peptide, IB-367, with activity against *Streptococcus mitis*, *Streptococcus sanguis*, *S. salivarius*, *S. aureus*, *Klebsiella* spp., *Escherichia coli* and *Pseudomonas* spp. are the best studied [122]. Peptide IB-367 passed phases I and II clinical trials and Intrabiotics started with phase III trials on treatment of oral mucositis [123]. Other examples of clinically tested broad-spectrum antimicrobial peptides are MSI-78, indolicidin, gramicidin, and polymyxin B [123]. Gramicidin S, CEMA, protegrin, indolicidin, and polyphemusin have antifungal activity and the latter three also have antiviral (including HIV) activity [123]. Peptide antibiotics are not classified as bacteriocins and they will not be discussed in this review.

Phase I clinical trials on the class II lantibiotic NVB302, aimed at the prevention of *C. difficile* infections, have been completed in 2012 (www.evaluatepharma.com). Two pharmaceutical companies (Oragenics and Intrexon) initiated a "lantibiotics program." The companies Astra and Merck commercialized nisin A for treatment of gastric *Helicobacter* infections and ulcers. Nisin A was incorporated in topical formulations [124] and a number of patents were filed for its use in the treatment of skin infections [125,126].

Continuous use of any antimicrobial increases the likelihood of bacteria developing resistance, it is therefore prudent to investigate combination therapies. A number of studies were performed with bacteriocins in combination with antibiotics. Nisin A inhibited the growth of *Pseudomonas aeruginosa* when used in combination with polymyxin E and clarithromycin [69]. The synergistic effect between colistin, and the bacteriocins nisin A and pediocin PA-1/AcH was demonstrated against *Salmonella cholerasuis*, *P. aeruginosa*, *Yersinia enterocolitica*, and *E. coli* [127]. The dosage of antibiotics may be lowered when used in combination with antibiotics [128] and may even broaden the spectrum of antibiotics [69,129,130]. Whether the combined use of bacteriocins and antibiotics will help to prevent the emergence of antibiotic-resistant strains remains to be proven.

7.5 DELIVERY SYSTEMS

Bacteriocins are, as most proteins and peptides, easily degraded by proteolytic enzymes in the oral cavity and the GIT. There is also no guarantee that intact bacteriocins will pass through the epithelial cells and, if they do enter the blood stream, it is almost certain that they will be recognized by the body's immune system. Although little research has been reported on the stability of bacteriocins *in vivo* and their protection from the immune system, a vast number of papers have been published on the protection of peptides in general. The first method experimented with was the inclusion of peptides into liposomes. Current methods are more focused on encapsulation with natural or synthetic polymers to form nanoparticles that are much more stable and smaller in size [131].

For uptake in the bloodstream after oral delivery it is important for the nanoparticles to cross the epithelium that lines the GIT. For this purpose, nanoparticles with a diameter of approximately 100nm are well suited. Their transport pathways across a model epithelial cell monolayer were recently studied by Chi and co-workers [132]. By selectively inhibiting specific pathways, the authors were able to determine the importance of individual transport mechanisms in the transcytosis of the nanoparticles. In earlier studies it was found that in nanoparticles that cross the epithelial cells, surface charge is important. In general, neutral or positively charged molecules pass much easier through pores in the gut wall than negatively charged groups. A 2006 review summarizes the role of physicochemical properties of nanoparticles in endocytosis and transcytosis mechanisms [133]. Crossing of biological barriers also depends on the tissue and blood circulation at the site of entrance [134,135].

The other important factor that needs to be taken into consideration is hydrophobicity. Nanoparticles with hydrophobic surfaces are rapidly recognized by macrophages of the mononuclear phagocytic system and are easily destroyed [136]. Coating with hydrophilic polymers usually increases circulation time and stability in blood [137]. Many studies have been devoted to the delivery of nanoparticles to the targeted site in the body. The general consensus is that once the nanoparticles are in the blood circulation, there are two mechanisms for delivery, i.e., passive targeting and active targeting. Passive targeting has mainly been investigated for delivery to tumor tissue. Use is made of the so-called Enhanced Permeability and Retention (EPR) effect. Intact epithelium that lines the blood vessels is impermeable for nanoparticles above a certain threshold size (diameter 6–10 nm). However, the vasculature inside tumor tissue is known to be "leaky," which results in penetration of nanoparticles in the diameter range of 10–200 nm [138]. The absence of a fully developed lymphatic system in tumor tissue further ensures that the nanoparticles are not removed from the tumor and as a consequence, they accumulate inside the tumor. Active targeting on the other hand can be directed to any type of tissue/cell although also here the majority of research has been dedicated to tumor tissue. Active targeting depends on specific recognition of overexpressed epitopes on the surface of the target cell. The nanoparticles are decorated with ligands that usually mimic natural recognition patterns. The processes of passive and active targeting to tumor cells have been reviewed before [139].

Several polymers have been used in the encapsulation of peptides. These include natural hydrogels such as collagen, gelatin, fibrin, hyaluronic acid, alginate, chitosan, and dextran, or synthetic polymers such as poly(D,L-lactide-*co*-glycolide) (PLGA), poly(lactic acid) (PLA), poly(ε-caprolactone) (PCL), poly(ethylene oxide) (PEO), poly(acrylic acid) (PAA), poly(*N*-isopropylacrylamide) (PNIPAAm), poly(vinyl alcohol) (PVA), and polyphosphazene (reviewed by [140]). PLGA is successfully used in the encapsulation of anticancer drugs, insulin, psychotic drugs (e.g., haloperidol) and hormones (e.g., estradiol and tetanus toxoid) [136,141]. PLGA hydrolyzes in the body to lactic acid and glycolic acid and causes minimal systemic toxicity [142]. PLA, which is also biodegradable, is used to encapsulate antipsychotic drugs (e.g., savoxepine), restenosis drugs (e.g., tyrphostins), hormones (e.g., progesterone), Oridonin (a natural diterpenoid), and proteins such as bovine serum albumin, BSA [136,143]. PCL is less biodegradable and is used in long-term implants, such as Tamoxifen and Taxol used in the treatment of cancer, and insulin used to treat diabetes [136]. Chitosan has been used in the encapsulation of insulin, cyclosporine A, and antihormonal drugs such as glycyrrhizin. Gelatin is used in the encapsulation of BSA, the anticancer drug paclitaxel, the anti-HIV drug didanosine, the antimalarial drug chloroquine phosphate and oligonucleotides [136].

An alternative approach to encapsulation is the attachment of peptides to carrier molecules. A classical method is attachment of peptides to polyethylene glycol (PEG), a process referred to as PEGylation [144–146]. Other methods include conjugation to fatty acids [147], substitution by peptoids [148,149] and incorporation of beta-peptides [150], fluorinated amino acids (Meng and Kumar, 2007), and acylation [151].

Salmaso et al. [152] succeeded in encapsulating nisin in PLA nanoparticles by using semicontinuous precipitation in compressed CO_2. Release of nisin from these nanoparticles depended on the salt concentration and the pH of the medium. Heunis et al. [153,154] were the first to electrospin a class II bacteriocin, plantaricin 423, into polymeric nanofibers prepared from PEO. Plantaricin 423 encapsulated in a PDLA-PEO scaffold was released in a slow and controlled manner and inhibited growth of *Lactobacillus sakei* and *Enterococcus faecium* in vitro [153]. Follow-up studies showed that nisin, electrospun into nanofibers of the same composition reduced bacterial cell numbers in a *S. aureus*-induced skin infection [155]. The antimicrobial spectrum of nisin was increased by the cospinning of silver nanoparticles and 2,3-dihydroxybezoic acid into PDLA-PEO nanofibers [156,157]. In a recent study, the bacteriocin subtilosin incorporated into PVA nanofibers inhibited the activity of the Herpes simplex virus type 1 [16]. Nanofibers of such small diameter are ideal for drug delivery and may be used to immobilize enzymes and release growth factors, or even applied in tissue engineering and the design of prostheses and wound dressings [158–164].

Modifications of conventional electrospinning are coaxial electrospinning, emulsion electrospining and electrospinning with dual spinnerets. Coaxial electrospinning uses a specialized needle consisting of two concentrical compartments to produce core shell structured nanofibers [165]. In coaxial electrospinning, a polymer solution in the outer part of the needle forms the shell and the solution in the inner part of the needle the core. The core material is typically a bioactive molecule (e.g., a bacteriocin, antibiotic, or growth factor). Core-shell structured nanofibers can also be produced via emulsion electrospinning [166]. In this case, an aqueous solution of the bacteriocin is emulsified in an organic polymer solution before electrospinning [158]. With this technique molecules in the dispersed aqueous phase are encapsulated.

Binding of bacteriocins to polymers is an exciting option, provided that the bacteriocin is released from the polymer in an active form. Maleic anhydride has been used in various polymer systems as an efficient compatibilizer in polymer blends. Maleic anhydride copolymerizes with various methacrylic and styrenic monomers to produce copolymers such as poly(styrene-*co*-maleic anhydride) (PSMA). The linking of hydroxyl and primary amine groups to PSMA renders it possible to attach bacteriocins to the structure and develop a drug carrier [167,168]. Patel and co-workers [169] linked the antiseptic agent Acriflavine (3, 6-diamino-10-methylacridinium) covalently to the surface of PSMA and poly(methyl methacrylate (MMA)-*co*-maleic anhydride (MAnh)) [170]. Acriflavine was released in the presence of a weak base and tested for antibacterial activity against *B. subtilis*. In another study, Ampicillin was bound to the anhydride groups of a matrix of poly(MMA-*co*-MAnh) via an amide bond [170]. The release of ampicillin from the polymeric carrier with different concentrations of MAnh was studied by testing for antimicrobial activity against *E. coli, B. subtilis*, and *S. aureus*. One of the major advantages of the construct is that the rate at which ampicillin was released could be controlled by the number of anhydride moieties in the copolymer.

Modifications of the PSMA drug carrier were described by Jeong and co-workers. The authors linked 4-aminobenzoic acid and 4-hydroxybenzoic acid, respectively, to PSMA [171] and later 4-aminophenol to PSMA [172]. All three polymer structures were antimicrobial towards *E. coli* and *S. aureus*. Although different results were recorded with the aminobenzoic acid and hydroxybenzoic conjugates, the aminophenol conjugate showed strong antibacterial activity against both species. It may be possible to anchor bacteriocins to the anhydride moieties in the form of amide and ester bonds. PSMA polymers are water soluble, but solubility depends on the styrene/maleic anhydride ratio and pH of the solution. Bacteriocins may be released from PSMA, but this will depend on the rate at which the amide or ester link between the peptide and PSMA is hydrolyzed.

Ignatova and co-workers [173] attached 5-amino-8-hydroxyquinoline, chlorhexidine (CHX), and poly(propylene glycol) monoamine covalently to the anhydride groups on the surface of PSMA nanofibers. Antimicrobial activity of the nanofibers was tested against *S. aureus, E. coli*, and *Candida albicans*.

Bi and coworkers (2011) used a colloidal emulsion of phytoglycogen octenyl succinate and oil to stabilize nisin. This prolonged the antimicrobial activity of nisin against *L. monocytogenes* with at least 40 days [174]. Similar results were reported when bacteriocin-like inhibitory substances (BLIS) and nisin were encapsulated into liposomes [175,176]. Bacteriocins produced by *L. salivarius, Streptococcus cricetus*, and *E. faecalis* encapsulated into liposomes were more effective than rifampicin when tested for activity against *M. tuberculosis* [56]. Bacteriocins produced by *S. salivarius* have been encapsulated into Bactoblis (PharmExtracta, Pontenure, Italy), a lozenge that releases the cells over 90 days.

7.6 **CONCLUSIONS**

Bacteriocins are natural antimicrobial peptides that have only recently been explored in the treatment and prevention of microbial infections. Compared to the vast number of bacteriocins described, little research has been devoted to the protection of bacteriocins from proteolytic enzymes and their delivery, at specific concentrations, to areas of infection. Rapid advancements in the field of nanotechnology provide many opportunities to develop target-specific drug delivery systems. Electrospinning to produce antimicrobial nanofibers is probably the most efficient and cost-effective method to protect bacteriocins from enzymatic degradation and provides the opportunity to design intelligent polymer-based drug delivery systems. Needless to say, all newly developed drug carriers need to be exposed to rigid cytotoxicity tests to determine their safety. We expect natural and synthetic polymers to play an increasing role in the fight against antibiotic-resistant pathogens.

REFERENCES

[1] Piper C, Cotter PD, Ross RP, Hill C. Discovery of medically significant lantibiotics. Curr Drug Discovery Technol 2009;6:1–18.

[2] Willey JM, van der Donk WA. Lantibiotics: peptides of diverse structure and function. Annu Rev Microbiol 2007;61:477–501.

[3] Cotter PD, Ross RP, Hill C. Bacteriocins - a viable alternative to antibiotics? Nature. Rev Microbiol 2013;11:95–105.

[4] Meindl K, Schmiederer T, Schneider K, Reicke A, Butz D, Keller S, et al. Labyrinthopeptins: a new class of carbacyclic lantibiotics. Angew Chem Int Ed Engl 2010;49:1151–4.

[5] Cotter PD, Hill C, Ross RP. Bacteriocins: developing innate immunity for food. Nat Rev Microbiol 2005;3:777–88.

[6] Diep DB, Skaugen M, Salehian Z, Holo H, Nes IF. Common mechanism of target cell recognition and immunity for class II bacteriocins. Proc Natl Acad Sci USA 2007;104:2384–9.

[7] Hsu ST, Breukink E, Tischenko E, Lutters MG, de Kruijff B, et al. The nisin-lipid II complex reveals a pyrophosphate cage that provides a blueprint for novel antibiotics. Nat Struct Mol Biol 2004;11:963–7.

[8] Bonelli RR, Schneider T, Sahl HG, Wiedemann I. Insights into *in vivo* activities of lantibiotics from gallidermin and epidermin mode-of-action studies. Antimicrob Agents Chemother 2006;50:1449–57.

[9] Hasper HE, Kramer NE, Smith JL, Hillman JD, Zachariah C, Kuipers OP, et al. An alternative bactericidal mechanism of action for lantibiotic peptides that target lipid II. Science 2006;313(5793):1636–7.

[10] Marki F, Hanni E, Fredenhagen A, van Oostrum J. Mode of action of the lanthionine-containing peptide antibiotics duramycin, duramycin B and C, and cinnamycin as indirect inhibitors of phospholipase A2. Biochem Pharmacol 1991;42:2027–35.

[11] Héchard Y, Pelletier C, Cenatiempo Y, Frère J. Analysis of sigma(54)-dependent genes in *Enterococcus faecalis*: a mannose PTS permease (EII(Man)) is involved in sensitivity to a bacteriocin, mesentericin Y105. Microbiology 2001;147:1575–80.

[12] Saeed S, Rasool RA, Ahmad S, Zaidi SZ, Rehmani S. Antiviral activity of *Staphylococcin* 188: a purified bacteriocin like inhibitory substance isolated from *Staphylococcus aureus* AB188. Res J Microbiol 2007;2:796–806.

[13] Serkedjieva J, Danova S, Ivanova I. Antiinfluenza virus activity of a bacteriocin produced by *Lactobacillus delbrueckii*. Appl Biochem Biotechnol 2000;88:122–9.

[14] Todorov SD, Wachsman MB, Knoetze H, Meincken M, Dicks LMT. An antibacterial and antiviral peptide produced by *Enterococcus mundtii* ST4V, isolated from soy beans. Int J Antim Agents 2005;25:508–13.

[15] Todorov SD, Wachsman MB, Tome E, Dousset X, Destro MT, Dicks LMT, et al. Characterisation of an antiviral pediocin-like bacteriocin produced by *Enterococcus faecium*. J Food Microbiol 2010;27:869–79.

[16] Torres N, Sutyak N, Xu S, Li J, Huang Q, Sinko P. Safety, formulation, and in vitro antiviral activity of the antimicrobial peptide subtilosin against herpes simplex virus type 1. Probiotics Antimicrob Proteins 2013;5:26–35.

[17] Wachsman MB, Castilla V, De Ruiz Holgado AP, de Torres RA, Sesma F, Coto CE. Enterocin CRL35 inhibits late stages of HSV-1 and HSV-2 replication in vitro. Antivir Res 2003;58:17–24.

[18] Wachsman MB, Farias ME, Takeda E, Sesma F, De Ruiz Hol-gado AP, de Torres RA, et al. Antiviral activity of enterocin CRL against herpes virus. Int J Antimicrob Agents 1999;12:293–9.

[19] Férir G, Petrova M, Andrei G, Huskens D, Hoorelbeke B. The lantibiotic peptide labyrinthopeptin A1 demonstrates broad anti-HIV and anti-HSV activity with potential for microbicidal applications. PLoS One 2013;8:e64010.

[20] Lehtoranta L, Pitkäranta A, Korpela R. Probiotics in respiratory virus infections. Eur J Clin Microbiol Infect Dis 2014;33:1289–302.

[21] Maeda N, Nakamura R, Hirose Y, Murosaki S, Yamamoto Y, Kase T, et al. Oral administration of heat-killed *Lactobacillus plantarum* L-137 enhances protection against influenza virus infection by stimulation of type I interferon production in mice. Int Immunopharmacol 2009;9:1122–5.

[22] Quintana WM, Torres NI, Wachsman MB, Sinko PJ, Castilla V, Chikindas M. Anti - herpes simplex virus type 2 activity of the antimicrobial peptide subtilosin. J Appl Microbiol 2014;117:1253–9.

[23] Silkin L, Hamza S, Kaufman S, Cobb SL, Vederas JC. Spermicidal bacteriocins:lacticin 3147 and subtilosin A. Bioorg Med Chem Lett 2008;18:3103–6.

[24] Sutyak KE, Anderson RA, Dover SE, Feathergill KA, Aroutcheva AA, et al. Spermicidal activity of the safe natural antimicrobial peptide subtilosin. Infect Dis Obstet Gynecol 2008;2008:540758.

[25] Aranha C, Gupta S, Reddy KV. Contraceptive efficacy of antimicrobial peptide nisin: *in vitro* and *in vivo* studies. Contraception 2004;69:333–8.

[26] Saar-Dover R, Bitler A, Nezer R, Shmuel-Galia L, Firon A, et al. D-alanylation of lipoteichoic acids confers resistance to cationic peptides in group B streptococcus by increasing the cell wall density. PLoS Pathog 2012;8:e1002891.

[27] Mazzotta AS, Montville TJ. Nisin induces changes in membrane fatty acid composition of *Listeria monocytogenes* nisin-resistant strains at 10°C and 30°C. J Appl Microbiol 1997;82:32–8.

[28] Ming X, Daeschel MA. Nisin resistance of foodborne bacteria and the specific resistance responses of *Listeria monocytogenes* Scott A. J Food Prot 1993;56:944–8.

[29] Crandall AD, Montville TJ. Nisin resistance in *Listeria monocytogenes* ATCC 700302 is a complex phenotype. Appl Environ Microbiol 1998;64:231–7.

[30] Peschel A, Otto M, Jack RW, Kalbacher H, Jung G, Götz F. Inactivation of the *dlt* operon in *Staphylococcus aureus* confers sensitivity to defensins, protegrins and other antimicrobial peptides. J Biol Chem 1999;274:8405–10.

[31] Kovacs M, Halfmann A, Fedtke I, Heintz M, Peschel A, et al. A functional dlt operon, encoding proteins required for incorporation of D-alanine in teichoic acids in gram-positive bacteria, confers resistance to cationic antimicrobial peptides in Streptococcus pneumoniae. J Bacteriol 2006;188:5797–805.

[32] Blake KL, Randall CP, O'Neill AJ. In vitro studies indicate a high resistance potential for the lantibiotic nisin in *Staphylococcus aureus* and define a genetic basis for nisin resistance. Antimicrob Agents Chemother 2011;55:2362–8.

[33] Collins B, Guinane CM, Cotter PD, Hill C, Ross RP. Assessing the contributions of the LiaS histidine kinase to the innate resistance of *Listeria monocytogenes* to nisin, cephalosporins, and disinfectants. Appl Environ Microbiol 2012;78:2923–9.

[34] Draper LA, Cotter PD, Hill C, Ross RP. Lantibiotic resistance. Microbiol Mol Biol Rev 2015;79:171–91.

[35] Robichon D, Gouin E, Débarbouillé M, Cossart P, Cenatiempo Y, Héchard Y. The *rpoN* (σ^{54}) gene from *Listeria monocytogenes* is involved in resistance to mesentericin Y105, an antibacterial peptide from *Leuconostoc mesenteroides*. J Bacteriol 1997;179:7591–4.

[36] Calvez S, Rincé A, Auffray Y, Prévost H, Drider D. Identification of new genes associated with intermediate resistance of *Enterococcus faecalis* to divercin V41, a pediocin-like bacteriocin. Microbiology 2007;153:1609–18.

[37] Kjos M, Oppegård C, Diep DB, Nes IF, Veening JW, et al. Sensitivity to the two-peptide bacteriocin lactococcin G is dependent on UppP, an enzyme involved in cell-wall synthesis. Mol Microbiol 2014;92:1177–87.

[38] Jarvis B. Resistance to nisin and production of nisin-inactivating enzymes by several *Bacillus* species. Microbiology 1967;47:33–48.

[39] Jarvis B, Farr J. Partial purification, specificity and mechanism of action of the nisin-inactivating enzyme from *Bacillus cereus*. Biochim Biophys Acta 1971;227:232–40.

[40] Van Kraaij C, de Vos WM, Siezen RJ, Kuipers OP. Lantibiotics: biosynthesis, mode of action and applications. Nat Prod Rep 1999;16:575–87.

[41] Fredenhagen A, Maerki F, Fendrich G, Maerki W, Gruner J, Van Oostrum J, et al. Duramycin B and C, two new lanthionine- containing antibiotics as inhibitors of phospholipase A2, and structural revision of duramycin and cinamycin Jung G, Sahl H, editors. Nisin and novel lantibiotics. Leiden, The Netherlands: ESCOM; 1991. p. 131–40.

[42] Shiba T, Wakamiya T, Fukase K, Ueki Y, Teshima T, Nishikawa M. Structure of the lanthionine peptides nisin, ancovenin and lanthiopeptin Jung G, Sahl HG, editors. Nisin and novel lantibiotics. Leiden, The Netherlands: ESCOM Science Publishers; 1991. p. 113–22.

[43] Jung G. Lantibiotics- ribosomally synthesized biologically active polypeptides containing sulfide bridges and α, β-didehydroamino acids. Ang Chem Intl Ed Engl 1991;30:1051–68.

[44] Jung G. Lantibiotics: a survey Jung G, Sahl HG, editors. Nisin and novel lantibiotics. Leiden, The Netherlands: ESCOM Science Publishers; 1991. p. 1–34.

[45] Kido Y, Hamakado T, Yoshida T, Anno M, Motoki Y, Wakamiya T, et al. Isolation and characterization of ancovenin, a new inhibitor of angiotensin I converting enzyme, produced by actinomycetes. J Antibiot 1983;36:1295–9.

[46] Kamarajan P, Hayami T, Matte B, Liu Y, Danciu T, et al. Nisin ZP, a Bacteriocin and Food Preservative, Inhibits Head and Neck Cancer Tumorigenesis and Prolongs Survival. PLoS One 2015;10:e0131008.

[47] Kaur S, Kaur S. Bacteriocins as potential anticancer agents. Front Pharmacol 2015;6:272.

[48] Howell TH, Fiorellini JP, Blackburn P, Projan SJ, de la Harpe J, Williams RC. The effect of a mouthrinse based on nisin, a bacteriocin, on developing plaque and gingivitis in beagle dogs. J Clin Periodontol 1993;20:335–9.

[49] Castiglione F, Cavaletti L, Losi D, Lazzarini A, Carrano L, et al. A novel lantibiotic acting on bacterial cell wall synthesis produced by the uncommon actinomycete *Planomonospora* sp. Biochemistry 2007;46:5884–95.

[50] Kruszewska D, Sahl HG, Bierbaum G, Pag U, Hynes SO, Ljungh A. Mersacidin eradicates methicillin-resistant *Staphylococcus aureus* (MRSA) in a mouse rhinitis model. J Antimicrob Chemother 2004;54:648–53.

[51] Ghobrial O, Derendorf H, Hillman JD. Pharmacokinetic and pharmacodynamics evaluation of the lantibiotic MU 1140. J Pharm Sci 2010;99:2521–8.

[52] Murinda SE, Rashid KA, Roberts RF. In vitro assessment of the cytotoxicity of nisin, pediocin, and selected colicins on simian virus 40-transfected human colon and Vero monkey kidney cells with trypan blue staining viability assays. J Food Prot 2003;66:847–53.

[53] Ingham A, Ford M, Moore RJ, Tizard M. The bacteriocin piscicolin 126 retains antilisterial activity *in vivo*. J Antimicrob Chemother 2003;51:1365–71.

[54] Oh S, Kim SH, Ko Y, Sim JH, Kim KS, Lee SH, et al. Effect of bacteriocin produced by *Lactococcus* sp. HY 449 on skin-inflammatory bacteria. Food Chem Toxicol 2006;44:1184–90.

[55] Belguesmia Y, Madi A, Sperandio D, Merieau A, Feuilloley M, Prévost H, et al. Growing insights into the safety of bacteriocins: the case of enterocin S37. Res Microbiol 2011;162:159–63.

[56] Sosunov V, Mischenko V, Eruslanov B, Svetoch E, Shakina Y, Stern N, et al. Antimycobacterial activity of bacteriocins and their complexes with liposomes. J Antimicrob Chemother 2007;59:919–25.

[57] Hertz C, Bono MR, Barros LF, Lagos R. Microcin E492, a channel-forming bacteriocin from *Klebsiella pneumonia*, induces apoptosis in some human cell lines. Proc Natl Acad Sci USA 2002;99:269–2701.

[58] Huang T, Geng H, Miyyapuram VR, Sit CS, Vederas JC, Nakano MM. Isolation of a variant of subtilosin A with hemolytic activity. J Bacteriol 2009;191:5690–6.

[59] Taylor JI, Hirsch A, Mattick ATR. The treatment of bovine streptococcal and staphylococcal mastitis with nisin. Vet Rec 1949;61:197–8.

[60] Broadbent JR, Chou YC, Gillies K, Kondo JK. Nisin inhibits several Gram-positive, mastitis-causing pathogens. J Dairy Sci 1989;72:3342–5.

[61] Ross RP, Galvin MG, McAuliffe O, Morgan SM, Ryan MP, Twomey DP, et al. Developing applications for lactococcal bacteriocins. Antonie van Leeuwenhoek 1999;76:337–46.

[62] Goldstein BP, Wei J, Greenberg K, Novick R. Activity of nisin against *Streptococcus pneumonia*, *in vitro*, and in a mouse infection model. J Antimicrob Chemother 1998;42:277–8.

[63] Delves-Broughton J, Blackburn RJ, Evans RJ, Hugenholtz J. Applications of the bacteriocin nisin. Antonie van Leeuwenhoek 1996;69:193–202.

[64] De Carvalho AA, Mantovani HC, Vanetti MC. Bactericidal effect of bovicin HC5 and nisin against *Clostridium tyrobutyricum* isolated from spoiled mango pulp. Lett Appl Microbiol 2007;45:68–74.

[65] Bartoloni A, Mantella A, Goldstein BP, Dei R, Benedetti M, Sbaragli S, et al. *In vitro* activity of nisin against clinical isolates of *Clostridium difficile*. J Chemother 2004;16:119–21.

[66] Kim TS, Hur JW, Yu MA, Cheigh CI, Kim KN, Hwang JK, et al. Antagonism of *Helicobacter pylori* by bacteriocins of lactic acid bacteria. J Food Prot 2003;66:3–12.

[67] Bower CK, McGuire J, Daeschel MA. Suppression of *Listeria monocytogenes* colonization following adsorption of nisin onto silica surfaces. Appl Environ Microbiol 1995;61:992–7.

[68] Bower CK, Parker JE, Higgins AZ, Oest ME, Wilson JT, Valentine BA, et al. Protein antimicrobial barriers to bacterial adhesion: *in vitro* and *in vivo* evaluation of nisin-treated implantable materials. Colloids Surf B, Biointerfaces 2002;25:81–90.

[69] Giacometti A, Cirioni O, Barchiesi F, Fortuna M, Sealise G. *In vitro* activity of cationic peptides alone and in combination with clinically used antimicrobial agents against *Pseudomonas aeruginosa*. J Antimicrob Chemother 1999;44:641–5.

[70] Sears PM, Smith BS, Stewart WK, Gonzalez RN. Evaluation of a nisin based germicidal formulation on teat skin of live cows. J Dairy Sci 1992;75:3185–90.

[71] de Kwaadsteniet M, Doeschate KT, Dicks LMT. Nisin F in the treatment of respiratory tract infections caused by *Staphylococcus aureus*. Lett Appl Microbiol 2009;48:65–70.

[72] Van Staden AD, Brand AM, Dicks LMT. Nisin F-loaded brushite bone cement prevented the growth of *Staphylococcus aureus in vivo*. J Appl Microbiol 2012;112:831–40.

[73] Brand AM, de Kwaadsteniet M, Dicks LMT. The ability of nisin F to control *Staphylococcus aureus* infection in the peritoneal cavity, as studied in mice. Lett Appl Microbiol 2010;51:645–9.

[74] Galvin M, Hill C, Ross RP. Lacticin 3147 displays activity in buffer against Gram-positive bacterial pathogens which appear insensitive in standard plate assays. Lett Appl Microbiol 1999;28:355–8.

[75] Zhang C, Occi J, Masurekar P, Barrett JF, Zink DL, et al. Isolation, structure, and antibacterial activity of philipimycin, a thiazolyl peptide discovered from Actinoplanes philippinensis MA7347. J Am Chem Soc 2008;130:12102–10.

[76] Ryan MP, Flynn J, Hill C, Ross RP, Meaney WJ. The natural food grade inhibitor, lacticin 3147, reduced the incidence of mastitis after experimental challenge with *Streptococcus dysgalactiae* in nonlactating dairy cows. J Dairy Sci 1999;82:2625–31.

[77] Tagg JR. Prevention of streptococcal pharyngitis by anti-*Streptococcus pyogenes* bacteriocin-like inhibitory substances (BLIS) produced by *Streptococcus salivarius*. Indian J Med Res 2004;119:13–16.

[78] Pieterse R, Todorov SD, Dicks LM. Mode of action and in vitro susceptibility of mastitis pathogens to macedocin ST91KM and preparation of a teat seal containing the bacteriocin. Braz J Microbiol 2010;41:133–45.

[79] Hillman JD, Brooks TA, Michalek SM, Harmon CC, Snoep JL, van Der Weijden CC. Construction and characterization of an effector strain of *Streptococcus mutans* for replacement therapy of dental caries. Infect Immun 2000;68:543–9.

[80] Mota-Meira M, Morency H, Lavoie MC. *In vivo* activity of mutacin B-Ny266. J Antimicrob Chemother 2005;56:869–71.

[81] Corr SC, Li Y, Riedel CU, O'Toole PW, Hill C, Gahan CGM. Bacteriocin production as a mechanism for the antiinfective activity of *Lactobacillus salivarius* UCC118. Proc Natl Acad Sci USA 2007;104:7617–21.

[82] Knoetze H, Todorov SD, Dicks LMT. A class IIa peptide from *Enterococcus mundtii* inhibits bacteria associated with otitis media. Int J Antimicrob Agents 2008;31:228–34.

[83] Naghmouchi K, Drider D, Kheadr E, Lacroix C, Prévost H, Fliss I. Multiple characterizations of *Listeria monocytogenes* sensitive and insensitive variants to divergicin M35, a new pediocin-like bacteriocin. J Appl Microbiol 2006;100:29–39.

[84] Dabour N, Zihler A, Kheadr E, Lacroix C, Fliss I. *In vivo* study on the effectiveness of pediocin PA-1 and *Pediococcus acidilactici* UL5 at inhibiting *Listeria monocytogenes*. Int J Food Microbiol 2009;133:225–33.

[85] Rihakova J, Cappelier JM, Hue I, Demnerova K, Fédérighi M, Prévost H, et al. *In vivo* activities of recombinant divercin V41 and its structural variants against *Listeria monocytogenes*. Antimicrob Agents Chemother 2010;54:563–4.

[86] Cushman DW, Cheung HS, Sabo EF, Ondetti MA. Development and design of specific inhibitors of angiotensin-converting enzyme. Am J Cardiol 1982;49:1390–4.

[87] Imig JD. ACE inhibition and bradykinin-mediated renal vascular responses: EDHF Involvement. Hypertension 2004;43:533–5.

[88] Skeggs LT, Kahn JR, Shumway NP. The preparation and function of the hypertension-converting enzume. J Exp Med 1956;103:295–9.

[89] Zhang R, Xu X, Chen T, Li L, Rao P. An assay for angiotensin-converting enzyme using capillary zone electrophoresis. Anal Biochem 2000;280:286–90.

[90] Fontana MB, de Bastos Mdo C, Brandelli A. Bacteriocins Pep5 and epidermin inhibit *Staphylococcus epidermidis* adhesion to catheders. Curr Microbiol 2006;52:350–3.

[91] Haste NM, Thienphrapa W, Tran DN, Loesgen S, Sun P, Nam S-J, et al. Activity of the thiopeptide antibiotic nosiheptide against contemporary strains of methicillin-resistant Staphylococcus aureus. The J Antibiot 2012;65:593–8.

[92] Lopez FE, Vincent PA, Zenoff AM, Salomón RA, Farías RN. Efficacy of microcin J25 in biomatrices and in a mouse model of Salmonella infection. J Antimicrob Chemother 2007;59:676–80.

[93] McGregor DP. Discovering and improving novel peptide therapeutics. Curr Opin Pharmacol 2008;8:616–9.

[94] Joerger RD. Alternatives to antibiotics: bacteriocins, antimicrobial peptides and bacteriophages. Poult Sci 2003;82:640–7.

[95] van Staden AD, Brand AM, Endo A, Dicks LMT. Nisin F, intraperitoneally injected, may have a stabilizing effect on the bacterial population in the gastrointestinal tract, as determined in a preliminary study with mice as model. Lett Appl Microbiol 2011;53:198–201.

[96] de Kwaadsteniet M, van Reenen CA, Dicks LMT. Evaluation of nisin F in the treatment of subcutaneous skin infections, as monitored by using a bioluminescent strain of *Staphylococcus aureus*. Probiotics Antimicro Prot 2010;2:61–5.

[97] Granger M, van Reenen CA, Dicks LMT. Effect of gastro-intestinal conditions on *Enterococcus mundtii* ST4SA and production of bacteriocin ST4SA, as recorded by real-time PCR. Int J Food Microbiol 2008;123:277–80.

[98] Ramiah K, Van Reenen CA, Dicks LMT. Expression of the mucus adhesion genes mub and mapA, adhesion-like factor EF-Tu and bacteriocin gene plaA of *Lactobacillus plantarum* 423 monitored with real-time PCR. Int J Food Microbiol 2007;116:405–9.

[99] Castiglione F, Lazzarini A, Carrano L, Corti E, Ciciliato I, Gastaldo L, et al. Determining the structure and mode of action of microbisporicin, a potent lantibiotic active against multiresistant pathogens. Chem Biol 2008;15:22–31.

[100] Piper C, Casey PG, Hill C, Cotter PD, Ross RP. The lantibiotic lacticin 3147 prevents systemic spread of *Staphylococcus aureus* in a murine infection model. Int J Microbiol 2012;2012:1–6.

[101] van Staden ADP, Heunis T, Smith C, Deane S, Dicks LMT. Efficacy of Lantibiotic Treatment of *Staphylococcus aureus*-induced Skin Infections, Monitored by *in vivo* Bioluminescent Imaging. Antimicrob Agents Chemother 2016;60 02938-15.

[102] Moll GN, Kuipers A, Rink R. Microbial engineering of dehydro-amino acids and lanthionines in non-lantibiotic peptides. Antonie Van Leeuwenhoek 2010;97:319–33.

[103] Kuipers A, Rink R, Moll GN. Translocation of a thioether-bridged azurin peptide fragment via the sec pathway in *Lactococcus lactis*. Appl Environ Microbiol 2009;75:3800–2.

[104] Kuipers A, Wierenga J, Rink R, Kluskens LD, Driessen AJ, Kuipers OP, et al. Sec-mediated transport of posttranslationally dehydrated peptides in *Lactococcus lactis*. Appl Environ Microbiol 2006;72:7626–33.

[105] Majchrzykiewicz JA, Lubelski J, Moll GN, Kuipers A, Bijlsma JJ, Kuipers OP, et al. Production of a class II two-component lantibiotic of *Streptococcus pneumoniae* using the class I nisin synthetic machinery and leader sequence. Antimicrob Agents Chemother 2010;54:1498–505.

[106] Bessalle R, Kapitkovsky A, Gorea A, Shalit I, Fridkin M. All-D-magainin: chirality, antimicrobial activity and proteolytic resistance. FEBS Lett 1990;274:151–5.

[107] Hong SY, Oh JE, Lee KH. Effect of D-amino acid substitution on the stability, the secondary structure, and the activity of membrane active peptide. Biochem Pharmacol 1999;58:1775–80.

[108] Li P, Roller PP. Cyclization strategies in peptide derived drug design. Curr Top Med Chem 2002;2:325–41.

[109] Rink R, Arkema-Meter A, Baudoin I, Post E, Kuipers A, Nelemans SA, et al. To protect peptide pharmaceuticals against peptidases. J Pharmacol Toxicol Methods 2010;61:210–8.

[110] Singh SB, Occi J, Jayasuriya H, Herath K, Motyl M, et al. Antibacterial evaluations of thiazomycin- a potent thiazolyl peptide antibiotic from Amycolatopsis fastidiosa. J Antibiot (Tokyo) 2006;60:565–71.

[111] Deleted in Review.

[112] Niu WW, Neu HC. Activity of mersacidin, a novel peptide, compared with that of vancomycin, tecoplanin, and datomycin. Antimicrob Agents Chemother 1991;35:988–1000.

[113] Kellner R, Jung G, Hörner T, Zähner H, Schnell N, Entian KD, et al. Gallidermin: a new lanthionine-containing polypeptide antibiotic. Eur J Biochem 1988;15:53–9.

[114] Brand AM, Smith C, Dicks LMT. Antimicrobial activity of nisin F, nisin A and Nisaplin® in the presence of serum and the effect of these lantibiotics on leukocyte functional capacity. Therapeutic properties of the lantibiotic nisin F. PhD Thesis. Stellenbosch: Stellenbosch University; 2013.

[115] Begde D, Bundale S, Mashitha P, Rudra J, Nashikkar N, Upadhyay A. Immunomodulatory efficacy of nisin - a bacterial lantibiotic peptide. J Pept Sci 2011;17:438–44.

[116] De Pablo MA, Gaforio JJ, Gallego AM, Ortega E, Gálvez AM, López AC. Evaluation of immunomodulatory effects of nisin-containing diets on mice. FEMS Immunol Med Microbiol 1999;24:35–42.

[117] Villamil L, Figueras A, Novoa B. Immunomodulatory effects of nisin in turbot (*Scophthalmus maximus* L.). Fish Shellfish Immunol 2003;14:157–69.

[118] Gupta SM, Aranha CC, Reddy KV. Evaluation of developmental toxicity of microbicide nisin in rats. Food Chem Toxicol 2008;46:598–603.

[119] Maqueda M, Gaèlvez A, Martènez-Bueno M, Guerra I, Valdivia E. Neutralizing antibodies against the peptide antibiotic AS-48: immunocytological studies. Antimicrob Agents Chemother 1993;37:148–51.

[120] Martínez MI, Rodrèguez JM, Suárez A, Martínez JM, Azcona JI, Hernández PE. Generation of polyclonal antibodies against a chemically synthesized N-terminal fragment of the bacteriocin pediocin. Lett Appl Microbiol 1997;24:488–92.

[121] Suárez AM, Rodrèguez JM, Hernaèndez PE, Azcona-Olivera JI. Generation of polyclonal antibodies against nisin: immunization strategies and immunoassay development. Appl Environ Microbiol 1996;62:2117–21.

[122] Mosca DA, Hurst MA, So W, Viajar BS, Fujii CA, Falla TJ. IB-367, a protegrin peptide with in vitro and *in vivo* activities against the microflora associated with oral mucositis. Antimicrob Agents Chemother 2000;44:1803–8.

[123] Hancock REW. Cationic antimicrobial peptides: towards clinical applications. Expert Opin Investig Drugs 2000;9:1723–9.

[124] Valenta C, Bernkop-Schnürch A, Rigler HP. The antistaphylococcal effect of nisin in a suitable vehicle: a potential therapy for atopic dermatitis in man. J Pharm Pharmacol 1996;48:988–91.

[125] Walsh SM, Shah AG, Mond JJ. Topical anti-infective formulations. Patent US20,040,192,581, 2004.

[126] Dawson MJ, Bargallo JC, Appleyard AN, Rudd BAM, Boakes S, Bierbaum G, et al. F3W variants of lantibiotic mersacidin and its use. Patent US7592308, 2009.

[127] Naghmouchi K, Baah J, Hober D, Jouy E, Rubrecht C, Sané F, et al. Synergistic effect between colistin and bacteriocins in controlling Gram-negative pathogens and their potential to reduce antibiotic toxicity in mammalian epithelial cells. Antimicrob Agents Chemother 2013;57:2719–25.

[128] Barriere SL. Bacterial resistance to beta-lactams, and its prevention with combination antimicrobial therapy. Pharmacother 1992;12:397–402.

[129] Steenbergen JN, Mohr JF, Thorne GM. Effects of daptomycin in combination with other antimicrobial agents: A review of in vitro and animal model studies. J Antimicrob Chemother 2009;64:1130–8.

[130] Wu YL, Scott EM, Po AL, Tariq VN. Ability of azlocillin and tobramycin in combination to delay or prevent resistance development in *Pseudomonas aeruginosa*. J Antimicrob Chemother 1999;44:389–92.

[131] Reis CR, Neufeld RJ, Ribeiro AJ, Veiga F. Nanoencapsulation II. Biomedical applications and current status of peptide and protein nanoparticulate delivery systems. Nanomed: Nanotechnol Biol Med 2006;2:53–65.

[132] Chai G-H, Hu F-Q, Sun J, Du Y-Z, You J, Yuan H. Transport pathways of solid lipid nanoparticles across Madin-Darby canine kidney epithelial cell monolayer. Mol Pharm 2014;11:3716–26.

[133] Des Rieux A, Fievez V, Garinot M, Schneider Y-V, Préat V. Nanoparticles as potential oral delivery systems for proteins and vaccines: a mechanistic approach. J Control Relase 2006;116:1–27.

[134] Brannon-Peppas L, Blanchette JO. Nanoparticle and targeted systems for cancer therapy. Adv Drug Delivery Rev 2004;56:1649–59.

[135] Feng SS. Nanoparticles of biodegradable polymers for new-concept chemotherapy. Expert Rev Med Devices 2004;1:115–25.

[136] Kumari A, Yadav SK, Yadav SC. Biodegradable polymeric nanoparticles based drug delivery systems. Colloids and Surfaces B: Biointerfaces 2010;75:1–18.

[137] Brigger I, Dubernet C, Couvreur P. Nanoparticles in cancer therapy and diagnosis. Adv Drug Delivery Rev 2002;54:631–51.

[138] Kobayashi H, Watanabe R, Choyke PL. Improving conventional enhanced permeability and retention effects; what is the appropriate target? Theranostics 2014;4:81–9.

[139] Wang M, Thanou M. Targeting nanoparticles to cancer. Pharmacol Res 2010;62:90–9.

[140] Lee KY, Yuk SH. Polymeric protein delivery systems. Prog Polym Sci 2007;32:669–97.

[141] Mundargi RC, Babu VR, Rangaswamy V, Patel P, Aminabhavi TM. Nano/micro technologies for delivering macromolecular therapeutics using poly(D,L-lactide-*co*-glycolide) and its derivatives. J Ctrl Rel 2008;125:193–209.

[142] Di Toro R, Betti V, Spampinato S. Biocompatibility and integrin-mediated adhesion of human osteoblasts to poly(DL-lactide-*co*-glycolide) copolymers. Eur J Pharm Sci 2004;21:161–9.

[143] Leo E, et al. *In vitro* evaluation of PLA nanoparticles containing a lipophilic drug in water-soluble or insoluble form. Int J Pharm 2004;278:133–41.

[144] Pasut G, Veronese FM. PEG conjugates in clinical development or use as anticancer agents: an overview. Adv Drug Delivery Rev 2009;61:1177–88.

[145] Veronese FM, Mero A. The impact of PEGylation on biological therapies. BioDrugs 2008;22:315–29.

[146] Jevsevar S, Kunstelj M, Porekar VG. PEGylation of therapeutic proteins. Biotechnol J 2010;5:113–28.

[147] Avrahami D, Shai Y. Conjugation of a magainin analogue with lipophilic acids controls hydrophobicity, solution assembly, and cell selectivity. Biochemistry 2002;41:2254–63.

[148] Wang P, Bang JK, Kim HJ, Kim JK, Kim Y, Shin SY. Antimicrobial specificity and mechanism of action of disulfide-removed linear analogs of the plant-derived Cys-rich antimicrobial peptide Ib-AMP1. Peptides 2009;30:2144–9.

[149] Chongsiriwatana NP, Patch JA, Czyzewski AM, Dohm MT, Ivankin A, Gidalevitz D, et al. Peptoids that mimic the structure, function, and mechanism of helical antimicrobial peptides. Proc Natl Acad Sci USA 2008;105:2794–9.

[150] Porter EA, Weisblum B, Gellman SH. Mimicry of host-defense peptides by unnatural oligomers: antimicrobial beta-peptides. J Am Chem Soc 2002;124:7324–30.

[151] Radzishevsky IS, Rotem S, Bourdetsky D, Navon-Venezia S, Carmeli Y, Mor A. Improved antimicrobial peptides based on acyl-lysine oligomers. Nat Biotechnol 2007;25:657–9.

[152] Salmaso S, Elvassore N, Bertucco A, Lante A, Caliceti P. Nisin-loaded poly-L-lactide nano-particles produced by CO_2 anti-solvent precipitation for sustained antimicrobial activity. Int J Pharm 2004;287:163–73.

[153] Heunis TDJ, Botes M, Dicks LMT. Encapsulation of *Lactobacillusplantarum* 423 and its bacteriocin in nanofibers. Probiotics Antimicrob Prot 2010;2:46–51.

[154] Heunis TDJ, Bshena O, Klumperman B, Dicks LMT. Release of bacteriocins from nanofibers prepared with combinations of poly (D, L-lactide) (PDLLA) and poly (ethylene oxide) (PEO). Int J Mol Sci 2011;12:2158–73.

[155] Heunis TDJ, Smith C, Dicks LMT. Evaluation of a nisin-eluting nanofiber scaffold to treat *Staphylococcus aureus*-induced skin infections in mice. Antimicrob Agents Chemother 2013;57:3928–35.

[156] Ahire JJ, Dicks LMT. Nisin incorporated with 2,3-Dihydroxybenzoic acid in nanofibers inhibits biofilm formation by a methicillin-resistant strain of *Staphylococcus aureus*. Probiotics Antimicro Prot 2015;7:52–9.

[157] Ahire JJ, Neveling DP, Dicks LMT. Co-spinning of silver nanoparticles with nisin increases antimicrobial spectrum of PDLLA: PEO nanofibers. Curr Microbiol 2015 doi.10.1007/s00284-015-0813-y.

[158] Maretschek S, Greiner A, Kissel T. Electrospun biodegradable nanofiber nonwovens for controlled release of proteins. J Controlled Release 2008;127:180–7.

[159] Porter JR, Henson A, Popat KC. Biodegradable poly (ε-caprolactone) nanowires for bone tissue engineering applications. Biomaterials 2009;30:780–8.

[160] Quaglia F. Bioinspired tissue engineering: the great promise of protein delivery technologies. Int J Pharm 2008;364:281–97.

[161] Chew SJ, Wen J, Yim EKF, Leong KW. Sustained release of proteins from electropsun biodegradable fibers. Biomacromol 2005;6:2017–24.

[162] Jiang H, Hu Y, Li Y, Zhao P, Zhu K, Chen W. A facile technique to prepare biodegradable coaxial electrospun nanofibers for controlled release of bioactive agents. J Controlled Release 2005;108:237–43.

[163] Kim TG, Lee DS, Park TG. Controlled protein release from electrospun biodegradable fiber mesh composed of poly (ε-caprolactone) and poly (ethylene oxide). Int J Pharm 2007;338:276–83.

[164] Liang D, Hsiao BJ, Chu B. Functional electrospun nanofibrous scaffolds for biomedical applications. Adv Drug Delivery Rev 2007;59:1392–412.

[165] Sun Z, Zussman E, Yarin AL, Wendorf JH, Greiner A. Compound core-shell polymer nanofibers by co-electrospinning. Adv Mater 2003;15:1929–32.

[166] Xu X, Zhuang X, Chen X, Wang X, Yang L, Jing X. Preparation of core-sheath composite nanofibers by emulsion electrospinning. Macromol Rapid Commun 2006;27:1637–42.

[167] Türk M, Rzayev ZM, Khalilova SA. Bioengineering functional copolymers. XIV. Synthesis and interaction of poly (*N*-isopropylacrylamide-*co*-3, 4-dihydro-2H-pyran-alt-maleic anhydride) s with SCLC cancer cells. Bioorg Med Chem 2010;18:7975–84.

[168] Ladaviere C, Delair T, Domard A, Pichot C, Mandrand B. Covalent immobilization of biological molecules to maleic anhydride and methyl vinyl ether copolymers—A physico-chemical approach. J Appl Polym Sci 1999;71:927–36.

[169] Patel H, Raval DA, Madamwar D, Patel SR. Polymeric prodrug: synthesis, release study and antimicrobial property of poly (styrene-*co*-maleic anhydride)-bound acriflavine. Angew Makromol Chem 1998;263:25–30.

[170] Patel H, Raval DA, Madamwar D, Sinha TJM. Polymeric prodrugs. Synthesis, release study and antimicrobial properties of polymer-bound acriflavine. Angew Makromol Chem 1997;245:1–8.

[171] Jeong JH, Byoun YS, Ko SB, Lee YS. Chemical modification of poly (styrene-alt-maleic anhydride) with antimicrobial 4-aminobenzoic acid and 4-hydroxybenzoic acid. J Ind Eng Chem 2001;7:310–5.

[172] Jeong JH, Byoun YS, Lee YS. Poly (styrene-alt-maleic anhydride)-4-aminophenol conjugate: synthesis and antibacterial activity. React Funct Polym 2002;50:257–63.

[173] Ignatova M, Stoilova O, Manolova N, Markova N, Rashkov I. Electrospun mats from styrene/maleic anhydride copolymers: modification with amines and assessment of antimicrobial activity. Macromol Biosci 2010;10:944–54.

[174] Bi L, Yang L, Bhunia A, Yao Y. Carbohydrate nanoparticle-mediated colloidal assembly for prolonged efficacy of bacteriocin against food pathogen. Biotech Bioeng 2011;108:1529–36.

[175] Malheiros P, Sant'Anna V, Barbosa M, Brandelli A, Franco B. Effect of liposome-encapsulated nisin and bacteriocin-like substance P34 on *Listeria monocytogenes* growth in Minas frescal cheese. Int J Food Microbiol 2012;156:272–7.

[176] Messi P, Guerrieri E, Bondi M. Bacteriocin-like substance (BLS) production in *Aermonas hydrophila* water isolates. FEMS Microbiol Lett 2003;220:121–5.

Graphene-Microbial Interactions

S. Szunerits[1] and R. Boukherroub[2]

*[1]Lille University of Science and Technology, Villeneuve d'Ascq, France,
[2]CNRS, Lille University of Science and Technology, Villeneuve d'Ascq, France*

CHAPTER OUTLINE

8.1 INTRODUCTION

Antibacterial molecules and materials are widely used in daily life to effectively protect public health and to cure acute bacterial infections. Antibiotics, metal ions, quaternary ammonium compounds, and others have been known to prevent bacteria attachment to materials' surfaces and their proliferation. The increase in the number of antibiotic-resistant bacteria, together with concerns about the toxicity of some of these materials, and costs, have motivated researchers to look for potential alternatives. Various nanomaterials [1] have been explored and proposed to meet these challenges, including silver nanoparticles (Ag NPs), titanium dioxide NPs (TiO_2 NPs), copper oxide NPs (CuO NPs) [2], zinc oxide (ZnO NPs) [3], calcium oxide NPs (CaO), magnesium oxide NPs (MgO NPs), and carbon-based materials [4] such as carbon nanotubes (CNTs) and diamond nanoparticles (NDs) [5–7] (Table 8.1). The antimicrobial character of Ag NPs has been known for a long time and Ag NPs are nowadays present in dressings for surgical wounds, in coating for medical devices as well as in lotions and gels to prevent bacteria and fungi contaminations. The mechanism of action of Ag NPs is well studied and is based on the attack of the respiratory chain and cell division that finally leads to cell death, while concomitantly releasing silver cations (Ag^+) enhances the bactericidal activity. Even though the antibacterial character of Ag NPs appears appealing, practical applications of Ag NPs are often hampered by their aggregation and subsequent loss of antibacterial activity [8]. In addition, concerns about the cytotoxicity of Ag NPs toward human cells have been voiced [9].

While CNTs were found to be cytotoxic to bacteria, they also exhibited strong toxicity to human cells unless their surface is functionalized [10]. CNTs are typically grown from carbon-containing gases in the presence

Table 8.1 Antimicrobial Nanomaterials

Nanomaterial	Mechanism	Application
Ag NPs	Ag^+ release, disruption of cell membrane and electron transport, DNA damage	Wound dressings, coatings of medical devices, portable water filters
ZnO NPs	Release of Zn^{2+}, H_2O_2 production	Antibacterial creams, mouthwash, coatings of medical devices
TiO_2 NPs	Production of ROS	Food sterilization agent
Au NPs	Interaction with cell membrane	Photothermal therapy
CuO NPs	Production of ROS	Wound dressings
CNTs	Membrane damage in bacteria due to an oxidative stress; aggregation between bacterial cells and carbon nanomaterials causes direct contact between the cells and carbon nanomaterials, which in turn leads to cell death	Currently only research based
NDs	Interactions with cellular components and interference with the permeability of the bacterial cell wall and/or membrane	Currently only research based

of catalytic nanoparticles, which remain in the nanostructures even after extensive purification procedures. These residual metallic impurities induce cytotoxicity within biological samples even at 50 ppm levels [11]. Most studies recently carried out demonstrate that the "unrolled" version of CNTs, graphene, free of metallic impurities, is a superior biocompatible material that allows the effective proliferation of human and mammalian cells with limited or no cytotoxicity. Such properties hold promise for potential applications of graphene-based materials in tissue engineering and implants, as well as for drug delivery. Several papers showed that graphene oxide (GO) can effectively promote the adhesion and proliferation of osteoblasts, kidney cells, embryonic cells, etc. Other reports indicated that cellular internalization of GO nanosheets added to culture media at concentrations of $50\,\mu g\,mL^{-1}$ causes 50% decrease in mammalian cell viability, suggesting that colloidal solutions of GO are mildly cytotoxic at higher concentrations [12]. The apparent low cytotoxicity of GO-based materials encouraged researchers to investigate their antimicrobial effects in solution and when deposited on surfaces. Indeed, a material with low mammalian cell cytotoxicity and efficient antimicrobial properties may be an ideal matrix for biomedical applications, helping to combat bacterial infections. The main research interests and objectives in recent years were thus to determine the antibacterial or bacteriostatic activity of graphene, graphene-based derivatives such as graphene oxide (GO) and reduced graphene oxide (rGO) and their nanocomposites. The importance of surface functional groups, size and shape of graphene was pointed out and will be discussed below. However, before detailed discussion about the antimicrobial properties of graphene nanostructures, some general information about the material itself and the different ways it can be produced are introduced.

8.2 GRAPHENE, GRAPHENE OXIDE (GO) AND REDUCED GRAPHENE OXIDE (rGO): PRODUCTION AND CHARACTERIZATION

Precisely speaking, graphene is the name given to a two-dimensional sheet of sp^2-hybridized carbon network with a carbon-carbon distance of 0.142 nm and an interlayer spacing of 0.34 nm [13,14]. The pristine form, as schematically depicted in Fig. 8.1A, contains complete sp^2 hybridized carbons with no defects; it consists of only a single layer of atoms and is essentially a very large polyaromatic hydrocarbon. In the literature, graphene may also refer to a multi-layered material (Fig. 8.1B) made up of many graphene layers stacked atop one another and held together by weak van der Waals forces and $\pi - \pi$ stacking interactions. Raman analysis

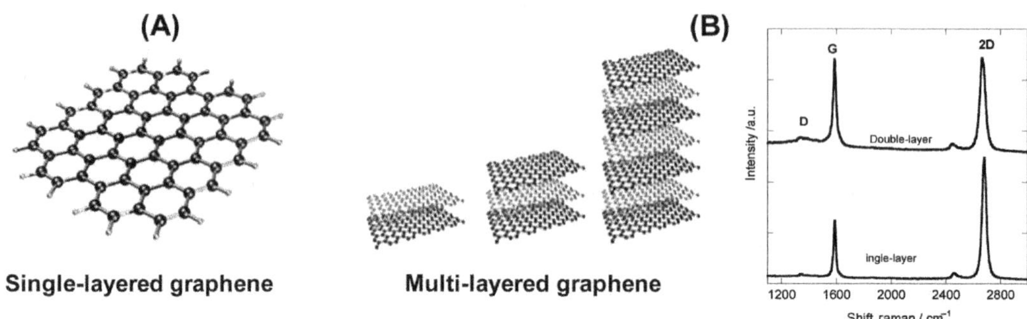

■ **FIGURE 8.1** (A) Schematic drawing of a single-layered pristine graphene; (B) From single to multilayered graphene nanostructures together with characteristic Raman spectra [21,22] to distinguish between them.

allows a distinction between single and multilayer characteristics of graphene layers as the 2D band is highly sensitive to stacking of the sheets [15–17]. While a single-layer graphene sheet presents a 2D band at $2679\,cm^{-1}$, for multi-layered graphene sheets the 2D band appears as a broadened peak with a $19\,cm^{-1}$ shift to higher wavenumbers.

While a graphene material of a few layers has some of the advantages of single-layered graphene such as high aspect ratio and surface area ($2630\,m^2\,g^{-1}$; for comparison graphite: $10\,m^2\,g^{-1}$ and carbon nanotubes: $1315\,m^2\,g^{-1}$) [18], yet it does not have the same optical and electrical properties such as remarkably high electron mobility at room temperature (measured experimentally to be $15.000\,cm^2\,V^{-1}\,s^{-1}$ with a theoretically intrinsic limit of $200.000\,cm^2\,V^{-1}\,s^{-1}$), a carrier density of $10^{12}\,cm^{-2}$ [19,20], or thermal conductivity in the range of $[(4.84–5.30) \times 10^3\,W\,m^{-1}\,K^{-1}]$.

The fabrication routes of graphene include mechanical exfoliation and unrolling of carbon nanotubes, chemical vapor deposition (CVD) and epitaxial growth. Mechanical exfoliation is often referred to as the "scotch-tape method," capable of separating a single graphene sheet from crystalline graphite (usually HOPG) through attachment to an adhesive tape (Fig. 8.2A) [13,14]. This method is an excellent technique to generate micrometer to submicrometer sized graphene sheets without the introduction of defects, and is ideally suited for the investigation of the physical properties of graphene. However, its low-yield is a serious limitation for any biomedical application for which larger quantities are needed.

Alternative methods for the synthesis of graphene by epitaxial growth and chemical vapor deposition (CVD) rely on a suitable substrate (SiC, Cu, Ni, Ir, Pt) for supporting the catalytic deposition of graphene [23]. Epitaxial growth of graphene was first demonstrated on a silicon carbide substrate [24].

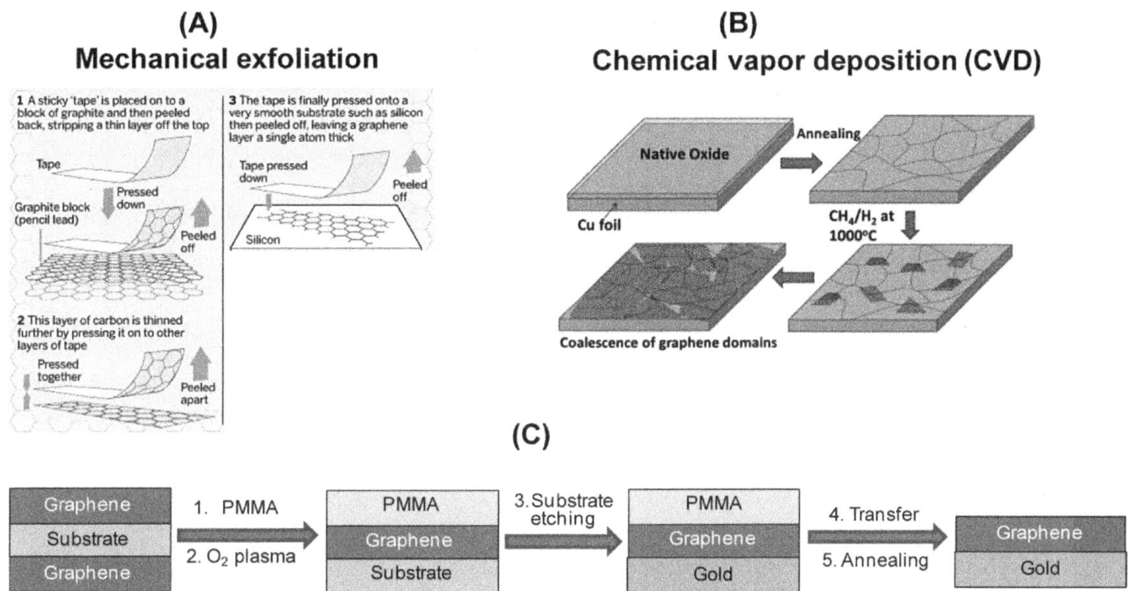

■ FIGURE 8.2 Fabrication processes of high-quality graphene: (A) mechanical exfoliation [27], (B) Formation of CVD graphene [28], (C) Wet chemical transfer procedure of CVD graphene to other substrates [22,25].

Temperatures higher than 1300°C and ultra-high vacuum are usually required to produce graphene. In the CVD process (Fig. 8.2B), nucleation and graphene growth occur after exposure of the appropriate substrate to a hydrocarbon gas such as methane or ethylene at low pressures and high temperatures. The condensed gas is converted to a carbon layer upon cooling of the substrate. Careful control of the experimental conditions allows the synthesis of single-layered graphene surfaces of high quality and defined size [22,25]. This method produces large-area graphene with high transparency and conductance, suitable for potential biomedical applications such as the direct growth of cells and for studies of bacteria–graphene interactions.

CVD grown graphene can be transferred onto various interfaces (gold, Ge, diamond, etc.) by wet chemical transfer process [22,25,26]. The method is based on the use of polymers such as polymethylmethacrylate (PMMA), which is spin-coated onto the graphene-coated substrate. The backside of the graphene can be removed by oxygen plasma treatment and the substrate with appropriate etchant solutions, before the graphene sheets are transferred onto the new substrate (Fig. 8.2C).

The most widely used approach and cost-effective technique for the production of larger scale quantities consists in chemical oxidation of graphite to graphite oxide followed by exfoliation through sonication to graphene oxide (GO), first developed by Hummers (Fig. 8.3A) [29,30]. It involves soaking of graphite in a solution of sulfuric acid (H_2SO_4) and potassium permanganate ($KMnO_4$) to produce graphite oxide [30]. Stirring or sonication of the formed graphite oxide is then performed to split apart the bulk material into single and few layers of GO. Aqueous suspensions of GO are relatively stable due to the hydrophilic nature of GO and its high negative surface charge (zeta potential around −40 mV). The surface of GO is indeed highly oxygenated, bearing hydroxyl, epoxide, diol, and carbonyl functional groups allowing GO to readily swell and disperse in water and many other organic solvents [31]. The presence of negatively charged oxygen species helps, in addition, to disperse GO as a single sheet by providing electrostatic repulsion. In contrast to CVD graphene, the presence of oxygen groups on the basal plane of GO disrupts the π-conjugation, and as a result, GO becomes an electrically insulating material.

The oxygen-containing functional groups on the GO surface can be easily removed through various reduction procedures using chemical (e.g., hydrazine monohydrate, sodium borohydride, hydroquinone), thermal,

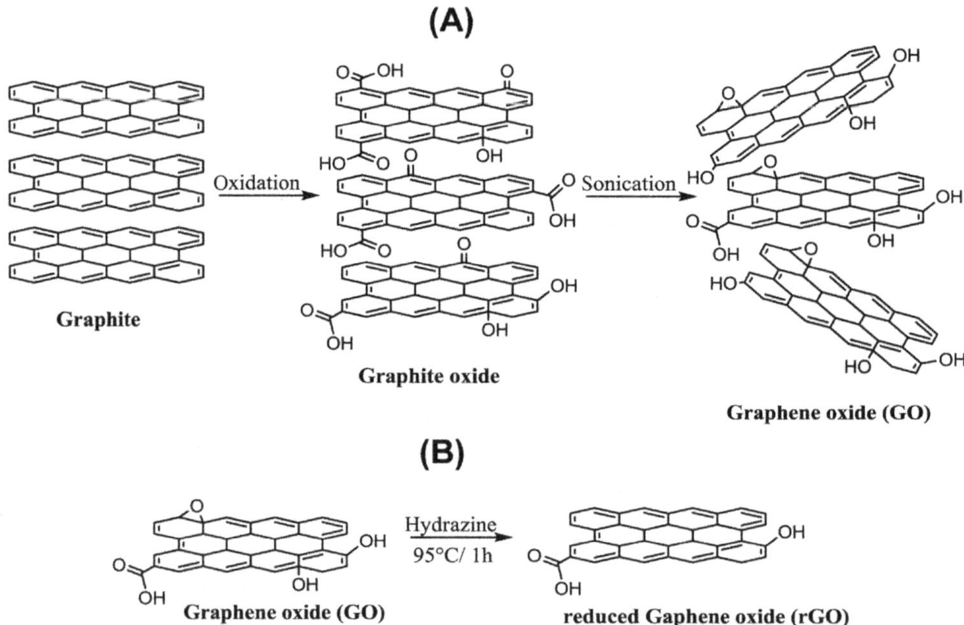

■ **FIGURE 8.3** Production of large quantities of graphene oxide (GO) and reduced graphene oxide (rGO) via chemical oxidation of graphite to graphite oxide followed by exfoliation through sonication to GO (A), and further reduction of GO to rGO (B) [30].

photochemical and other means to yield reduced graphene oxide (rGO) (Fig. 8.3B) [23,30,32–47]. Chemical treatment with hydrazine monohydrate is still the most widely used process to regenerate $C = C$ bonds, mainly due to its strong reduction ability to eliminate most oxygen-containing functional groups of GO, and its ability to yield stable rGO aqueous dispersions [46]. This method has, however, some significant drawbacks including the highly toxic nature of hydrazine and the presence of a significant level of defects, reducing considerably the conductivity of rGO. To avoid using hydrazine, many environmentally friendly and highly efficient reductants have been developed in the last years including vitamin C [48,49], amino acids [49], reducing sugars [50], alcohols [41], dopamine [44,51], etc. The degree of reduction and thus the conductivity of rGO depends strongly on the reduction method used. Any of these rGO nanostructures are often incorrectly referred as graphene.

In terms of the potential interaction with bacteria, the reduced conductivity of GO and rGO is not necessarily detrimental. The presence of different surface functions has the additional advantage of tuning the interaction with bacteria through the incorporation of other chemical functions or molecules. At present, the field is rather limited to some examples dealing with GO or rGO modified with silver nanoparticles [52,53], positively charged polymers such as poly(L-lysine) (PLL) [54] or chitosan [55], and surfactants such as TWEEN 20 [56] (Fig. 8.4).

In order to get an overview of some of the past and present findings on the antibacterial and bacteriostatic properties of graphene, GO, rGO, and graphene-based nanocomposites, in the following the most important works in this field will be discussed. It will become clear that the few studies conducted in this area have led to conflicting results in some cases [12,53,58]. These differences are linked to the rather complex and versatile structures of graphene and its derivatives. Currently, it is evident that the orientation of graphene on surfaces, the presence of sharp edges, the size of the graphene nanocomposite as well as the presence of functional molecules and chemical groups are strongly influencing the antimicrobial activity of these materials for both Gram-negative and Gram-positive bacteria. Ivanova and coworkers lately confirmed that the density of the edges of graphene was one of the principle parameters that contribute to the antibacterial behavior of graphene nanosheets films [58]. It is hoped that the following discussion will help understand in more depth the different parameters that require special attention in order to reach optimal antimicrobial characteristics and/or inhibition of bacterial attachment to surfaces. It is indeed important to make a distinction between a graphene surface that inhibits bacterial attachment and a surface that effectively destroys bacteria on contact. Both

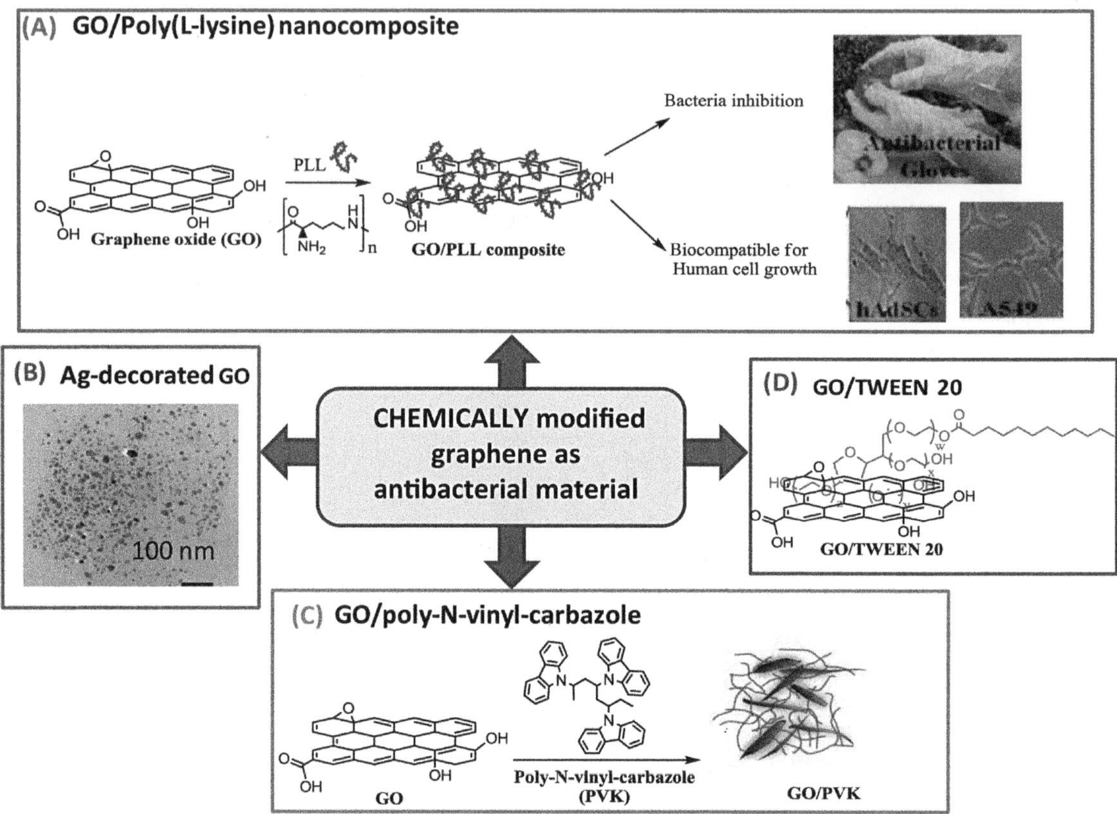

■ **FIGURE 8.4** Surface-functionalized GO and rGO for antibacterial studies. (A) A GO/Poly-L-lysine (PLL) composite to inhibit bacteria and support human cells; [54] (B) TEM image of Ag-decorated GO; [53] (C) Inhibition of biofilm formation using poly-N-vinyl-carbazole (PVK) modified GO; [57] (D) Free standing paper composed of a TWEEN/GO composite [56].

types of surfaces are often called "antibacterial," while the latter should more correctly be classified as bactericidal.

8.3 INHIBITION OF BACTERIAL ATTACHMENT AND ANTIBACTERIAL ACTIVITY OF GRAPHENE-COATED SURFACES

The adhesion of bacteria to surfaces is a very complex process and depends on a multitude of parameters. The physical surface structures as well as the chemistry of the substrate are critical factors influencing the interactions with bacteria and the build-up of biofilms. Bacteria have in general greater affinity for hydrophobic surfaces and an effective strategy for inhibiting pathogen attachment to surfaces is by making them hydrophilic using materials such as polyethylene glycol (PEG). In order to enhance bacteria attachment to the surface, the water molecules have to be displaced. At high ionic strength, the PEG chains are dehydrated and their efficiency for preventing bacterial adhesion is strongly depressed [59]. The other parameter influencing bacterial adhesion is the presence of surface charge. In the case of graphene, the surface is of intermediate wettability (water contact angle around 80°C) with a varying level of charge depending on the method of production. Surface roughness is also affecting bacteria attachment and rough surfaces cause the bacterial cell wall to rupture under its own weight upon contact. In the case of graphene, the sharp-edges would fulfill this criterion. However, the basal planes can be considered atomically smooth. The interaction of bacteria with graphene is thus highly dependent on the orientation of the surface relative to the bacteria [58].

Graphene surfaces can be prepared in a number of ways using GO and rGO suspensions as starting materials and forming "paper" like surfaces through application of a vacuum filtration protocol [12], electrophoretic deposition of GO [60], Langmuir-Blodgett technique [61], self-drying upon solvent evaporation in the presence of chitosan as matrix [55], or by simple drop- or spin-casting the solutions onto planar surfaces (Fig. 8.5) [53,54]. Considering that CVD derived graphene films can be transferred to any surface of interest, the antibacterial and bactericidal properties of such surfaces become of interest for practical applications [26].

Amongst the first studies on the interaction of bacteria with graphene are those by Akhavan and Ghaderi [60] and Hu et al. [12]. GO nanowells (GONWs) were deposited onto a stainless steel substrate in such a manner that a significant amount of edges were exposed [60]. From the SEM image in Fig. 8.5D, it can be seen that single and/or multilayered GO sheets could be deposited using this approach in high density and random orientation, with some of them almost perpendicular to the surface of the substrate. These sheets provide extremely sharp edges on the surface and should effectively interact with bacteria. Reduction of the GONWs by hydrazine vapor did not change the morphology, but resulted in a significant reduction of the C-O, C=O and O=C-OH functions as evidenced by the C_{1s} XPS core level spectra of both graphene structures (Fig. 8.6A). To investigate bacterial toxicity of the GONWs and rGNWs, *Escherichia coli* and *Staphylococcus aureus* were used as models for Gram-negative and Gram-positive bacteria with bare stainless steel as control. A substantial loss in cell viability for both bacteria strains was observed (Fig. 8.6A). The GONWs exhibited stronger antibacterial activity against *S. aureus* than *E. coli* due to higher resistance of *E. coli* bacteria against the direct

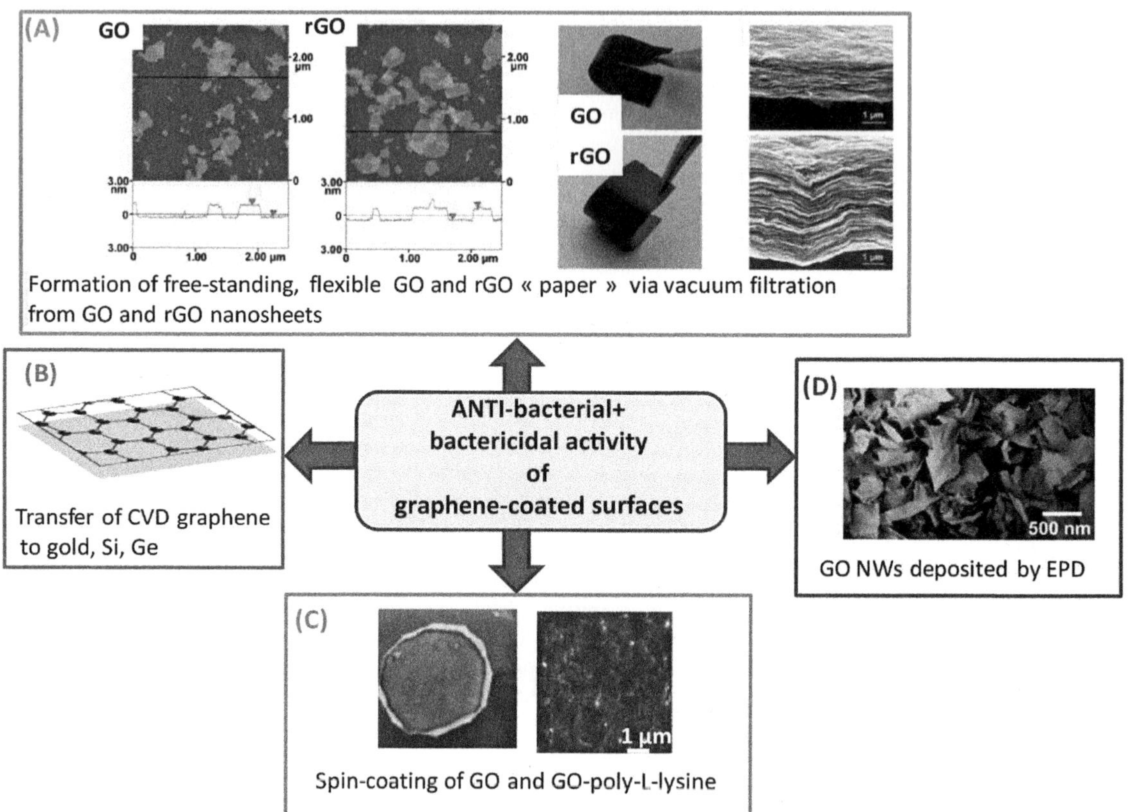

■ **FIGURE 8.5** Different strategies for the formation of graphene-coated substrates used for the determination of their antibacterial and bactericidal properties: (A) AFM images of GO and rGO sheets together with photographs of free-standing flexible GO and rGO paper produced via vacuum filtration; [12] (B) Transfer of CVD grown graphene nanosheets to different solid interfaces; [26] (C) Formation of GO films on polyvinylidene floride (PVDF) membranes [53] or composites such as GO/Poly(L-lysine) [54] interfaces by spin-coating; (D) Electrophoretic deposition of GO nanowalls (GONWs) on stainless steel [60].

interaction with the edge of the nanowalls. Indeed, as a Gram-positive bacterium, *S. aureus* is constituted of a peptidoglycan layer with a thickness ranging from 20 to 80 nm without an outer membrane, while the Gram-negative *E. coli* has a much thinner peptidoglycan layer (1–8 nm) and possesses an additional outer membrane. The reduced nanowalls exhibited stronger toxicity than the GONWs. It was suggested that rGO has sharper edges than the oxidized counterpart, leading to increased potency [60].

At the same time, Hu *et al.* investigated the interaction of *E. coli* with graphene-based paper prepared by vacuum filtration (Fig. 8.5A) [12]. In contrast to the results of Akhavan and Ghaderi [60], photographs of *E. coli* grown on GO and rGO paper overnight showed a limited number of *E. coli*

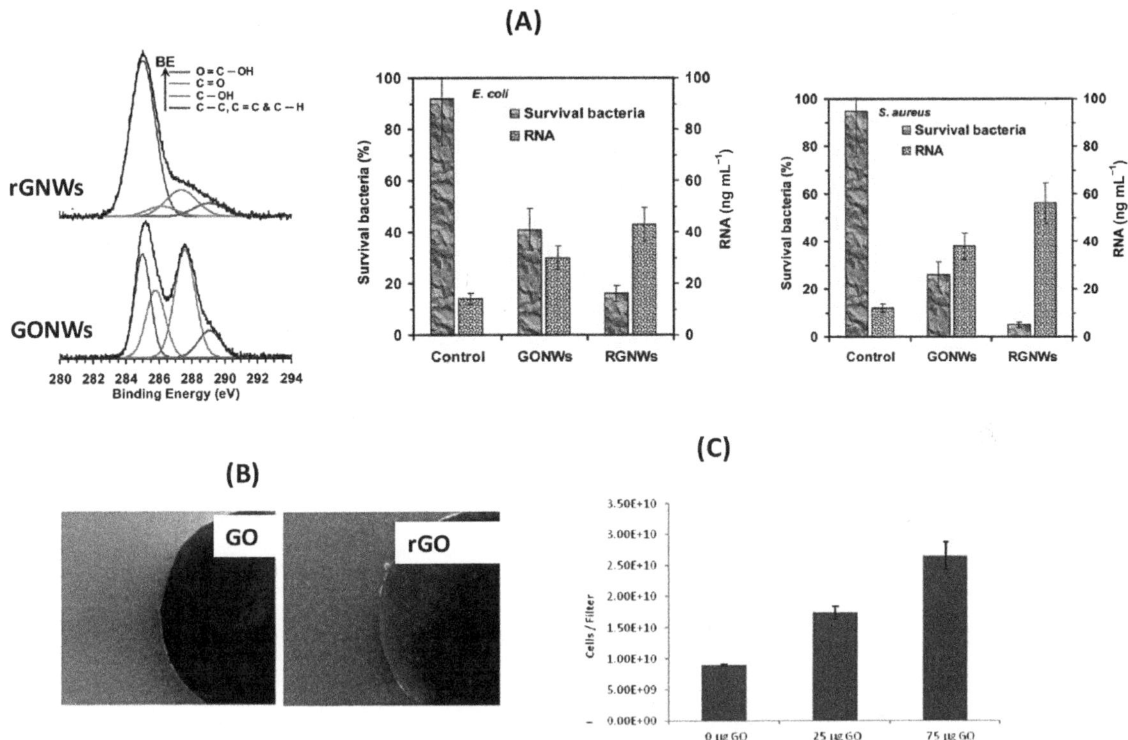

■ **FIGURE 8.6**   (A) Peak deconvolution of C_{1s} core level XPS spectra of GONWs and rGNWs together with cytotoxicity of both materials to *E. coli* and *S. aureus*; [60] (B) Photographs of *E. coli* grown on GO and rGO paper; [12] (C) Bacterial growth assessed by quantitative real-time PCR in PVDF filters coated with GO (0, 25 and 75 µg) when inoculated with *E. coli* at 37°C foe 18 h [53].

colonies on rGO paper, while no colonies at all were found on GO (Fig. 8.6B). SEM studies confirmed that *E. coli* cells on the graphene paper lost the membrane integrity, which is responsible for the bacterial killing effect. One hypothesis is that, in this case, GO and rGO lay relatively flat on the surface with only a few edges oriented away from the interface. The effect of size on the antimicrobial activity of GO surface coatings was highlighted recently by Elimelche et al., who showed a 4-fold increase in antimicrobial activity when the GO sheet area decreased from 0.65 to $0.01 \, \mu m^2$ [62].

Ruiz et al. demonstrated that GO deposited on polymer filters has no antibacterial properties and even supports *E. coli* growth (Fig. 8.6C). It is likely that this preparation results in GO lying flat on the substrate with few exposed edges [53]. The antibacterial activity of structurally flat graphene is thus an interesting avenue to investigate.

We have investigated recently the adhesion behavior of two different *E. coli* strains, UTI89 and LF82, to gold interfaces modified with CVD graphene. As seen from SEM images in Fig. 8.7A, no morphological changes or membrane damage of the bacteria occurred, demonstrating that such interfaces possess no antibacterial properties. Adherent and invasive *E. coli* such as LF82 bacteria can on the contrary easily proliferate into biofilms (Fig. 8.7A). The antibacterial character of such interfaces is, however, related to the electronic properties of the substrate [26]. Large-area graphene films on Cu and Ge inhibited bacterial growth and resulted in dead bacteria, while the proliferation of both bacteria cannot be restricted by graphene films on SiO_2 (Fig. 8.7B). The easy transfer of electrons was believed to be the underlying reason, as for both Cu and Ge, the graphene-on-substrate junctions can act as an electron pump (Fig. 8.7B) and

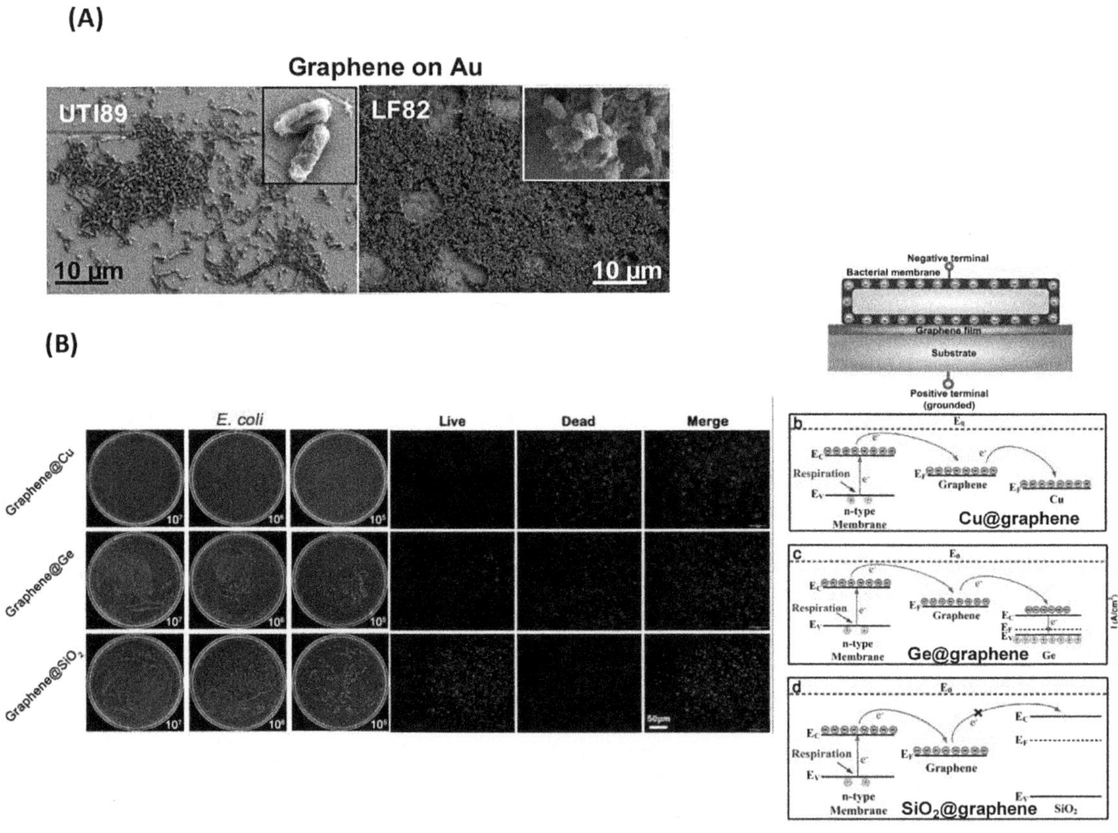

■ **FIGURE 8.7**   (A) SEM images of *E. coli* UTI89 and LF82 seeded on gold coated with graphene (24 h at 37°C) at low and high magnifications; (B) photographs of *E. coli* colonies on different substrates coated with graphene together with fluorescence images showing the viability of the bacteria after 24 h incubation, visualized by staining with a LIVE/DEAD bacterial viability Kit (live bacteria are green, dead ones are read); proposed mechanism of observed phenomenon [26].

electrons are steadily pumped away from the microbial membrane producing oxidative stress to the membrane.

Next to pure GO and rGO, composite surfaces have also been the subject of investigation [53,54,56,63,64]. Poly(L-lysine) (PLL)-modified GO enriched with anionic carboxylic acid groups to favor the electrostatic binding to cationic PLL was shown to inhibit bacteria adhesion and proliferation, while supporting human cell growth (Fig. 8.4) [54]. The efficiency of polyvinyl-*N*-carbazole modified GO to inhibit the build-up of biofilms was demonstrated by Mejeas [57]. Not only were no biofilms formed, the GO material also inactivated significant proportions of bacteria with greater loss of viability to Gram-positive bacteria. Park et al. found that free-standing paper, composed of TWEEN/graphene, prevented the binding of bacteria due to the presence of the surfactant [56]. The integration of silver particles onto GO paper proved to induce antibacterial properties [53]. An efficient way to tailor the physico-chemical properties of graphene film and its antibacterial properties was proposed by Kim et al. [64]. It is based on combining colloidal suspensions of rGO nanosheets and exfoliated layered titanate nanosheets for the fabrication of freestanding hybrid films comprised of stacked and overlapped nanosheets. The intimate incorporation of layered titanate nanosheets onto the graphene films gave rise to an increased mechanical strength, increased surface roughness and hydrophilicity in addition to exhibiting complete sterilization effects to *E. coli* O157:H7 cells in 15 min [64].

8.4   INTERACTION OF GO AND rGO WITH BACTERIA IN SUSPENSION

Numerous studies investigated the interaction of bacteria with GO and rGO nanosheets in suspension [53–55,62,65]. These studies revealed the loss of bacterial cell viability in a concentration and time-dependent manner, with greater loss of viability for Gram-positive bacteria. Membrane rupture was proposed to be the dominant process underlying the toxicity mechanism. An important work in this respect is that of Liu et al., who attempted to shed some light on the antibacterial activity of a family of related material dispersions including graphite, graphite oxide, GO and rGO using *E. coli* as a model [26].

Under otherwise similar conditions and incubation times, GO dispersions exhibited the highest antibacterial activity, followed by rGO, reduced graphite and graphite (Fig. 8.8A). SEM images of *E. coli* cells before and after incubation with GO for 2 h showed that the bacteria are flattened, have lost their cell integrity and are wrapped in a thin layer of GO. In contrast, *E. coli* cells were embedded in large rGO aggregates. This suggested

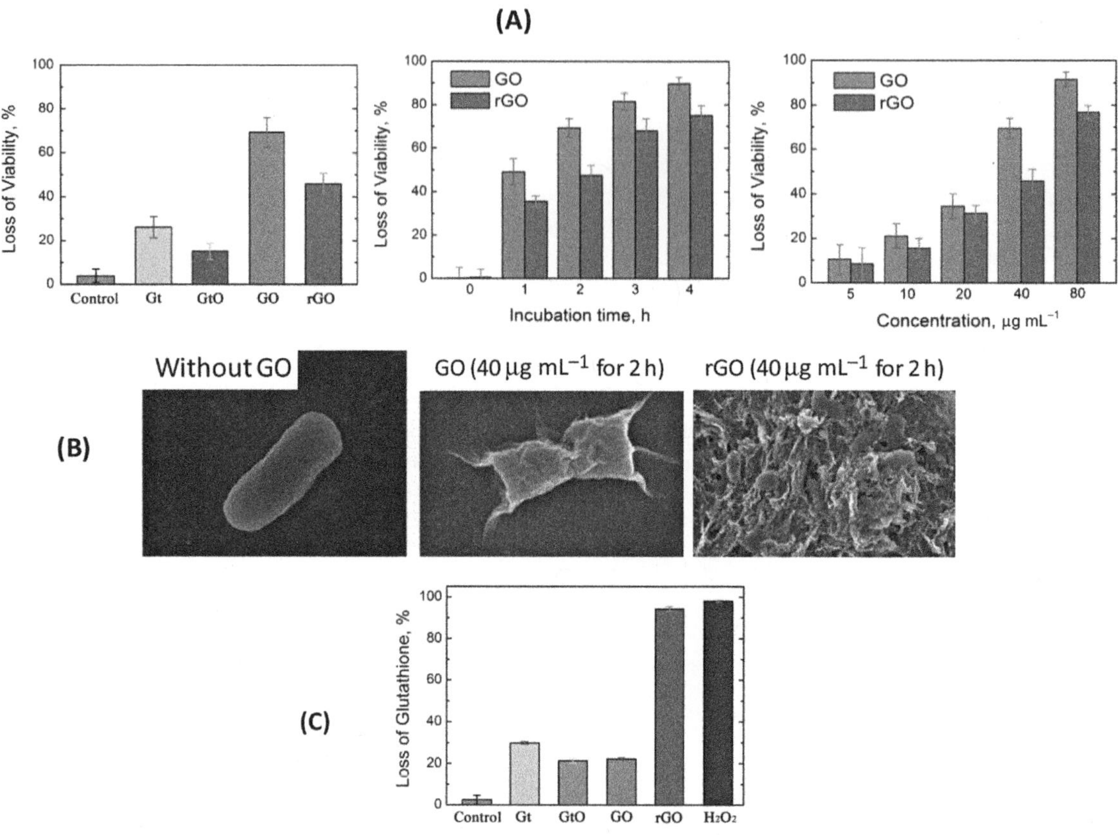

■ **FIGURE 8.8** (A) Viability of bacterial cells after incubation with different graphene-related nanomaterials such as graphite (Gt), graphite oxide (GtO), GO and rGO suspensions at different concentrations and times; (B) SEM images of *E. coli* without graphene-based materials and after incubation with GO and rGO dispersions (40 µg mL⁻¹) for 2 h; (C) loss of glutathione after in vitro incubation with graphite (Gt), graphite oxide (GtO), GO and rGO suspensions [67].

that aggregation/dispersion and the lateral dimensions of graphene–based materials play an important role in their antibacterial avidities [66,67]. While no superoxide radical anion ($O_2{}^{\circ-}$) induced reactive oxygen species production was detected, the materials could oxidize glutathione, which serves as a redox state mediator in bacteria (Fig. 8.8A). Conductive rGO and graphite showed higher oxidative capacities than the insulating GO. These results suggest that antimicrobial actions are contributed by both membrane and oxidative stress and a three-step antimicrobial mechanism was proposed, including (1) cell deposition on graphene materials, (2) membrane stress caused by the direct contact with sharp nanosheets, and (3) superoxide anion-independent oxidation [26]. The underlying molecular mechanism was recently investigated by Tu et al. [68]. They demonstrated experimentally and theoretically that graphene nanosheets can penetrate into cell membranes and extract large amounts of phospholipids because of the strong dispersion interaction between graphene and lipid molecules, these being the molecular bases for the graphene's antibacterial activity and cytotoxicity [68].

The potential of wrapping *E. coli* by graphene nanosheets was also investigated by Akhavan et al. [69]. They found that the bacteria entrapped within the aggregated sheets were biologically disconnected from their environment, and did not proliferate. However, after removing the sheets from the surface of the microorganisms by sonication, the bacteria were reactivated and proliferated. This indicates that they were alive within the aggregated graphene sheets at least for 24 h [69]. They also suggested that the interaction of GO with *E. coli* under acidic conditions leads to the formation of rGO [70].

Currently, the vast majority of the literature is focused on the antibacterial properties of *E. coli* with some on *S. aureus* [60] and other species such as *Pseudomonas aeruginosa, Enterococcus faceclis, Bacillus subtilis*, and *Salmonella typhimurium* [71,72]. In the case of *P. aeruginosa* [71], vaguely better antibacterial activity was observed for rGO. Most importantly, exposure to GO and rGO induced a significant production of $O_2^{\circ-}$, in contrast to the work on *E. coli* [67]. Krishnamoorthy et al. determined the minimum inhibitory concentration (MIC) of rGO nanosheets against different pathogens [72]. MICs between 1 and $8\,\mu g\,mL^{-1}$ were obtained, while a standard drug such as kanamycin showed more elevated MICs between 64 and $128\,\mu g\,mL^{-1}$, with the lower MIC values being for Gram-negative bacteria. This is in direct contrast to other studies, suggesting that the secondary cell membrane of Gram-negative bacteria provides better resistance to membrane-induced damage.

8.5 ANTIBACTERIAL ACTIVITY OF NANOPARTICLE DECORATED GO AND rGO

Recently, the antibacterial properties of graphene-based nanocomposites decorated with nanoparticles have attracted considerable interest. By growing Ag NPs on the surface of GO, several groups using different approaches have synthesized GO-Ag NPs and rGO-Ag NPs nanocomposites with significantly improved antibacterial activities when compared to GO and Ag NPs only [73–82]. The enhanced antimicrobial activity of GO-Ag NPs nanocomposites is attributed to the presence of Ag NPs, whereas GO provides a high surface area for cellular interaction and deposition. Additionally, the high concentration of oxygenated groups confers to GO-Ag NPs sheets the versatility to bind to solid surface via covalent and noncovalent interactions, thus providing multiple opportunities to develop multifunctional materials with antimicrobial activity. Elimelech and coworkers recently reported an interesting approach to produce antimicrobial electrospun mates (Fig. 8.9A) [83]. Poly(lactide-*co*-glycolide) (PLGA) and chitosan were electrospun to yield cylindrical polymer fibers, and were further functionalized with GO/Ag NPs via chemical reaction between the carboxyl groups of GO and the primary amine functional groups of the PLGA-chitosan fibers. Upon direct contact with bacterial cells, Gram-negative and Grampositive bacteria were effectively inactivated [83]. The importance of the ratio of Ag NPs to GO on the antibacterial activity of the nanocomposite was highlighted by Tang et al. [75]. Optimized GO/Ag NPs composites showed strong antibacterial activities at low concentrations ($2.5\,\mu g\,mL^{-1}$). In the case of *E. coli*, the antibacterial effect of the GO-Ag NPs was shown to be based on the disruption of the bacterial cell wall integrity, whereas in the case of *S. aureus* cell division inhibition was observed [75]. The deposition of Ag nanowires onto GO was recently proposed as an alternative antibacterial material where bacterial death is due to damage of the cell membrane integrity and DNA, RNA, and protein leakage due to oxidative damage of the bacteria through the release of reactive oxygen species brought about by the release of silver ions [84].

The integration of gold nanostructures onto graphene-based materials has become another strategy for the treatment and ablation of bacteria [85–88]. Das and coworkers showed that rGO nanosheets when decorated with Au NPs possess high antibacterial activity; bacterial cell death is caused by the leakage of sugars and proteins from the cell membrane once in contact with the rGO-Au NPs material [88]. Gold nanorods coated with pegylated reduced graphene oxide nanoparticles (rGO-PEG-Au NRs) and further functionalized with multimeric heptyl α-D-mannoside probes allowed for the selective killing of uropathogenic *E. coli* UTI89, taking advantage of the excellent light absorption properties of rGO-PEG and Au NRs particles in the near-infrared (NIR) (Fig. 8.9B). Full photothermal ablation of Gramnegative pathogens was achieved in shorter irradiation times (10 min) [86].

More recently, other nanoparticles such as copper [89], magnetic nanoparticles [90–94] as well as semiconductor [95–97] nanoparticles have been integrated onto graphene nanosheets and the antibacterial properties of the resulting nanocomposites have been investigated. The integration of magnetic properties onto graphene-based matrices allowed the preparation of antibacterial materials with the additional possibility of magnetically removing bacteria from solutions by means of an external magnetic field (Fig. 8.9C) [90–92].

Photocatalytic inactivation under solar irradiation using titanium oxide (TiO_2) is considered as another promising alternative for the removal of pathogens from water [94,98]. Upon excitation of TiO_2 by light, the photon energy generates electron-hole pairs on the TiO_2 surface, which can react with oxygen and water molecules to form hydroxyl radicals, responsible for the disinfection of bacteria [99]. The reduction of bacteria and fungi in the presence of GO-TiO_2 nanocomposites on cotton was investigated by Karimi et al. [98]. As can be seen from Fig. 8.9D, raw cotton and even GO impregnated cotton provided suitable environments for bacterial growth. However, in the case of cotton impregnated with GO-TiO_2,

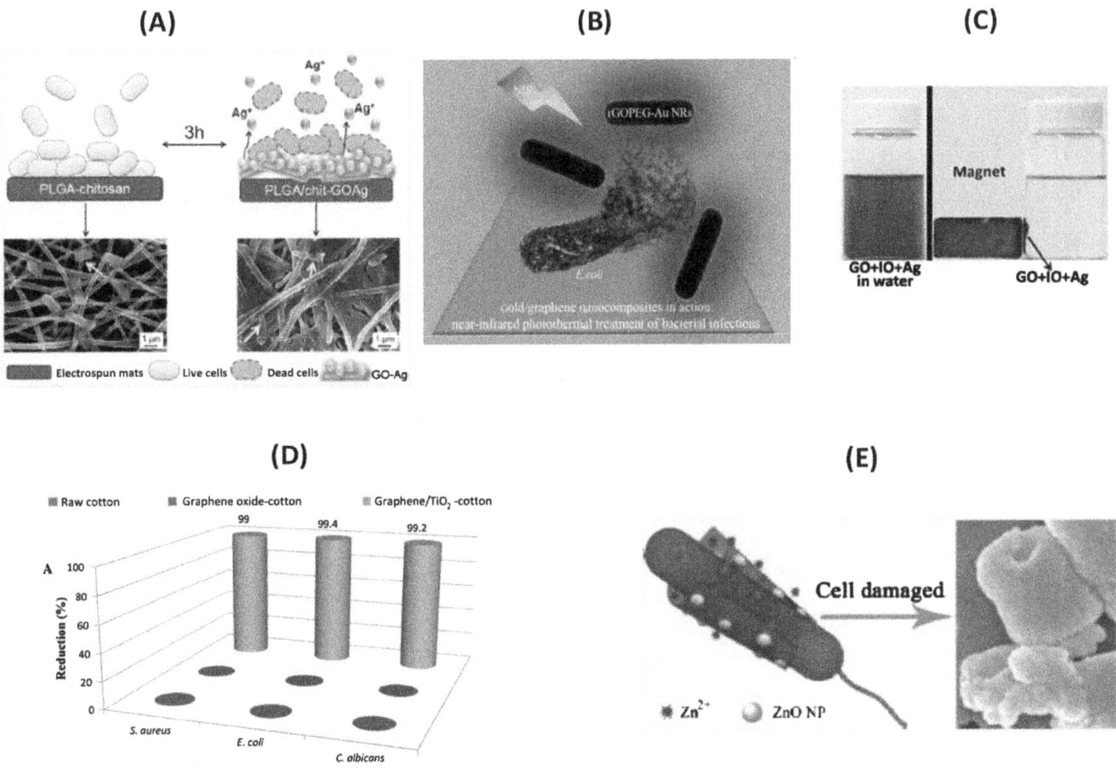

■ **FIGURE 8.9** (A) Antibacterial electrospun nanofiber mats of Poly(lactide-co-glycolide) (PLGA) and chitosan functionalized with GO/Au NPs [83]; (B) photothermal ablation of *E. coli* using rGO-coated gold nanorods [86]; (C) magnetic GO for the removal of pathogens from solution [92]; (D) Antimicrobial efficiency of cotton, cotton treated with GO (5%) and cotton treated with GO-TiO_2 [98]; (E) antibacterial activity of GO-ZnO NPs originating from high zinc concentration localized around the bacteria [96].

excellent antibacterial and antifungal activities were observed [98]. The antibacterial properties of GO-TiO$_2$ modified with magnetic particles was recently investigated and excellent antimicrobial properties were obtained [94]. After 30 min of solar irradiation, the bacterial inactivation rates nearly achieved 100%. The low cost of GO and TiO$_2$ and the simplicity of manufacturing magnetic GO-TiO$_2$ make it a promising candidate in water disinfection treatment.

Like Ag NPs, ZnO has been widely used as an antibacterial agent. Release of zinc ions from ZnO was suggested as one of the primary antibacterial mechanisms of ZnO NPs. Moreover, the penetration and disorganization of the bacterial membrane upon contact with ZnO NPs is believed to contribute in addition to the antibacterial ability or ZnO NPs. However, the easy aggregation of ZnO NPs hampers their antibacterial activity. Wang et al. prepared GO-ZnO NPs by a facile one-pot reaction where the 4 nm sized ZnO NPs are homogenously anchored onto the GO sheets (Fig. 8.9E) [96]. GO helped the dispersion of ZnO NPs, slowed down the dissolution of ZnO, and acted as storage site for dissolved zinc ions, thus enabling the intimate contact of *E. coli* with ZnO NPs and zinc ions [96].

8.6 **CONCLUSION AND PERSPECTIVES**

Graphene and its derivatives have opened up new roads for research with high expectations as long-lasting antimicrobial materials. Size, shape, structural defects as well as surface functional groups of graphene and its derivatives all strongly influence the biological response and define the antibacterial activity of the materials. The most relevant material properties for the antibacterial effect of graphene include the orientation of the graphene nanosheets on surfaces and the lateral dimension of graphene nanosheets in suspensions, which are strongly correlated with their aggregation. Larger GO sheets in solution exhibit stronger antibacterial activity than smaller ones, in a time and a concentration dependent manner. Large GO sheets more easily cover cells, resulting in cell viability loss upon wrapping. Well-dispersed GO sheets display stronger antibacterial activity among several graphene-family nanomaterials. The effective mode of action of graphene nanostructures on bacteria is strongly material dependent and ranges from reactive oxygen species (ROS) production for the destruction of the bacteria strain and cell division inhibition, to the ability to penetrate into cell membranes and extract large amounts of phospholipids from them together with a reduction in metabolic activity or bacteria cell entrapment. While *E. coli* is mostly used as a model, there is a lack of the repetitive use of other pathogenic species and phenotypes to illustrate the bactericidal properties of graphene nanostructures. Indeed, most of their action has still to be validated on other bacteria strains. Even though some studies on *P. aeruginosa*, *E. faceclis*, *B. subtilis*, or *S. typhimurium* are emerging, a deeper understanding is still required in order to draw conclusions about the general antibacterial ability of graphene. Kanchanapally et al. showed for example recently that a 3D porous membrane comprising GO and nisin, an antimicrobial peptide, is capable of effectively identifying, separating and disinfecting water from methicillin-resistant *S. aureus* (MRS) [100]. Xie et al. investigated the influence of graphene on gut microbiota [101]. A detailed report on the interaction of graphene family materials with foodborne bacterial pathogens (*Listeria monocytogenes*, *Salmonellea enteric*) has been given by Chwalibog and coworkers [102]. However, the targeting of a specific pathogen group is also still in its infant shoes. While some of the recent examples have shown that graphene-based nanostructures can bypass the bacterial resistance barriers, it becomes ultimately necessary to stimulate clinical conditions to evaluate the practical applications of these new nanomaterials.

The integration of GO and rGO nanosheets into hydrogels [103] and polymers [104–107] is another current field of intense investigation, which will help advance the formulation of highly efficient, low cost, broad spectrum antibacterial materials with applications in water purification and in the biomedical field. Moreover, the combination of metal/metal oxide nanoparticles with GO or rGO offers an interesting alternative for developing efficient antibacterial materials. These nanocomposites often exhibit higher antibacterial activity than the metal/metal oxide nanoparticles or GO/rGO taken separately. GO/rGO enhances the stability of the nanoparticles against aggregation and dissolution.

An important current development is also the use of the optical properties of graphene. Light absorbed by the nanostructures can be efficiently converted into localized heat energy and used for the hyperthermic destruction of pathogens [86,108]. The strong optical absorption across the NIR spectrum of water-soluble polyethylene glycol modified rGO coupled with high chemical and thermal stability provided rapid temperature rise and an efficient way of heating. Wu et al. showed that magnetic rGO functionalized with glutaraldehyde efficiently captures bacteria and upon NIR irradiation for 10 min kills up to 99% of both Gram-positive and Gram-negative bacteria [108]. We have recently studied the possibility of specifically killing *E. coli* pathogens using gold nanoparticles wrapped with pegylated rGO and loaded with multimetric heptyl α-D-mannoside probes [86]. The presence of multiple copies of mannose on the photothermal agent led to a strong specific adherence of uropathogenic *E. coli* to the graphene-coated nanostructures and selective photothermal killing. This currently offers a unique biocompatible method for the ablation of pathogens with the opening of probably a new possibility for clinical treatments of patients with urinary infections.

ACKNOWLEDGMENTS

Financial support from the Centre National de la Recherche Scientifique (CNRS), the University Lille 1, the Nord-Pas-de Calais region, the CPER Photonics4Society, the Institut Universitaire de France (IUF), and the EU union through the Marie Sklodowska-Curie action (H2O2-MSCA-RISE-2015, PANG-690836) is acknowledged.

REFERENCES

[1] Huh AJ, Kwon YJ. "Nanoantibiotics": a new paradigm for treating infectious diseases using nanomaterials in the antibiotics resistant era. J Control Release 2011;156:128–45.

[2] Applerot G, Lellouche J, Lipovsky A, Nitzan Y, Lubart R, Gedanken A, et al. Understanding the antibacterial mechanism of CuO nanoparticles: revealing the route of induced oxidative stress. Small 2012;8:3326–37.

[3] Bondarenko O, Juganson K, Ivask A, Kasemets K, Mortimer M, Kahru A. Toxicity of Ag, CuO and ZnO nanoparticles to selected environmentally relevant test organisms and mammalian cells in vitro: a critical review. Arch Toxicol 2013;87:1181–200.

[4] Dizaj SM, Mennati A, Jafari S, Khezri K, Adibkia K. Antimicrobial activity of carbon-based nanoparticles. Adv Pharm Bull 2015;5:19–23.

[5] Beranova J, Seydlova G, Kozak H, Potocky S, Konopasek I, Kromka A. Antibacterial behavior of diamond nanoparticles against *Escherichia coli*. Phys Stat Sol B 2012;12:2581–4.

[6] Khanal M, Raks V, Issa R, Chernyshenko V, Barras A, Garcia Fernandes JM, et al. Selective antimicrobial and antibiofilm disrupting properties of functionalized diamond nanoparticles against Escherichia coli and Staphylococcus aureus. Part Part Syst Charact 2015;32:822–30.

[7] Wehling J, Dringer R, Zare RN, Maas M, Rezwan K. Bactericidal activity of partially oxidized nanodiamonds. ACS Nano 2014;8:6475–83.

[8] Wong KKY, Liu X. Silver nanoparticles—the real "silver bullet" in clinical medicine? Med Chem Commun 2010;1:125–31.

[9] Kong H, Jang J. Antibacterial properties of novel poly(methyl methacrylate) nanofiber containing silver nanoparticles. Langmuir 2008;24:2051–6.

[10] Schipper ML, Nakayama-Ratchford N, Davis CR, Kam NWS, Chu P, Liu Z, et al. A pilot toxicology study of single-walled carbon nanotubes in a small sample of mice. Nat Nanotechnol 2008;3:216–21.

[11] Pumera M, Miyahara Y. What amount of metallic impurities in carbon nanotubes is small enough not to dominate their redox properties? Nanoscale 2009;1:260–5.

[12] Hu W, Peng C, Luo W, Lv M, Li X, Li D, et al. Graphene-based antibacterial paper. ACS Nano 2010;4:4317–23.

[13] Novoselov KS, Geim AK, Morozov SV, Jiang D, Zhang Y, Dubonos SV, et al. Electric field effect in atomically thin carbon films. Science 2004;306:666–9.

[14] Novoselov KS, Jiang D, Schedin F, Booth TJ, Khotkevich SV, Morozov SV, et al. Two-dimensional atomic crystals. Proc Natl Acad Sci USA 2005;102:10451–3.

[15] Akhavan E, Ghaderi E, Esfandiar A. Wrapping bacteria by graphene nanosheets for isolation from environment, reactivation by sonication, and inactivation by near-infrared irradiation. J Phys Chem B 2011;115:6279–88.

[16] Ferrari AC, Meyer JC, Casiraghi C, Lazzeri M, Mauri F, Piscanec S, et al. Raman spectrum of graphene and graphene layers. Phys Rev Lett 2006;97:187401.

[17] Kim KS, Zhao Y, Jang H, Lee SY, Kim JM, Kim KS, et al. Large-scale pattern growth of graphene films for stretchable transparent electrodes. Nature 2009;457:706–10.

[18] Pumera M, Smid B, Veltruska K. Influence of nitric acid treatment of carbon nanotubes on their physico-chemical properties. J Nanosci Nanotechnol 2009;9:2671.

[19] Geim AK, Novoselov KS. The rise of graphene. Nat Mater 2007;6:183.

[20] Chen J, Juang C, Xiao S, Ishigami M, Fuhrer MS. Intrinsic and extrinsic performance limits of graphene devices on SiO_2. Nat Nanotechnol 2008;3:206–9.

[21] Graf D, Molitor F, Ensslin K, Stampfer C, Jungen A, Hierold C. Spatially resolved Raman spectroscopy of single- and few-layer graphene. Nano Lett 2007;7:238–42.

[22] Zagorodko O, Spadavecchia J, Yanguas Serrano A, Larroulet I, Pesquera A, Zurutuza A, et al. Highly sensitive detection of DNA hybridization on commercialized graphene-coated surface plasmon resonance interfaces. Anal Chem 2014;86:11211–6.

[23] Allen MJ, Tung VC, Kaner RB. Honeycomb carbon: a review of graphene. Chem Rev 2010;110:132–45.

[24] Berger C, Song Z, Li X, Wu X, Brown N, Naud C, et al. Electronic confinement and coherence in patterned epitaxial graphene. Science 2006;26:1191–6.

[25] Li W, Cai W, An J, Kim S, Nah J, Yang D, et al. Large-area synthesis of high-quality and uniform graphene films on copper foils. Science 2009;324:1312–4.

[26] Li J, Wang G, Zhu H, Zhang M, Zheng X, Di Z, et al. Antibacterial activity of large-area monolayer graphene film manipulated by charge transfer. Sci Rep 2014;4:4359.

[27] Novoselov KS, Fal'ko VI, Colombo L, Gellert PR, Schwab MG, Kim K. A roadmap for graphene. Nature 2012;490:192–200.

[28] Kumar A, Lee CH. Synthesis and biomedical applications of graphene: present and future trends Aliofkhazraei M, editor. Nanotechnology and nanomaterials "advances in graphene science", INTECH; 2013. Chapter 3. <http://dx.doi.org/10.5772/55728 > .

[29] Brodie BC. On the atomic weight of graphite. Philos Trans R Soc London 1859;149:249–59.

[30] Hummers WS, Offerman JRE. Preparation of graphite oxide. J Am Chem Soc 1958;80:1339.

[31] Paredes JI, Villar-Rodil S, Martinez-Alonso A, Tascon JMD. Graphene oxide dispersions in organic solvents. Langmuir 2008;24:10560.

[32] Dreyer DR, Park SJ, Bielawski CW, Ruoff RS. The chemistry of graphene oxide. Chem Soc Rev 2010;39:228.

[33] Fellahi O, Das MR, Coffinier Y, Szunerits S, Hadjersi T, Maamache M, et al. Silicon nanowire arrays-induced graphene oxide reduction under UV irradiation. Nanoscale 2011;3:4662.

[34] Stankovich S, Dikin DA, Piner RD, Kohlhaas KA, Kleinhammes A, Jia Y, et al. Synthesis of graphene-based nanosheets via chemical reduction of exfoliated graphite oxide. Carbon 2007;45:1558.

[35] Shin HJ, Kim KK, Benayad A, Yoon SM, Park HK, Jung IS, et al. Efficient reduction of graphite oxide by sodium borohydride and its effect on electrical conductance. Adv Funct Mater 2009;19:1987.

[36] Park H, Ruoff RS. Chemical methods for the production of graphenes. Nat Nanotechnol 2009;4:217.

[37] Amarnath CA, Hong CE, Kim NH, Ku BC, Kuila T, Lee JH. Efficient synthesis of graphene sheets using pyrrole as a reducing agent. Carbon 2011;49:3497.

[38] Fan X, Peng W, Li Y, Wang S, Zhang G, Zhang F. Deoxygenation of exfoliated graphite oxide under alkaline conditions: a green route to graphene preparation. Adv Mater 2008;20:4490.

[39] Fan ZJ, Kai W, Yan J, Wei T, Zhi LJ, Feng J, et al. Facile synthesis of graphene nanosheets via Fe reduction of exfoliated graphite oxide. ACS Nano 2011;5:191.

[40] Moon IK, Lee JH, Ruoff RS, Lee H. Reduced graphene oxide by chemical graphitization. Nat Commun 2010;73:1.

[41] Dreyer DR, Murali S, Zhu S-E, Ruoff RS, Bielawski RS. Reduction of graphite oxide using alcohols. J Mater Chem 2011;21:3443.

[42] Compton OC, Jain B, Dikin DA, Abouimrane A, Amine K, Nguyen ST. Chemically active reduced graphene oxide with tunable C/O ratios. ACS Nano 2011;5:4380.

[43] Kaminska I, Barras A, Coffinier Y, Lisowski W, Niedziolka-Jonsson J, Woisel P, et al. Preparation of a responsive carbohydrate-coated biointerface based on graphene/azido-terminated tetrathiafulvalene nanohybrid material. ACS Appl Mater Interfaces 2012;4:5386.

[44] Kaminska I, Das MR, Coffinier Y, Niedziolka-Jonsson J, Sobczak J, Woisel P, et al. Reduction and functionalization of graphene oxide sheets using biomimetic dopamine derivatives in one step. ACS Appl Mater Interfaces 2012;4:1016.

[45] Kaminska I, Das MR, Coffinier Y, Niedziolka-Jonsson J, Woisel P, Opallo M, et al. Preparation of graphene/tetrathiafulvalene nanocomposite switchable surfaces. Chem Commun 2012;48:1221.

[46] Li D, Muller MB, Gilje SK, Kaner RB, Wallace GG. Processable aqueous dispersions of graphene nanosheets. Nat Nanotechnol 2008;3:101.

[47] Tung VC, Allen MJ, Yang Y, Kaner RB. High-throughput solution processing of large-scale graphene. Nat Nanotechnol 2009;4:25.

[48] Fernandez-Merino MJ, Guardia L, Paredes JI, Villar-Rodil S, Solis-Fernandez P, Martinez-Alonso AT, et al. Vitamin C is an ideal substitute for hydrazine in the reduction of graphene oxide suspensions. J Phys Chem C 2010;114:6426.

[49] Gao J, Liu F, Liu YL, Ma N, Wang ZQ, Zhang X. Environment-friendly method to produce graphene that employs vitamin C and amino acid. Chem Mater 2010;22:2213.

[50] Zhu CZ, Guo SJ, Fang YX, Dong SJ. Reducing sugar: new functional molecules for the green synthesis of graphene nanosheets. ACS Nano 2010;4:2429–37.

[51] Xu LQ, Yang WJ, Neoh K-G, Kang E-T, Fu GD. Dopamine-induced reduction and functionalization of graphene oxide nanosheets. Macromolecules 2010;43:8336.

[52] Das MR, Sarma RK, Saikia R, Kake VS, Shelke MV, Sengupta P. Synthesis of silver nanoparticles in an aqueous suspension of graphene oxide sheets and its antimicrobial activity. Colloid Surf B 2011;83:16–22.

[53] Ruiz ON, Shiral Fernando KA, Wang B, Brown NA, Luo PG, McNamara ND, et al. Graphene oxide: a nonspecific enhancer of cellular growth. ACS Nano 2011;5:8100–7.

[54] Some S, Ho S-M, Dua P, Hwang E, Hun Shin Y, Yoo H, et al. Dual functions of highly potent graphene derivative poly-L-lysine composites to inhibit bacteria and support human cells. ACS Nano 2012;6:7151–61.

[55] Sreeprasad TS, Shihabudheen Maliyekkal M, Deepti K, Chaudhari K, Lourdu Xavier P, Pradeep T. Transparent, luminescent, antibacterial and patternable film forming composites of graphene oxide/reduced graphene oxide. ACS Appl Mater Interfaces 2011;3:2643–54.

[56] Park S, Mohanty N, Suk JW, Nagaraja A, An J, Pinr R, et al. Biocompatible, robust free-standing paper composed of a TWEEN/graphene composite. Adv Mater 2010;22:1736–40.

[57] Mejías Carpio IE, Santos CM, Wei X, Rodrigues DF. Toxicity of a polymer-graphene oxide composite against bacterial planktonic cells, biofilms, and mammalian cells. Nanoscale 2012;7:4746–56.

[58] Pham VTH, Khanh Truong V, Quinn MDJ, Notley SM, Guo Y, Baulin VA, et al. Graphene induces formation of pores that kill spherical and rod-shaped bacteria. ACS Nano 2015;9:8458–67.

[59] Liu G, Chen Y, Zhang G, Yang S. Protein resistance of (ethylene oxide)$_n$ monolayers at the air/water interface: effects of packing density and chain length. Phys Chem Chem Phys 2007;9:6073–82.

[60] Akhavan O, Ghaderi E. Toxicity of graphene and graphene oxide nanowalls against bacteria. ACS Nano 2010;4:5731–6.

[61] Mangadlao JD, Santos CM, Felipe MJL, de Leon ACC, Rodrigues DF, Advincula RC. On the antibacterial mechanism of graphene oxide (GO) Langmuir-Blodgett films. Chem Commun 2015;51:2886–9.

[62] Perreault F, de Faria AF, Nejati S, Elimelech M. Antimicrobial properties of graphene oxide nanosheets: why size matters. ACS Nano 2015;9:7226–36.

[63] Duan L, Wang Y, Zhang Y, Liu J. Graphene immobilized enzyme/polyethersulfone mixed matrix membrane: enhanced antibacterial, permeable and mechanical properties. Appl Surf Sci 2015;355:436–45.

[64] Kim Y, Park S, Kim H, Park S, Ruoff RS, Hwang S-J. Strongly-coupled freestanding hybrid films of graphene and layered titanate nanosheets: an effective way to tailor the physicochemical and antibacterial properties of graphene film. Adv Funct Mater 2013;24:2288–94.

[65] He J, Zhu X, Qi Z, Wang C, Mao X, Zhu C, et al. Killing dental pathogens using antibacterial graphene oxide. ACS Appl Mater Interfaces 2015;7:5605–11.

[66] Liu S, Ming H, Zeng TH, Wu R, Jiang R, Wei J, et al. Lateral dimension-dependent antibacterial activity of graphene oxide sheets. Langmuir 2012;28:12364–72.

[67] Liu S, Zeng TH, Hofmann M, Burcombe E, Wei J, Jiang R, et al. Antibacterial activity of graphite, graphite oxide, graphene oxide, and reduced graphene oxide: membrane and oxidative stress. ACS Nano 2011;5:6971–80.

[68] Tu Y, Lv M, Xiu P, Huynh T, Zhang M, Castelli M, et al. Destructive extraction of phospholipids from *Escherichia coli* membranes by graphene nanosheets. Nat Nanotechnol 2013;8:594.

[69] Akhavan O, Ghaderi E, Esfandiar A. Wrapping bacteria by graphene nanosheets for isolation from environment, reactivation by sonication, and inactivation by near-infrared irradiation. J Phys Chem B 2011;115:6279–88.

[70] Akhavan O, Ghaderi E. *Escherichia coli* bacteria reduce graphene oxide to bactericidal graphene in a self-limiting manner. Carbon 2012;50(5):1853–60.

[71] Gurunathan S, Han JW, Dayem AA, Eppakayala V, Kim J-H. Oxidative stress-mediated antibacterial activity of graphene oxide and reduced graphene oxide in *Pseudomonas aeruginosa*. Int J Nanomed 2012;7:5901–14.

[72] Krishnamoorthy K, Veerapandian M, Zhang L-H, Yun K, Kim SJ. Antibacterial efficiency of graphene nanosheets against pathogenic bacteria via lipid peroxidation. J Phys Chem C 2012;116:17280–7.

[73] Das MR. Synthesis of silver nanoparticles in an aqueous suspension of graphene oxide sheets and its antimicrobial activity. Colloids Surf B 2011;83:16–22.

[74] Zhang D, Liu X, Wang X. Green synthesis of graphene oxide sheets decorated by silver nanoprisms and their anti-bacterial properties. J Inorg Biochem 2011;105:1181–6.

[75] Tang J, Chen Q, Xu LQ, Zhang S, Feng L, Cheng L, et al. Graphene oxide-silver nanocomposite as a highly effective antibacterial agent with species-specific mechanisms. ACS Appl Mater Interfaces 2013;5:3867–74.

[76] Chook S, Chia C, Zakaria S, Ayob M, Chee K, Huang N, et al. Antibacterial performance of Ag nanoparticles and AgGO nanocomposites prepared via rapid microwave-assisted synthesis method. Nanoscale Res Lett 2012;7:1–7.

[77] Pasricha R, Gupta S, Srivastava AK. A facile and novel synthesis of Ag-graphene-based nanocomposites. Small 2009;5:2253–9.

[78] Tang XZ, Cao Z, Zhang HB, Liu J, Yu ZZ. Growth of silver nanocrystals on graphene by simultaneous reduction of graphene oxide and silver ions with a rapid and efficient one-step approach. Chem Commun 2011;47:3084–6.

[79] Shao W, Liu X, Min H, GDong G, Feng Q, Zuo S. Preparation, characterization, and antibacterial activity of silver nanoparticle-decorated graphene oxide nanocomposite. ACS Appl Mater Interfaces 2015;7:6966–73.

[80] Chook SW, Chia CH, Zakaria S, Ayob MK, Huang NM, Neoh HM, et al. Antibacterial hybrid cellulose–graphene oxide nanocomposite immobilized with silver nanoparticles. RSC Adv 2015;5:26263–8.

[81] Zhang P, Wang H, Zhang X, Xu W, Li Y, Li Q, et al. Graphene film doped with silver nanoparticles: self-assembly formation, structural characterizations, antibacterial ability, and biocompatibility. Biomater Sci 2015;3:852–60.

[82] Chen X, Huang X, Zheng C, Liu YL, Xu T, Liu J. Preparation of different sized nano-silver loaded on functionalized graphene oxide with highly effective antibacterial properties. J Mater Chem B 2015;3:7020.

[83] de Faria AF, Perreault F, Shaulsky E, Arias Chavez LH, Elimelech M. Antimicrobial electrospun biopolymer nanofiber mats functionalized with graphene oxide-silver nanocomposites. ACS Appl Mater Interfaces 2015;7:12751–9.

[84] Cui J, Liu Y. Preparation of graphene oxide with silver nanowires to enhance antibacterial properties and cell compatibility. RSC Adv 2015;5:85748–55.

[85] Lin D, Qin T, Wang Y, Sun X, Chen L. Graphene oxide wrapped SERS Tags: Multifunctional platforms toward optical labeling, photothermal ablation of bacteria, and the monitoring of killing effect. ACS Appl Mater Interfaces 2014;6:1320–329.

[86] Turcheniuk K, Hage C-H, Spadavecchia J, Serrano AY, Larroulet I, Pesquera A, et al. Plasmonic photothermal destruction of uropathogenic *E. coli* with reduced graphene oxide and core/shell nanocomposites of gold nanorods/reduced graphene oxide.. J Mater Chem B 2015;3:375–86.

[87] Turchniuk K, Boukherroub R, Szunerits S. Gold–graphene nanocomposites for sensing and biomedical applications. J Mater Chem B 2015;3:4301–24.

[88] Hussain N, Gogoi A, Sarma RK, Sharma P, Barras A, Boukherroub R, et al. Reduced graphene oxide nanosheets decorated with Au nanoparticles as an effective bactericide: investigation of biocompatibility and leakage of sugars and proteins. ChemPlusChem 2014;79:1774–84.

[89] Ouyang Y, Cai X, Shi Q, Liu L, Wan D, Tan S, et al. Poly-l-lysine-modified reduced graphene oxide stabilizes the copper nanoparticles with higher water-solubility and long-term additively antibacterial activity. Colloid Surf B 2013;107:107–14.

[90] Ma S, Zhan S, Jia Y, Zhou Q. Highly efficient antibacterial and Pb(II) removal effects of Ag-CoFe$_2$O$_4$-GO nanocomposite. ACS Appl Mater Interfaces 2015;7:10576–86.

[91] Chella S, Kollu P, Komarala EVPR, Doshi S, Saranya M, Felix S, et al. Solvothermal synthesis of MnFe$_2$O$_4$-graphene composite—investigation of its adsorption and antimicrobial properties. Appl Surf Sci 2015;327:27–36.

[92] Moosavi R, Ramanathan S, Lee YY, Ling KCS, Afkhami A, Archunan G, et al. Synthesis of antibacterial and magnetic nanocomposites by decorating graphene oxide surface with metal nanoparticles. RSC Adv 2015;5:76442–50.

[93] Santosh C, LKollu P, Doshi S, Sharma M, Bahadur D, Vanchinathan MT, et al. Adsorption, photodegradation and antibacterial study of graphene–Fe$_3$O$_4$ nanocomposite for multipurpose water purification application. RSC Adv 2014;4:28300–8.

[94] Chang Y-N, Ou X-M, Zeng G-M, Gong J-L, Deng C-H, Jiang Y, et al. Synthesis of magnetic graphene oxide–TiO$_2$ and their antibacterial properties under solar irradiation. Appl Surf Sci 2015;343:1–10.

[95] Yu S, Liu J, Zhu W, Hu Z-T, Lim T-T, Yan X. Facile room-temperature synthesis of carboxylated graphene oxide-copper sulfide nanocomposite with high photodegradation and disinfection activities under solar light irradiation. Sci Rep 2015;5:16369.

[96] Wang Y-W, Cao A, Jiang Y, Zhang X, Liu J-H, Liu Y. Superior antibacterial activity of zinc oxide/graphene oxide composites originating from high zinc concentration localized around bacteria. ACS Appl Mater Interfaces 2014;6:2791–8.

[97] Zhong L, Yun K. Graphene oxide-modified ZnO particles: synthesis, characterization, and antibacterial properties. Int J Nanomed 2015;10:79–92.

[98] Karimi L, Yazdanshena ME, Khajavi R, Rashidi A, Mirjalili M. Using graphene/TiO$_2$ nanocomposite as a new route for preparation of electroconductive, self-cleaning, antibacterial and antifungal cotton fabric without toxicity. Cellulose 2014;21:3813–27.

[99] Markowska-Szczupak A, Ulfig K, Morawski AW. The application of titanium dioxide for deactivation of bioparticulates: an overview. Catal Today 2011;169:249–57.

[100] Kanchanapally R, Nellore BPV, Sinha SS, Pedraza F, Jones SJ, Pramanik A, et al. Antimicrobial peptide-conjugated graphene oxide membrane for efficient removal and effective killing of multiple drug resistant bacteria. RSC Adv 2015;5:18881–7.

[101] Xie Y, Wu B, Zhang X-X, Yin J, Mao L, Hu M. Influences of graphene on microbial community and antibiotic resistance genes in mouse gut as determined by high-throughput sequencing. Chemosphere 2016;144:1306–12.

[102] Kurantowicz N, Sawosz E, Jaworski S, Kutwin M, Strojny B, Wierzbicki M, et al. Interaction of graphene family materials with *Listeria monocytogenes* and *Salmonella enterica*. Nanoscale Res Lett 2015;10:23.

[103] Wang X, Liu Z, Ye X, Hu K, Zhong H, Yuan X, et al. A facile one-pot method to two kinds of graphene oxide-based hydrogels with broad-spectrum antimicrobial properties. Chem Eng J 2015;260:331–7.

[104] Li P, Sun S, Dong A, Hao Y, Shi S, Sun Z, et al. Developing of a novel antibacterial agent by functionalization of graphene oxide with guanidine polymer with enhanced antibacterial activity. Appl Surf Sci 2015;355:446–52.

[105] Tang X-Z, Mu C, Zhu W, Yan X, Hu X, Yang J. On the dispersion systems of graphene-like two-dimensional materials: from fundamental laws to engineering guidelines. Carbon 2016;98:432–40.

[106] Nanda SS, An SSA, Yi DK. Oxidative stress and antibacterial properties of a graphene oxide-cystamine nanohybrid. Int J Nanomed 2015;10:549–56.

[107] Cai X, Lin M, Tan S, Mai W, Zhang Y, Liang Z, et al. The use of polyethyleneimine-modified reduced graphene oxide as a substrate for silver nanoparticles to produce a material with lower cytotoxicity and long-term antibacterial activity. Carbon 2012;50:3407–15.

[108] Wu M-C, Deokar AR, Liao J-H, Shih P-Y, Ling Y-C. Graphene-based photothermal agent for rapid and effective killing of bacteria. ACS Nano 2013;7:1281–90.

Index

Note: Page numbers followed by "*f*" and "*t*" refer to figures and tables, respectively.

Q